Biodiversity
New Leads for the Pharmaceutical and Agrochemical Industries

Edited by

Stephen K. Wrigley
TerraGen Discovery (UK) Ltd, Slough, Berkshire, UK

Martin A. Hayes
*GlaxoWellcome Research and Development,
Stevenage, Hertfordshire, UK*

Robert Thomas
Biotics Ltd, Guildford, Surrey, UK

Ewan J.T. Chrystal
AstraZeneca Agrochemicals, Jeallotts Hill, Berkshire, UK

Neville Nicholson
SmithKline Beecham, Harlow, Essex, UK

RS•C
ROYAL SOCIETY OF CHEMISTRY

The proceedings of the International Meeting Biodiversity: A Source of New Leads for the Pharmaceutical and Agrochemical Industries held on 5–8 September 1999 at the University of St Andrews.

Special Publication No. 257

ISBN 0-85404-830-8

A catalogue record for this book is available from the British Library

Published by The Royal Society of Chemistry,
Thomas Graham House, Science Park, Milton Road,
Cambridge CB4 0WF, UK

For further information see our web site at www.rsc.org

Printed and bound by MPG Books Ltd, Bodmin, Cornwall

Preface

The present volume is based on the proceedings of the international conference entitled *Biodiversity: A Source of New Leads for the Pharmaceutical and Agrochemical Industries* held at the University of St. Andrews in September 1999. This was organised by the Biotechnology Group of the Industrial Division of The Royal Society of Chemistry, as was the preceding meeting on *Phytochemical Diversity* held at the University of Sussex in 1996.[1] Whereas the primary topic of the earlier conference was industrially useful plant products, the current volume is of wider scope and includes substances derived from microbial, marine and plant sources.

Recent developments in genetic engineering techniques are opening up the possibility of transferring the gene clusters responsible for the biosynthesis of individual metabolites from any phylum to more convenient production sources, such as microorganisms or crop plants. At the same time, advances in disease-related high throughput screening and the combinatorial synthesis of structural analogues have rekindled interest in examining the widest possible sources of new products. Despite the extensive exploration of natural products throughout the past century, only a minority of species of plants, microbes or other organisms have received more than superficial screening. Furthermore, our understanding of structure–activity relationships is still limited to the extent that whenever a new disease target of pharmaceutical or agrochemical interest is developed, it is usually necessary to start screening afresh.

This book is divided into six sections comprising 21 chapters covering a wide range of natural product topics. These are based on the individual lectures given by senior scientists from major pharmaceutical and agrochemical companies and from academic and government laboratories based in Australia, Brazil, Canada, the UK and the USA.

In the first section, the opening chapters review the historical development and future potential of microbial secondary metabolites and examine novel concepts and approaches to exploring the still largely untapped genetic resources of global biodiversity. This section includes a timely review of past achievements and new directions in drug discovery at the US National Cancer Institute. It also contains a discussion of how best to deploy natural products alongside synthetic and combinatorial chemical libraries in modern high-throughput drug discovery screens and a comparison of the screening productivity of natural products *vis-à-vis* compounds prepared by combinatorial chemistry.

Since the development of penicillin and the subsequent launch of the antibiotic era, the screening of microorganisms, particularly filamentous bacteria and fungi, has dominated natural products discovery. The exploration of microbial sources of a variety of useful new chemical entities forms the subject matter of the second section. Topics covered include current research on drugs of both bacterial and fungal origin, ranging from clinically-useful antifungal products to inhibitors of signal transduction and lipoprotein-associated phospholipase.

The third section is devoted to investigations of the rapidly expanding area of marine products as new sources of anticancer drugs and other disease-related activities. A number of marine metabolites have been shown to exhibit intrinsically useful medicinal properties *per se* or to provide valuable sources of new drug development leads. Tropical species of marine invertebrates are described which accumulate high concentrations of copper and zinc through the formation of complexes with cyclic peptide metabolites. In a review of the cytotoxic dolastatins, it is demonstrated that although originally isolated from the Indian Ocean seahare, these are actually produced by cyanobacteria and enter the macroorganism *via* a dietary route.

Historically, the earliest natural products to be investigated were plant constituents, the utilisation of which provided the principal focus of the previous conference on *Phytochemical Diversity*. Special attention is devoted in the fourth section of this volume to the exploration of regional flora, ranging from the temperate species of the United Kingdom to the tropical rain forests of Brazil and Northern Australia. Specific topics addressed in the opening chapter are water soluble bioactive alkaloids, whereas other chapters examine more general aspects, including traditional plant uses, phytomedicines and ecology-based bio-prospecting for new leads.

Two chapters in section five describe different aspects of the biosynthesis of polyketides, which are numerically the most abundant and structurally diverse class of natural products. The first chapter reports *in vivo* and *in vitro* studies aimed at enhancing our understanding of the basic pathways of polyketide assembly with a view to producing novel compounds. In the second chapter, a consistent difference in the modes of cyclisation of the fused ring polyketides of fungi and streptomycetes is described, which provides the basis for a new biosynthetic classification of these metabolites.

The sixth and final section is devoted to the chemical synthesis of natural products and their derivatives. Target molecules include the strobilurin fungicides produced by wood-rotting fungi and the events are described which led to the discovery of the successful synthetic analogue azoxystrobin, following the synthesis of *ca.* 1400 separate analogues. The final two chapters report a new strategy for the synthesis of the Prelog-Djerassi lactonic acid, which additionally yielded unexpected products, and also the preparation of candidate herbicides and aminoacyl tRNA synthase inhibitors modelled on known microbial metabolites.

This compendium includes the majority of the St Andrews conference lectures, but not the poster presentations. While some aspects of new lead discovery have inevitably received more attention than have others, the editors have endeavoured to minimise the duplication of subject matter and to provide a balanced coverage of key topics.

It is a particular pleasure to acknowledge the collective contributions of the individual authors and the much valued sponsorship received from the Tony and Angela Fish Bequest and the AstraZeneca Agrochemicals, GlaxoWellcome Research and Development and SmithKline Beecham companies and also the assistance of the staff of the Publications Section of The Royal Society of Chemistry in the preparation of this volume.

May 2000   Robert Thomas

[1]*Phytochemical Diversity: A Source of New Industrial Products*, S.K. Wrigley, M.A. Hayes, R. Thomas and E.J.T. Chrystal, (eds.), Royal Society of Chemistry, Cambridge, 1997.

Contents

1 Natural Products – History, Diversity and Discovery

MICROBIAL NATURAL PRODUCTS: A PAST WITH A FUTURE

Arnold L. Demain

Fermentation Microbiology Laboratory, Department of Biology, Massachusetts Institute of Technology, Cambridge, Massachusetts 02139, USA

For over 50 years, antibiotics have served us well in combating infectious bacteria and fungi. The recently increased development of resistance to older antibacterial and antifungal drugs is being challenged by (i) newly discovered antibiotics (e.g., pneumocandins), (ii) new semi-synthetic versions of old antibiotics (e.g., glycylcyclines), (iii) older underutilized antibiotics (e.g., teicoplanin), and (iv) new derivatives of previously undeveloped narrow-spectrum antibiotics (e.g., streptogramins, everninomycin). Many of these products are in late stage clinical testing at the moment. In addition, many antibiotics are used commercially, or are potentially useful, in medicine or agriculture for activities other than their antibiotic action. They are used as antitumor agents, enzyme inhibitors including powerful hypocholesterolemic agents, immunosuppressive agents, and anti-migraine agents, in medicine. Agricultural products include bioherbicides, antiparasitic agents, bioinsecticides and growth promotants for animals (especially ruminants) and plants. A number of these products were first discovered as mycotoxins, or as antibiotics which failed in their development as such. Combinatorial chemistry will accelerate the discovery of new derivatives of natural products. It will join structure-function drug design, semi-synthesis, and recombinant DNA technology as techniques complementing the screening of natural products.

1 INTRODUCTION

Natural products have been an overwhelming success in our society. The doubling of our life span in the twentieth century has been attributed to the use of plant and microbial secondary metabolites.[1] These have reduced pain and suffering, and revolutionized medicine by allowing for the transplantation of organs. Natural products are the most important anticancer and anti-infective agents. Over 60% of approved and pre-NDA candidates are either natural products or related to them, not including biologicals such as vaccines and monoclonal antibodies.[2] Almost half of the best selling pharmaceuticals are natural or are related to natural products. Often, the natural molecule has not been used itself but served as a lead molecule for manipulation by chemical or genetic means.

Secondary metabolism has evolved in nature in response to needs and challenges of the natural environment. Nature has continually carried out its own version of combinatorial chemistry[1] for the period of over 3 billion years during which bacteria have inhabited the earth.[3] During that time, there has been an evolutionary process in progress in which producers of secondary metabolites evolve according to their local environments. If the metabolites are useful to the organism, the biosynthetic genes are retained and genetic modifications further improve the process. Combinatorial chemistry practiced by nature is much more sophisticated than combinatorial chemistry in the laboratory, yielding exotic structures rich in stereochemistry, concatenated rings and reactive functional groups.[1] As a result, an amazing variety and number of products have been found in nature. The total number of natural products produced by plants has been estimated to be over 500,000.[4] About 100,000 secondary metabolites of molecular weight less than 2500 have been characterized, mainly produced by microbes and plants;[5] some 50,000 are from microorganisms.[6,7] The enormous diversity existing in secondary metabolism can be illustrated by the following two examples.

(i) 22,000 terpenoids have been described from living organisms.[8] They are all produced from hydroxymethylglutaryl coenzyme A *via* mevalonate. Their structures contain repeats of the five-carbon isoprene unit unless subsequently modified. Their functions include intercellular communication in animals (steroid hormones) and in plants (gibberellins), aroma (volatile terpenoids), pigments (carotenoids) and sexual hormones in animals and fungi. There are 86 known gibberellins, of which 26 are produced by fungi and the rest by plants.[9]

(ii) 10,000 polyketides are known, most of which are produced by bacteria and fungi.[10] These include antibiotics (e.g., erythromycin, tetracyclines, rifamycins), antitumor agents (e.g., doxorubicin, daunorubicin, enediynes), immunosuppressants (e.g., FK 506, rapamycin), antiparasitic agents (e.g., avermectins), antifungals (e.g., amphotericin, griseofulvin), cardiovascular agents (e.g., lovastatin, pravastatin) and veterinary products (e.g., monensin, tylosin).

Natural product research is at its highest level now due to unmet needs, remarkable diversity of structures and activities, utility as biochemical probes, novel and sensitive assay methods, improvements in isolation, purification and characterization, and new production methods.[11] Many new products have been made by genetic methods involving modification or exchange of genes between organisms to create hybrid molecules; the technique is known as combinatorial biosynthesis.[12,13]

The enormous diversity of microorganisms is a factor that must be kept in mind for future drug development. Only a minor proportion of bacteria and fungi have been cultured or examined for secondary metabolite production. For example, only 5% of the total number of fungal species have been described. Of those described (69,000), only 16% (11,500) have been cultured.[14] It has been estimated that 1 gram of soil contains 1000 to 10,000 species of undiscovered prokaryotes![15] The extensively used concept of isolation of microbial strains from different geographical and climatic locations around the world still gathers support.[16]

2 MEDICAL ANTIBIOTICS

The selective action exerted on pathogenic bacteria and fungi by microbial secondary metabolites ushered in the antibiotic era and for fifty years, we have benefited from this remarkable property of these "wonder drugs". The successes were so impressive that these antibiotics were virtually the only drugs utilized for chemotherapy against pathogenic microorganisms. Antibiotics are defined as low molecular weight organic natural products (secondary metabolites; idiolites) made by microorganisms which are active at low concentration against other microorganisms. Of the 12,000 antibiotics known in 1995, 55% were produced by filamentous bacteria (=actinomycetes) of the genus *Streptomyces*, 11% from other actinomycetes, 12% from non-filamentous bacteria and 22% from filamentous fungi.[7,17] New bioactive products from microbes continue to be discovered at an amazing pace: 200 to 300 per year in the late 1970s increasing to 500 per year by 1997. The world market for antimicrobials involves some 150-300 products, either natural, semi-synthetic or synthetic, and includes cephalosporins (45%), penicillins (15%), quinolones (11%), tetracyclines (6%) macrolides (5%), aminoglycosides, ansamycins, glycopeptides and polyenes.[17]

About 30 years ago, the difficulty and high cost of isolating novel structures and agents with new modes of action for such uses became apparent and the field looked like it might enter a phase of decline. This is understandable because the chance of finding useful antibiotics from microbes is very low, i.e. one in 10,000 to 100,000 compounds.[18-20] Similar data have been observed with plants.[21] Indeed, the number of anti-infective investigational new drugs (INDs) declined by 50% from the 1960s to the late 1980s.[22] By the technique of semi-synthesis, chemists had been improving antibiotics for many years but despite success, new screening techniques were sorely needed in 1970 to isolate new bioactive molecules from nature. Many felt that the golden era of antibiotic discovery was over, but this was far from the truth. Due mainly to the development of novel target-directed screening procedures, important new antibiotics appeared on the scene and became commercial successes in the 1970s and 1980s. These included cephamycins (e.g., cefoxitin), fosfomycin, carbapenems (e.g., thienamycin), monobactams (e.g., aztreonam), glycopeptides (e.g., vancomycin, teicoplanin), aminoglycosides (e.g., amikacin, sisomicin) as well as semisynthetic versions of cephalosporins and macrolides. Thienamycin is the most potent and broadest in spectrum of all antibiotics known today. Although a β-lactam, it is not a member of the penicillins or cephalosporins; rather it is a carbapenem. It is active against aerobic and anaerobic bacteria, both Gram-positive and Gram-negative including *Pseudomonas*. This novel structure was isolated in Spain from a new soil species which was named *Streptomyces cattleya*.[23] Interestingly, this culture also produces penicillin N and cephamycin C. Despite a low degree of production and instability of the molecule, the structure of thienamycin was elucidated in 1976. The main differences between the conventional β-lactams and thienamycin are the possession of a carbon atom instead of sulfur in the ring attached to the β-lactam ring and the *trans* configuration of the hydrogen atoms of the β-lactam ring. In addition to its broad spectrum and high potency, the molecule possesses resistance to plasmid and chromosomal β-lactamases and activity against pathogens resistant to penicillins, cephalosporins and aminoglycosides. Its chemical instability is due to a type of self-destruction, i.e., the β-lactam ring of one molecule of thienamycin is broken *via* aminolysis by the amine group of a second thienamycin molecule. To eliminate this problem, chemists created the semi-synthetic carbapenem, imipenem, by adding a formididoyl group to the side chain amine. Not only was imipenem found to be more stable than thienamycin, but also twice as active! The mode of action of imipenem (and thienamycin) is inhibition of cell-wall peptidoglycan synthesis and it is bactericidal.

In humans, imipenem was found to be metabolized by an enzyme in the kidney, renal dehydropeptidase-I, which acts as a β-lactamase. Since the enzyme appears to serve no essential role in human metabolism, scientists were able to develop a synthetic competitive inhibitor, cilastatin, which they then used with imipenem to produce the combination drug, primaxin (Tienam). Primaxin was introduced into medical practice in 1985.

The unique activity of thienamycin is due to several factors: (i) it permeates the Gram-negative outer cell membrane through porin channels at 10-20 times the rate of classical β-lactams; (ii) it is not destroyed by the β-lactamases of the periplasmic space; (iii) it binds to and inhibits all penicillin-binding proteins (PBPs) but is principally active against PBPs-2 and-1b rather than PBP-3 which is attacked by conventional penicillins and cephalosporins. Sequential inhibition of PBPs-2 and-1b converts the cells of the pathogen to non-growing spheroplasts which rapidly die. On the other hand, inhibition of PBP-3 blocks septum formation resulting in long viable filaments which take longer to die; (iv) after removal of thienamycin, there is a long delay before regrowth of any unkilled spheroplasts. With conventional β-lactams, unkilled filaments (which contain 20 or more cells per unit) septate and regrow immediately.

It is quite interesting that despite the very high potency of thienamycin, it took over 25 years of worldwide screening before it was discovered. Today, we know that carbapenems are not rare. Many members of this family have been isolated in laboratories all over the world although none equals thienamycin in potency and spectrum of activity. The answer probably lies in its very low level of production by wild strains and its instability. Conventional screening procedures evidently missed this activity and only after the development of specific and sensitive modern assay procedures was it found. Thienamycin was discovered by a sensitive mode of action screen, the details of which have never been revealed.

Microbiologists knew in 1970 that technology had not yet won the war against infectious microoganisms due to resistance development in pathogenic microbes. Indeed, it may never win the war and we may have to be satisfied to merely stay one step ahead of the pathogens for a long time to come; thus, the search for new drugs must not be stopped. New antibiotics are continually needed because of (i) the development of resistant pathogens, (ii) the evolution of new diseases (e.g., AIDS, Hanta virus, Ebola virus, *Cryptospiridium,* Legionnaire's disease, Lyme disease, *Escherichia coli* 0157:H7), (iii) the existence of naturally resistant bacteria (e.g., *Pseudomonas aeruginosa* causing fatal wound infections, burn infections and chronic and fatal infections of lungs in cystic fibrosis patients), *Stenotrophomonas maltophilia, Enterococcus faecium, Burkholderia cepacia* and *Acinetobacter baumanni,*[24] and (iv) the toxicity of some of the current compounds.[17] The resistant bacteria are generally uninhibited by all commercial antibiotics.[25] Other organisms exist which are not normally virulent but do infect immunocompromised patients.[26] Semi-synthetic tetracyclines, e.g. glycylcyclines, are being developed for use against tetracycline-resistant bacteria.[27] Exploitation of old but underutilized antibiotics is occurring. In recent years, there has been great concern about resistance development among Gram-positive pathogens. Clinical isolates of penicillin-resistant *Streptococcus pneumoniae*, the most common cause of bacterial pneumonia, increased in the US from 1987 to 1992 by 60-fold.[28] Methicillin-resistant *Staphylococcus* infections increased to an alarming extent throughout the world.[29] At present, vancomycin is the molecule of choice to treat infections caused by such organisms; however, resistance is developing to this glycopeptide antibiotic, especially in the case of nosocomial vancomycin-resistant *Enterococcus* infections. Fortunately, some vancomycin-resistant enterococci are treatable by the related underutilized glycopeptide antibiotic, teicoplanin.[30] Furthermore, a number of "old" compounds are now available for development which were not previously developed because of their narrow antibacterial spectrum which was restricted to Gram-positive bacteria. At that time (in the 1970's and 1980's), breadth of spectrum was the commercial goal but today, a major aim is to inhibit resistant Gram-positive pathogens. About 90% of natural antibiotics fail to inhibit Gram-negative organisms such as *Escherichia coli.*[31] The reasons include its outer membrane which contains (i) narrow porin channels which retard the entry into the cell of even small hydrophilic compounds and (ii) a lipopolysaccharide moiety which slows down the trans-membrane diffusion of lipophilic antibiotics.[32] Furthermore, Gram-negative bacteria often possess a multiple-drug efflux pump which eliminates many antibiotics from the cells.[33]

One group of useful narrow-spectrum compounds are the streptogramins which are synergistic pairs of antibiotics made by single microbial strains. The pairs are constituted by a (Group A) polyunsaturated macrolactone containing an unusual oxazole ring and a dienylamide fragment and a (Group B) cyclic hexadepsipeptide possessing a 3-hydroxypicolinoyl exocyclic fragment. Such streptogramins include virginiamycin and pristinamycin.[34] Pristinamycin, made by *Streptomyces pristinaespiralis*, is a mixture of a cyclodepsipeptide (pristinamycin I) and a polyunsaturated macrolactone (pristinamycin II). Although the natural streptogramins are poorly water-soluble and cannot be used intravenously, new derivatives have been made by semi-synthesis and mutational biosynthesis. Synercid (=RP59500) is a mixture of two water-soluble semi-synthetic streptogramins, quinupristin (=RP57669) and dalfopristin (=RP54476), is in the approval phase of development for resistant bacterial infections.[35] The two Synercid components synergistically (100-fold) inhibit protein synthesis and are active against vancomycin-resistant enterococci and methicillin-resistant staphylococci.[36] Another product, called RPR106972, is being developed for oral treatment of community-acquired infections and is in phase II clinical trials. It is a co-crystalline association of two minor natural products, pristinamycin I$_B$ and pristinamycin II$_B$, produced by *Streptomyces pristinaespiralis*. Another potentially useful compound is a new everninomycin derivative called ziracin (=Sch27899) which is being developed for drug-resistant Gram-positive infections.[37] Everninomycin was discovered in 1979, but this narrow spectrum molecule, containing oligosaccharide and aromatic moieties, was never developed. New screens for further discovery include those for inhibitors of bacterial signal peptidases,[38] non-β-lactam inhibitors of β-lactamase, inhibitors of lipid A biosynthesis and inhibitors of tRNA sythetases.[39]

A new potential for antibiotics is the treatment of ulcers caused by *Helicobacter pylori*. Hundreds of millions of people throughout the world are infected with this Gram-negative bacterium which causes most gastric and duodenal ulcers. Broths of an actinomycete, *Amycolata* sp., were reported to contain eight novel quiniolones which inhibit *Helicobacter pylori* at 0.1 ng/ml while not inhibiting other bacteria.[40]

Fungal infections are a real problem today, having doubled from the 1980s to the 1990s, with bloodstream infections increasing 5-fold and an observed mortality of 55%.[41] There is an increasing incidence of candidiasis, cryptococcosis and aspergillosis especially in AIDS patients.[42] Aspergillosis failure rates exceed 60%. Fungal infections occur often after transplant operations: 5% for kidney, 15-35% for heart and lung, up to 40% for liver transplants, usually (80%) by *Candida* and *Aspergillus* spp.[43] Pulmonary aspergillosis is the main factor involved in death of recipients of bone marrow transplants and *Pneumocystis carinii* is the number one cause of death in patients with AIDS from Europe and North America.[44] Current treatments include the synthetic azoles (e.g., fluconazole and flucytosine) or the natural polyenes (e.g., amphotericin B). However usage is becoming limited by resistance development to the azoles and toxicity of the polyenes. A possible answer to this problem are the echinocandins produced by *Aspergillus nidulans* var. *echinulatus*[45] and the structurally-related pneumocandins produced by *Zalerion arboricola*.[46] These cyclic lipopeptides inhibit (1,3)-β-glucan synthase and thus biosynthesis of the 1,3 glucan layer of the *Candida albicans* cell wall; they are relatively non-toxic. Semi-synthetic versions of echinocandin and pneumocandin are in phase II/III clinical tests. The pneumocandin clinical entry is L-743,872 (=MK-0991) which has been improved in water solubility, antifungal spectrum and potency.[47] The echinocandin candidate is LY303366, which has been improved in spectrum and bioavailability. The spectra of these compounds include many *Candida* species such as *Candida albicans* and many aspergilli including *Aspergillus fumigatus*. In contrast to the currently used azoles which are fungistatic, these are fungicidal.

3 AGRICULTURAL ANTIBIOTICS (FUNGICIDES)

A number of synthetic fungicides have been removed from the agricultural market due to safety and environmental problems. The post-harvest diseases which were their targets cause losses of 25 to 50 % of harvested fruits and vegetables. Fortunately, a number of agricultural fungicides modeled after natural products are in use. The β-methylacrylates were designed based on the structure of the antifungal oudemansins and strobilurins produced by basidiomycetes such as *Oudemansiella mucida* and *Strobiluris tenacellus* respectively.[48-50] These compounds inhibit mitochondrial respiration of fungi by specifically binding to cytochrome b.[51] Another commercial fungicide, Fenpiclonil (=Beret[R]) is a semisynthetic version of the *Pseudomonas* antibiotic, pyrrolnitrin.[52]

Polyoxins are peptide-nucleosides that are substrate analogs of UDP-N-acetylglucosamine which is the essential building block of chitin, this polymer making up 1% of the yeast cell wall. Polyoxins have been used in agriculture for many years and act by inhibiting chitin synthetase.

4   NOVEL ACTIVITIES OF ANTIBIOTICS AND OTHER SECONDARY METABOLITES

An important concept for the further development of natural products is that compounds which possess antibiotic activity also possess other activities. Some of these activities had been quietly exploited in the past, and it became clear in the 1980s that such broadening of scope should be further expanded. Thus, a broad screening of antibiotically-active molecules for antagonistic activity against organisms other than microorganisms, as well as for activities useful for pharmacological or agricultural applications, was proposed in order to yield new and useful lives for "failed antibiotics" (for a review, see Demain[53]) . This resulted in the development of a large number of simple *in vitro* laboratory tests [e.g., enzyme inhibition screens[54, 55]] to detect, isolate and purify useful compounds. Fortunately, we entered into a new era in which microbial metabolites were applied to diseases heretofore only treated with synthetic compounds, i.e. diseases not caused by bacteria and fungi, and huge successes were achieved.

4.1 Anticancer drugs

Most of the important antitumor compounds used for chemotherapy of tumors are microbially-produced antibiotics.[56, 57] These include actinomycin D, mitomycin, bleomycins and the anthracyclines, daunorubicin and doxorubicin. The recent successful molecule, taxol (=paclitaxel), was discovered in plants[58] but also is a fungal metabolite.[59] It is approved for breast and ovarian cancer and is the only antitumor drug known to act by blocking depolymerization of microtubules. In addition, taxol promotes tubulin polymerization and inhibits rapidly dividing mammalian cancer cells.[60] It also inhibits fungi such as *Pythium*, *Phytopthora* and *Aphanomyces* spp. by the same mechanism.[61]

A provocative recent finding is that neomycin inhibits human angiogenin-induced angiogenesis in human endothelial cells.[62] The mechanism appears to involve neomycin's ability to inhibit phospholipase C. Amazingly, other aminoglycosides (gentamicin, streptomycin, kanamycin, amikacin and paromomycin) are inactive even though paromomycin differs from neomycin by merely having -OH at position 6 of the glucose ring instead of -NH$_2$.

4.2 Enzyme inhibitors

4.2.1 *Hypocholesterolemic agents*. The extremely successful hypocholesterolemic agents of the statin family have antifungal activities especially against yeasts. Brown *et al.* discovered the first member of this group, compactin (ML-236B; mevastatin) as an antibiotic product of *Penicillium brevicompactum*.[63] All members of the group are substituted hexahydronaphthalene lactones. Independently, Endo *et al.* discovered compactin in broths of *Penicillium citrinum* as an inhibitor of 3-hydroxy-3-methylglutaryl coenzyme A reductase, the regulatory and rate-limiting enzyme of cholesterol biosynthesis.[64] Later, Endo[65] and Alberts *et al.*[66] independently discovered the more active methylated form of compactin known as lovastatin (monacolin K; mevinolin) in broths of *Monascus ruber* and *Aspergillus terreus* respectively. The statins inhibit *de novo* production of cholesterol in the liver, the major source of blood cholesterol. High blood cholesterol leads to atherosclerosis, which is a causal factor in many types of coronary heart disease, a leading cause of human death. Lovastatin was approved by the FDA in 1987 when clinical tests in humans showed a lowering of total blood cholesterol of 18 to 34%, a 19-39% decrease in low-density lipoprotein cholesterol ("bad cholesterol") and a slight increase in high-density lipoprotein cholesterol ("good cholesterol"). This compound and its derivatives have been a huge commercial and clinical success. One such successful derivative is pravastatin which is produced by hydroxylation of compactin using actinomycetes.[67, 68]

4.2.2 *Other enzyme inhibitors*. In addition to the enzyme inhibitors used to lower cholesterol, others have also succeeded in the world of medicine. These include clavulanic acid, desferal and acarbose. Clavulanic acid is a β-lactam with poor antibiotic activity, produced by *Streptomyces clavuligerus*.[69] It is an inhibitor of penicillinase, and is thus included with penicillins in combination therapy of penicillin-resistant bacterial infections.[70] Desferal is a siderophore produced by *Streptomyces pilosus*.[71] Its high level of metal binding activity has led to its use in iron-overload diseases (hemochromatosis), and aluminum overload in kidney dialysis patients. Acarbose is used as an inhibitor of intestinal α-glucosidase in diabetes and hyperlipoproteinemia. It is produced by *Actinoplanes* sp. SE50.[72] Additional enzyme inhibitors include bialaphos (see Herbicides below), polyoxin and nikkomycin [see Insecticides below and Agricultural Antibiotics (Fungicides) above], and echinocandins and pneumocandins (see Medical Antibiotics above).

4.3 Immunosuppressants

Cyclosporin A was originally discovered as a narrow spectrum antifungal peptide produced by the mold, *Tolypocladium inflatum*.[73] Discovery of its immunosuppressive activity led to its use in heart, liver and kidney transplants and to the overwhelming success of the organ transplant field. Although cyclosporin A had been the only product on the market for many years, two other products, produced by actinomycetes, are providing new opportunities. These are FK-506 (=tacrolimus)[74] and rapamycin,[75] both polyketide narrow spectrum antifungal agents, which are 100-fold more potent that cyclosporin as immunosuppressants and less toxic. FK-506 was approved for use a few years ago and rapamycin awaits FDA approval after completing clinical trials. Studies on the mode of action of these immunosuppressive agents have markedly expanded current knowledge of T-cell activation and proliferation.[76] Rapamycin, FK-506 and cyclosporin A all act by a new mechanism of drug action, i.e., interaction with an intracellular protein (an immunophilin) thus forming a novel complex which selectively disrupts signal transduction events of lymphocyte activation.[77] By binding to its immunophilin (FKBP), rapamycin inhibits a unique growth regulation path utilized by lymphocytes in responding to several cytokines. A previously unknown protein called mTOR (a member of the family of lipid/protein kinases) is part of the rapamycin-sensitive signal transduction pathway.

A very old broad-spectrum compound, mycophenolic acid, first discovered in 1896 and never commercialized as an antibiotic, has recently been developed as a new immunosuppressant. Its morpholinoethylester is the commercial immunosuppressant molecule.[78] Mycophenolic acid was isolated in the crystalline state in 1896 by Gosio and was the first pure compound to show antibiotic activity. This product of *Penicillium glaucum* was shown to inhibit the growth of *Bacillus anthracis* and was then forgotten, until rediscovered in 1913 and given its name.[79] Before being developed into an approved immunosuppressant, mycophenolic acid was used to treat psoriasis.

4.4 Antiparasitic Agents

One of the major economic diseases of poultry is coccidiosis caused by species of the parasitic protozoan, *Eimeria*. For years, this disease was treated using only synthetic chemicals and indeed only synthetic compounds were screened for coccidiostat activity. Although they were generally effective, resistance developed rapidly in the coccidia and new chemical modifications of the existing coccidiostats had to be made. Surprisingly, a parenterally toxic and narrow spectrum antibiotic, monensin, was found to have extreme potency against coccidia.[80] Even with this finding, there were grave doubts that the fermentation process for this polyether compound could be improved to the point where monensin would become economical for use on the farm. However, the power of industrial genetics and biochemical engineering is so great that there is almost no limit to the improvements possible in a fermentation process. As a result, the polyether antibiotics, especially monensin (produced by *Streptomyces cinnamonensins*), lasalocid (produced by *Streptomyces lasaliensis*) and salinomycin (produced by *Streptomyces albus*)[81] dominate the commercial coccidiostat field today.

An interesting sidelight of the monensin story is the discovery of its further use as a growth promotant in ruminants. For years, synthetic chemicals had been screened for activity in cattle and sheep diets to eliminate the wasteful methane production and to increase volatile fatty acid formation (especially propionate) in the rumen, which would improve feed efficiency. Although the concept was sound, no useful products resulted. Experimentation with monensin showed that polyethers have this activity and these compounds are used widely today for this purpose.[81]

Another major agricultural problem has been the infection of farm animals by worms. The predominant type of screening effort over the years was the testing of synthetic compounds against nematodes and some commercial products did result. Certain antibiotics had also been shown to possess antihelmintic activity against nematodes or cestodes but these failed to compete with the synthetic compounds. By examining microbial broths for antihelmintic activity, US investigators found a non-toxic fermentation broth which killed the intestinal nematode, *Nematosporoides dubius*, in mice. The *Streptomyces avermitilis* culture, which was isolated by Omura and coworkers at the Kitasato Institute in Japan,[82] produced a family of secondary metabolites having both antihelmintic and insecticidal activities which were named "avermectins". They are disaccharide derivatives of macrocyclic lactones with exceptional activity against parasites, i.e., at least ten times   higher than any synthetic antihelmintic agent known. Despite   their   macrolide   structure, avermectins lack activity against bacteria and fungi, do not inhibit protein synthesis nor are they ionophores; instead they interfere with neurotransmission in many invertebrates. They have activity against both nematode and arthropod parasites in sheep, cattle, dogs, horses, and swine. A semi-synthetic derivative, 22, 23-dihydroavermectin B_1 (Ivermectin) is one thousand times more active than the synthetic antihelmintic, thiobenzole, and is a commercial veterinary product. An excellent review of avermectin biosynthesis have been published by Ikeda and Omura.[83] The avermectins are closely related to the milbemycins, a group of non-glycosidated macrolides produced by *Streptomyces hygroscopicus* subsp. *aureolacrimosus.*[84] These compounds possess activity against worms and insects. A new avermectin, called Doramectin (=cyclohexylavermectin B_1), was recently developed by the technique of mutational biosynthesis.[85] Indeed it was the first commercially successful example of mutational biosynthesis. The natural avermectins contain C-25 side chains of 2-methyl butyryl ("a" components) or isobutyryl ("b" components). Branched-chain 2 keto acid dehydrogenase was eliminated from the parent culture by knocking out gene *bkd*.[86] This enzyme normally supplies the 2-methylbutyryl-CoA and isobutyryl-CoA starter units of avermectins from isoleucine and valine respectively.[87, 88] The mutant produced no avermectin unless fed isobutyric acid or (S)-2-methylbutyric acid. Upon feeding other fatty acids, novel avermectins were made.[89] More than 800 fatty acids were tested, yielding over 60 avermectins including cyclohexyl B_1 avermectin (Doramectin), which resulted from incorporation of cyclohexane carboxylic acid. Doramectin is claimed to have some commercial advantages over Ivermectin.[90]

A fortunate fallout of the avermectin work was the finding that Ivermectin has activity against the black fly vector of human onchocerciasis ("river blindness"). It interferes with transmission of the filarial nematode, *Onchocerca volvulus*, to the human population. Since 40 million people are affected by this disease, the decision by Merck to supply ivermectin free of charge to the World Health Organization, for use in humans in the tropics, was met with great enthusiasm and hope for conquering this parasitic disease.

4.5 Herbicides

The agricultural use of synthetic chemicals as herbicides has worried many environmentalists. Two widely used herbicides, alachlor (Lasso) and atrazine (AAtrex) have been reported to cause cancer in long-term animal tests. To fill the void, antibiotics have been considered for use as agricultural herbicides. One such herbicide, bialaphos, which is active against broad-leaved weeds, was developed in Japan.[91] This *Streptomyces viridochromogenes* product is N-{4[hydroxy(methyl)phosphinoyl]homoalanyl}alanyl-alanine. It was first discovered by Zahner's group in Germany as a broad-spectrum antibiotic active against bacteria and *Botrytis cinerea.*[92] Its hydrolysis product, DL-homoalanin-4-yl(methyl) phosphinic acid (=DL-phosphinothricin) is a glutamine synthetase inhibitor which is marketed as glufosinate (=Basta). Although sold as the racemate, only the L-form is active. Of great interest to environmentalists is that bialaphos is easily degraded in the environment, having a half-life only two hours.[93]

4.6 Mycotoxins

It is difficult to conceive that even poisons can be harnessed as medically-useful drugs, yet this is the case with the ergot alkaloids. These mycotoxins (i.e., toxins produced by molds) were responsible for fatal poisoning of humans and animals (ergotism) throughout the ages after consumption of bread   made from grain contaminated with species of *Claviceps*. The ergot alkaloids   were responsible in the Middle Ages for the disease known in Europe as "Holy Fire" or "St. Antony's Fire". This widespread epidemic disease produced gangrene, cramps, convulsions and hallucinations. These early names of the disease relate to the care of patients by the monks of the Antoniter Brotherhood. A major epidemic occurred in the USSR during the famine of 1926-1927. It is amazing but true that these "poisons" are now used for angina pectoris, hypertonia, migraine headache, cerebral circulatory disorder, uterine contraction, hypertension, serotonin-related disturbances, inhibition of prolactin release in agaloactorrhoea, reduction in bleeding after childbirth, and for prevention of implantation in early pregnancy.[79] Among their physiological activities are the inhibition of action of adrenalin, noradrenalin and serotonin and the contraction of smooth muscles of the uterus. Some of the ergot alkaloids possess antibiotic activity.

With the above in mind, one wonders how many potentially useful compounds lie in that huge mountain of secondary metabolites which we call mycotoxins. Ergot alkaloids are not the only successful example. The genus *Gibberella* is responsible for two others, zearelanone and gibberellins. Zearelanone, produced by *Gibberella zeae* (syn. *Fusarium graminearum*),[94] is an estrogen and its reduced derivative zeranol is used as an anabolic agent in cattle and sheep, increasing both growth and feed efficiency. Gibberellic acid, a member of the phytotoxic mycotoxin group known as the gibberellins, is produced by *Gibberella fujikuroi* and is the cause of the "foolish rice seedling" disease of rice.[95] In this disease, the plant grows so abnormally fast that it virtually dies of exhaustion. Gibberellins have been successfully exploited in regulating the growth of plants. They are used to speed up barley malting, improve malt quality, increase the yield of vegetables and cut in half the time required to obtain lettuce and sugar beet seed crops. It is not known whether the gibberellins have any antibiotic activity.

4.7 Insecticides

Bioinsecticides are only a small part of the insecticide field but their market is increasing.[96] They cost somewhat more than chemical insecticides and have narrow specificity and short life in the field. By 1984, fourteen agents had been approved by the Environmental Protection Agency (EPA) (6 bacterial, 4 viral, 3 fungal and 1 protozoan). The bioinsecticide field has been dominated by a non-antibiotic product, the protein crystal of *Bacillus thuringiensis*. The selective toxicity of this protein (i.e., the delta-endotoxin; BT toxin) against insects of the order *Lepidoptera* has been exploited successfully in agriculture for many years. Indeed the fantastic selectivity of its toxicity against these insects has limited its commercial success because agriculture has been spoiled by broad-spectrum chemical insecticides. However, the world is becoming wary of the ecological damage done by many synthetic chemicals and the resistance which is developing to them in insects. On the other hand, the insecticidal toxin of *Bacillus thuringiensis* has not disturbed the environment and only little resistance has developed.[97] BT toxin has played a major role in the development of agricultural biotechnology in that its gene, when transferred to the plant, renders them resistant to certain insects. Other intriguing applications include the activities of different strains of *Bacillus thuringiensis* and other species of bacilli such as *Bacillus sphaericus* and *Bacillus popillae* against other insects.[98]

With regard to smaller metabolites, the nikkomycin antibotics are of considerable interest.[99] These are nucleoside analogues, structurally related to the polyoxins which are used as agricultural antifungal agents. Since these compounds function as inhibitors of chitin synthetase and since chitin is an important structural material for insects, the nikkomycins   have   potent   insecticidal   activity.   The   macrotetrolide, tetranactin,[100] has been in use   since 1974 as a mitocide for plants.

An exciting new development is the discovery of a new family of bioinsecticides, the spinosyns. These are sold today as Spinosad. They are non-toxic, environmentally friendly tetracyclic macrolides produced by *Saccharopolyspora spinosa* with activity against insects of the orders Coleoptera, Diptera, Hymenoptera, Isoptera, Lepidoptera, Siphonoptera and Thysanoptera but without antibacterial activity.[101]

5 CLOSING COMMENTS

There has been a major change in the field of discovery and application of secondary metabolites over the past twenty years. This change is characterized by the broadening of the scope of the search. No longer are microbial sources looked upon solely as potential solutions for microbial diseases. This change in screening philosophy has been followed by ingenious applications of molecular biology to detect receptor antagonists and agonists and other agents inhibiting or enhancing cellular activities on a molecular level.[102]

Lessons that have been learned are as follows: (i) even an antibiotic screen can exploit the antibiotic activity of a molecule of interest for non-antibiotic application in that the screening procedure might be used as a simple assay during purification and drug metabolism studies; (ii) if the candidate product does not fulfill its promise for the targeted application, the presence of antibiotic activity may indicate wide bioactivity and suggest high priority status for broad-based pharmacological, molecular, microbiological and agricultural testing of the compound or mixture of compounds; (iii) companies which maintain a library of previously discovered antibiotics which never quite made it in the world of clinical antibiotics should have these compounds or mixtures screened broadly for non-antibiotic application; and (iv) deeper examination of known mycotoxins could lead to the development of new drugs or agents for agriculture or even medicine.

Some people in the field feel that combinatorial chemistry will replace natural product screening for discovery of new drugs. Some companies are even dropping their natural product programs to support combinatorial chemistry efforts. Such moves are ill-conceived since the role of combinatorial chemistry, like those of structure-function drug design and recombinant DNA technology two and three decades ago, is that of complementing and assisting natural product discovery and development, not replacing them. The main factors for the discovery of useful compounds of the future are commitment, ingenuity, ability to exploit nature's biodiversity and to devise simple *in vitro* high-throughput screening procedures for desirable activities.

References

1 G. L. Verdine, *Nature*,1996, **384 Suppl**, 11.
2 G. M. Cragg, D. J. Newman and K. M. Snader, *J. Nat. Prod.*, 1997, **60**, 52.
3 H. D. Holland, *Science*, 1998, **275**, 38.
4 R. Mendelson and M. J. Balick, *Econ. Bot.*, 1995, **49**, 223.
5 C. A. Roessner and A. I. Scott, *Ann. Rev. Microbiol.*, 1996, **50**, 467.
6 W. Fenical and P. R. Jensen, in: *Marine Biotechnology I: Pharmaceutical and Bioactive Natural Products*, D. H. Attaway and O. R. Zaborsky, Eds., Plenum, NY, USA, 1993.
7 J. Berdy, Proc. 9th Internat Symp Biol Actinomycetes, Part 1, 3-23. Allerton Press Inc., NY, USA, 1995.
8 J. D. Connolly and R. A.   Hill, *Dictionary of Terpenoids*, Chapman & Hall, London, UK, 1991.

9   L. N. Mander, *Chem. Rev.*, 1992, **92**, 573.

10   R. L. Rawls, *Chem. & Eng. News*, 1998, **76** (10), 29.

11   A. M. Clark, *Pharmaceut. Res.*, 1996, **13**, 1133.

12   C. R. Hutchinson, in: *Biotechnology of Antibiotics*, 2nd ed., W. Strohl, Ed., pp 683-702. Marcel Dekker, NY, USA, 1997.

13   J. McAlpine, in: *Natural Products II: New Technologies to Increase Efficiency and Speed*, D. M. Sapienza and L. M. Savage, Eds., pp 251-278. Internat. Bus. Comm., Southborough, USA, 1998.

14   D. L. Hawksworth, *Mycol. Res.*, **95**, 641.

15   V. Torsvik, R. Sorheim and J. Goksoyr, *J. Ind. Microbiol.*, 1996, **17**, 170.

16   C. Möller, G. Weber and M. M. Dreyfuss, *J. Ind. Microbiol.*, 1996, **17**, 359.

17   W. Strohl, in: *Biotechnology of Antibiotics*, 2nd ed., W. Strohl, Ed., pp 1-47, Marcel Dekker, NY, USA, 1997.

18   I. D. Fleming, L. J. Nisbet and S. J. Brewer, in: *Bioactive Microbial Products: Search and Discovery*, J. D. Bu'lock , L. J. Nisbet and D. J. Winstanley , Eds., pp 107-130, Academic Press, London, UK, 1982.

19   H. B. Woodruff, S. Hernandez and E. D. Stapley, *Hindustan Antibiot. Bull.*, 1979, **21**, 71.

20   H. B. Woodruff and L. E. McDaniel, *Soc. Gen. Microbiol. Symp.*, 1958, **8**, 29.

21   C. G. Jones and R. D. Firn, *Phil. Trans. R. Soc.*, 1991, **333**, 273.

22   J. DiMasi, M. Seibring and L. Lasagna, *Clin. Pharmacol. Ther.*, 1994, **55**, 609.

23   F. S. Kahan, F. M. Kahan, R. T. Goegelman, S. A. Currie, M. Jackson, E. O. Stapley, T. W. Miller, A. K. Miller, D. Hendlin, S. Mochales, S. Hernandez, H. B. Woodruff and J. Birnbaum, *J. Antibiot.*, 1979, **32**, 1.

24   F. C. Tenover and J. M. Hughes, *JAMA*, 1996, **275**, 300.

25   C. Stephens and L. Shapiro, *Chem. & Biol.*, 1997, **4**, 637.

26   A. Morris, J. D. Kellner and D. E. Low, *Curr. Opin. Microbiol.*, 1998, **1**, 524.

27   P.-E. Sum, F.-W. Sum and S. J. Projan, *Curr. Pharm. Des.*, 1998, **4**, 119.

28   R. Breiman, J. Butler, F. Tenover, J. Elliot and R. Facklam, *JAMA*, 1994, **271**, 1831.

29   D. A. Goldman, R. A. Weinstein, R. P. Wenzel, O. C. Tablan, R. J. Duma, R. P. Gaynes, J. Schlosser and W. J. Martone, *JAMA*, 1996, **275**, 234.

30   R.C. Moellering, 37th *Intersci. Conf. Antimicrob. Agents Chemother.*, Toronto, Canada, 1997.

31   M. Vaara, *Antimicrob. Agents. Chemother.*, 1993, **37**, 2255.

32   H. Nikaido, *J. Bacteriol.*, 1996, **178**, 5853.

33   K. Lewis, *TIBS*, 1994, **19**, 119.

34   J. C. Barriere, N. Berthaud, D. Beyer, S. Dutka-Malen, J. M. Paris and J. F. Desnottes, *Curr. Pharm. Des.*, 1998, **4**, 155.

35   T. Nichterlein, M. Kretschmar and H. Hof, *J. Chemother.*, 1996, **8**, 107.

36   S. C. Stinson, *Chem. & Eng. News*, 1996, **74**, (39), 75.

37   P. Stead, *Drug Discov. Today*, 1997, **2**, 256.

38   M. T. Black and G. Bruton, *Curr. Pharmaceut. Design*, 1998, **4**,133.

39   K. Bush, *Curr. Opin. Chem. Biol.*, 1997, **1**, 169.

40   K. A. Dekker, T. Inagaki, E. Nomura, T. Gootz, L. H. Huang, W. E. Kohlbrenner, Y. Matsunaga, P. R. McGuirk, T. Sakakibara, S. Sakemi, Y. Suzuki, Y. Yamauchi, Y. Kojima and N. Kojima, Abstr P22, 5th *Internat. Conf. Biotechnol. Microb. Prods.: Novel Pharmacol. Agrobiol. Act.*, Williamsburg, USA, 1997.

41 F. A. Tally, P. A. Wendler and F. Houman, Abstr. S5, *5ᵗʰ Internat. Conf. Biotechnol. Microb. Prods.: Novel Pharmacol. Agrobiol. Act.*, Williamsburg, USA, 1997.

42 T. C. White, *ASM News*, 1997, **63**, 427.

43 B. P. Alexander and J. R. Perfect, *Drugs*, 1997, **54**, 657.

44 N. H. Georgopapadakou, *Curr. Opin. Microbiol.*, 1998, **1**. 547.

45 M. Debono, *Exp. Opin. Invest. Drugs*, 1994, **3**, 821.

46 R. E. Schwartz, D. F. Sesin, H. Joshua, K. E. Wilson, A. J. Kempf, K. A. Goklen, D. Kuehner, P. Gailliot, C. Gleason, R. White, E. Inamine, G. Bills, P. Salmon and L. Zitano, *J. Antibiot.*, 1992, **45**, 1853.

48 T. Anke, F. Oberwinkler, W. Steglich and G. Schramm, *J. Antibiot.*, 1977, **30**, 806.

49 T. Anke, H. J. Hecht, G. Schramm and W. Steglich, *J. Antibiot.*, 1979, **32**, 1112.

50 A. Miller, *Chem. Ind.*, 1997, (1): 7.

51 J. M. Clough, *Nat. Prod. Rep.*, 1993, **10**, 565.

52 S. B. Rees and K. A. Powell, in *Dev. Ind. Microbiol.*, C. Nash, J. Hunter-Cevera, R. Cooper, P. E. Eveleigh and R. Hamill, Eds., **32**, pp 73-78, Wm. C Brown, Dubuque, USA, 1992.

53 A. L. Demain, *Science*, 1983, **219**, 709.

54 H. Umezawa, *Enzyme Inhibitors of Microbial Origin*, University Park Press, Baltimore, USA, 1972.

55 H. Umezawa, *Annu. Rev. Microbiol.*, 1982, **36**, 75.

56 T. Oki and A. Yoshimoto, in: *Annual Reports on Fermentation Processes*, D. Perlman, Ed., **3**, pp 215-251, Academic Press, NY, USA, 1979.

57 M. Tomasz, *Curr. Biol.*, 1995, **2**, 575.

58 M. E. Wall and M. C. Wani, *Cancer Res.*, 1995, **55**, 753.

59 A. Stierle, G. Strobel and D. Stierle, *Science*, 1993, **260**, 214.

60 J. J. Manfredi and S. B. Horowitz, *Pharmacol. Ther.*, 1984, **25**, 83.

61 G. A. Strobel and D. M. Long, *ASM News*, 1998, **64**, 263.

62 G. F. Hu, *Proc. Natl. Acad. Sci. USA*, 1998, **95**, 9791.

63 A. G. Brown, T. C. Smale, T. J. King, R. Hasenkamp and R. H. Thompson, *J. Chem. Soc. Perkin Trans. I,* 1976, 1165.

64 A. Endo, M. Kuroda and Y. Tsujita, *J. Antibiot.*, 1976, **29**, 1346.

65 A. Endo, *J. Antibiot.*, 1979, **32**, 852.

66 A. W. Alberts, J. Chen, G. Kuron, V. Hunt, J. Huff, C. Hoffman, J. Rothrock, M. Lopez, H. Joshua, E. Harris, A. Patchett, R. Monaghan, S. Currie, E. Stapley, G. Albers-Schonberg, O. Hensens, J. Hirshfield, K. Hoogsteen, J. Liesch and J. Springer, *Proc. Natl. Acad. Sci. USA*, 1980, **77**, 3957.

67 N. Serizawa and T. Matsuoka, *Biochim. Biophys. Acta*, 1991, **1084**, 35.

68 Y. Peng and A. L. Demain, *J. Indust. Microbiol. Biotechnol.*, 1998, **20**, 373.

69 J. Birnbaum, E. O. Stapley, A. K. Miller, H. Wallick, D. Hendlin and H. B. Woodruff, *J. Antimicrob. Chemother.*, 1978, **4** Suppl **B**, 15.

70 A. G. Brown, *Drug Design Deliv.*, 1986, **1**, 1.

71 G. Winkelmann, in: *Biotechnology*, H. J. Rehm and G. Reed, Eds., **4**, pp 215-243, VCH, Weinheim, Germany, 1986.

72 E. Truscheit, W. Frommer, B. Junge, L. Müller, D. D. Schmidt and W. Wingender, *Angew. Chem. Intl. Ed. Engl.*, 1981, **20**, 744.

73 J. F. Borel, C. Feurer, H. U. Gabler and H. Stahelin, *Agents & Actions*, 1976, **6**, 468.

74 T. Kino, H. Hatanaka, M. Hashimoto, M. Nishiyama, T. Goto, M. Okuhara, M. Kohsaka, H. Aoki and H. Imanaka, *J. Antibiot.*, 1987, **40**, 1249.

75 D. Vezina, A. Kudelski and S. N. Sehgal, *J. Antibiot.*, 1975, **28**, 721.

76 J. Liu, *TIPS*, 1993, **14**, 182.

77 F. J. Dumont and Q. Su, *Life Sci.*, 1996, **58**, 373.

78 W. A. Lee, L. Gu, A. R. Kikszal, N. Chu, K. Leung and P. H. Nelson, *Pharmaceut. Res.*, 1990, **7**, 161.

79 R. Bentley, *Persp. Biol. & Med.*, 1997, **40**, 364.

80 M. E. Haney Jr. and M. M. Hoehn, *Antimicrob. Agents Chemother.*, 1968, **1967**, 349.

81 J. W. Westley, *Adv. Appl. Microbiol.*, 1977, **22**, 177.

82 E.O. Stapley, in *Trends in Antibiotic Research*, H. Umezawa, A.L. Demain, R. Hata and C.R. Hutchinson, Eds., pp 154-170, Jpn. Antibiot. Res. Assoc., Tokyo, Japan, 1982.

83 H. Ikeda and S. Omura, *Chem. Rev.*, 1997, **97**, 2591.

84 H. Mishima, J. Ide, S. Muramatsu and M. Ono, *J. Antibiot.*, 1983, **36**, 980.

85 H.I.A. McArthur, in *Developments in Industrial Microbiology-BMP '97*, C.R. Hutchinson and J. McAlpine, Eds., pp 43-48, Soc. Ind. Microbiol., Fairfax, USA, 1998.

86 E. W. Hafner, B. W. Holley, K. S. Holdom, S. E. Lee, R. G. Wax, D. Beck, H. McArthur and W. C. Wernau, *J. Antibiot.*, 1991, **44**, 349.

87 T. Chen, B. Arison, V. Gullo and E. Inamine, *J. Indust. Microbiol.*, 1989, **4**, 231.

88 D.J. MacNeil, J.L. Occi, K.M. Gewain and T. MacNeil, *Ann. NY Acad. Sci.*, 1994, **721**, 123.

89 C. Dutton, S. Gibson, A. Goudie, K. Holdom, M. Pacey, J. Ruddock, J. D. Bu'Lock and M. K. Richards, *J. Antibiot.*, 1991, **44**, 357.

90 A. C. Goudie, N. A. Evans, K. A. F. Gration, B. F. Bishop, S. P. Gibson, K. S. Holdom, B. Kaye, S. R. Wicks, D. Lewis, A. J. Weatherley, C. I. Bruce, A. Herbert and D. J. Seymour, *Vet. Parasitol.*, 1993, **49**, 5.

91 Y. Kondo, T. Shomura, Y. Ogawa, T. Tsuruoka, H. Watanabe, K. Totukawa, T. Suzuki, C. Moriyama, J. Yoshida, S. Inouye and T. Nida, *Sci. Rep. Meiji Seika*, 1973, **13**, 34.

92 E. Bayer, K. H. Gugel, K. Hagele, H. Hagenmaier, S. Jessipow, W. A. Konig and H. Zahner, *Helv. Chim. Acta*, 1972, **55**, 224.

93 K. Tachibana, in: *Pesticide Science and Biotechnology*, R. Greenhalgh and T. R. Roberts, Eds., pp146-148, Blackwell, Oxford, UK, 1987.

94 P. Hidy, R. S. Baldwin, R. L. Greasham, C. L. Keith and J. R. McMullen, *Adv. Appl. Microbiol.*, 1977, **22**, 59.

95 E. G. Jefferys, *Adv. Appl. Microbiol.* 1970, 13, 283.

96 A. Klausner, *Bio/Technology*, 1984, **2**, 408.

97 A. I. Aronson, W. Beckman and P. Dunn, *Microbiol. Revs.*, 1986, **50**, 1.

98 L. K. Miller, A. J. Lingg and L. A. Bulla Jr., *Science*, 1983, **219**, 715.

99 G. U. Brillinger, *Arch. Microbiol.*, 1979, **121**, 71.

100 H. Oishi, T. Sugawa, T. Okutomi, K. Suzuki, T. Hayashi, M. Sawada and K. Ando, *J. Antibiot.*, 1970, **23**, 105.

101 H.A. Kirst, K.H. Michel, J.S. Mynderase, E.H. Chio, R.C. Yao, W.M. Nakasukasa, L.D. Boeck, J.L. Occlowitx, J.W. Paschal, B. Deeter and G.D. Thompson, in *Synthesis and Chemistry of Agrochemicals III*, D.R. Baker, J.G. Fenyes and J.J. Steffens, Eds., pp 214-225, ACS Symp. Ser. 504, ACS, Washington, USA, 1992.

102 H. Tanaka and S. Omura, in: *Biotechnology,* 2nd ed, H.J. Rehm, G. Reed, H. Kleinkauf and H. von Doehren, Eds.,   7, pp 107-132, VCH, Weinheim, Germany, 1997.

NEW CONCEPTS AND APPROACHES TO BIODIVERSITY

D.F.Marshall and J.R. Hillman

Scottish Crop Research Institute, Invergowrie, Dundee, DD2 5DA

1 INTRODUCTION

Biodiversity has become a word with a multiplicity of meanings that can be applied at many levels of scale from global to local. It is a term that can be adapted to describe the range of species in an ecosystem or to the levels of genetic diversity that are contained within a species.[1]

Biodiversity can essentially be thought of as the balance between the rate of generation of new variation (by mutational and recombinational processes as well as speciation) *versus* the rate of loss through local or global extinction of genetic variation within species or of species themselves. Though most would agree with the notion that the maintenance of biodiversity is a 'good thing', it is difficult to quantify biodiversity or even to identify an optimum biodiversity level for a given habitat or to identify the optimum level of genetic diversity within a particular species. Nevertheless there is now growing concern about the increasing human impact on the rate of loss of biodiversity and the realisation that, without modern technologies, species that go extinct are lost forever.[2,3]

The biological variation that we see is the manifestation of the expression of the range of genes that are continually evolving in the global ecosystem, together with their interactions with the changing biotic and physical environments in which they are found. Much of the current public perception of the biodiversity issue is focused on 'visible' biodiversity in terms of higher plant and animal species-richness. In reality, the biological survival of an individual species is uniquely dependent on the genetic variation that it contains and the microbial environment is a key, often overlooked, element of any ecosystem.

In this article we shall focus for simplicity on the biodiversity of crop plants, though we are aware that even a consideration of the impacts of crop plant biodiversity cannot be treated in isolation from either their pests and pathogens or the other species in their agricultural or natural ecosystems. As well as their role in food chains, human nutrition, industrial feedstock and biomedicine, higher plants can be viewed as providing much of the physical substance of the living landscape. In viewing these issues from a British perspective most, if not all, of our visible landscape is the product of thousands of years of human impact. Even here, the adoption of appropriate actions (e.g. a viable Biodiversity Action Plan (BAP)) is required to mediate the impacts of modern agriculture and forestry as well as other human activities.

Rapidly developing modern molecular technologies help deal with the biodiversity issue. Traditional methodologies such as species inventories can yield valuable information for the first stage of biodiversity analysis. PCR-based DNA technologies provide rapid screening tools to obtain relatively objective measurements of genetic diversity within species, and can, for example, estimate biodiversity in the soil for microbial species yet to be identified. For crop plants, molecular marker technologies have enabled us to track the increasing loss of genetic diversity from wild ancestors, through land races to the advanced varieties of modern agricultural monocultures.[4] Such biodiversity audits have brought the dual benefits of highlighting the dangers of reliance on a dramatically limited set of crop genotypes as well as identifying genes of value to advanced agriculture that have been left behind in the wild gene pool by the bottle-neck of domestication.[5, 6]

Clearly, the availability of a broad gene pool for crop species must be maintained through appropriate *ex situ* and *in situ* strategies. New transgenic techniques[7] mean that the breadth of the available gene pool can now transcend not only species but, subject to ethical and social considerations, even taxonomic phyla. By ensuring that the biodiversity resources of our planet are sustained, the genetic wealth of, not only our crop plants, but even other under-utilised or even undiscovered species, will be available to provide food and other biological resources for future generations. The rapidly developing science of genomics and the post-genomics sciences will provide new ways to catalogue and sympathetically exploit our biodiversity heritage as well as helping us retain or even recover a natural environment that is compatible with a high quality of human life.

2 THE ECONOMIC VALUE OF PLANT BIODIVERSITY

If we consider the role of plants in agriculture we find that, of the estimated 300,000 to 500,000 species of higher plants, only a very small minority are directly exploited as food. Even if we take an extremely liberal view of what constitutes a food species we find that only around 7,000 species (of an estimated total of approximately 30,000 edible species) have ever been exploited as food by man. The situation is even more constrained if we take an objective account of the true impact of these species. The generally held view is that some 30 species are largely responsible for feeding the world. Wheat, rice and maize provide more than half of all plant derived dietary energy input at a global level, though a number of other species have a considerable impact at a regional or local basis.

Given the importance of such a relatively few species for world food supplies, the maintenance of genetic diversity within these species is of crucial importance for global food security. This diversity provides the basic resource for breeding programmes for the development of cultivars that are resistant to current and future pest and pathogen strains or are adapted to environmental stresses such as drought or salinity. A major problem exists, however, in that there is no simple direct measure of relevant genetic diversity within a crop species that is globally applicable. For example, in future we may require genetic diversity for resistance to pest and pathogen strains that are yet to appear in agriculture. How can we identify such diversity now to ensure that changes in land use and agricultural practice do not lead to the loss of relevant resistance genes?

One potential route to achieve this is to establish and maintain genetic resource collections for all key crop species. Currently this operation is underway on a global scale based on *ex situ* (i.e. genebank) or *in situ* (i.e. based on farm or 'wild' reserve) collections.[8] The successful operation of such genetic conservation strategies is uniquely dependent on international co-operation and agreements and is characterised by many notable success stories.[8] There are, however, many operational difficulties in practice. The survival of genetic resource collections requires significant long-term investment. In particular the maintenance of stored seed collections through cycles of regeneration and the management of *in situ* reserves are both labour intensive and require to be undertaken in the context of a lack of simple scale or baseline for genetic diversity. The major difficulty is that we have no simple concept of how much diversity is *enough*. This is partially the result of the complexity of plant genomes. A typical plant species may have as many as 50,000 genes each with many possible allelic states. In practice, the only absolute measure of genetic diversity for agriculturally relevant traits is based on the traits themselves, which for genetic resource collections requires direct evaluation in large scale agronomic trials over many sites and seasons or screens against known races of pests and pathogens. This difficulty has led to the adoption of a growing range of molecular technologies to quantify or partition genetic diversity in crop plants. This approach has, until relatively recently, been based on the exploitation of various classes of 'neutral' genetic markers such as isozymes, RFLPs (Restriction Fragment Length Polymorphisms), AFLPs (Amplified Fragment Length Polymorphisms) and SSRs (Simple Sequence Repeats).[9, 10] These technologies are all based on gel electrophoresis and allow scientists to quantify genetic diversity at a series of sample points in the genome. Their use, in combination with the theories of population genetics, has led to the development of improved strategies for both collection and maintenance of plant germplasm.

There is, however, still a significant problem in that such 'neutral' marker systems, though giving value information about the overall genetic architecture of diversity, do not allow us to gain detailed information about the actual genes that determine key traits. Plant molecular biology has begun to make significant progress towards the characterisation of agriculturally relevant genes in crop plants. This is particularly true in the case of pest and disease resistance.[11, 12] Nevertheless, we have only begun the process of identifying which of the 40 -50,000 genes in the plant are of primary importance in crop improvement.

3 THE IMPACT OF THE NEW 'OMICS' TECHNOLOGIES

The new science of genomics and the imminent availability of the complete sequence of the *Arabidopsis* and rice genomes are changing the way we approach many genetic problems in crop plants. Together with the growing collections of ESTs (or Expressed Sequence Tags) for most of the world's major crop species this whole genome level sequence will underpin new high-throughput technologies in crop diversity analysis and improvement. A particularly valuable feature is the commonality of these new 'omics' technologies across all living organisms. This enables agriculture to benefit from technology advances that are been driven by the large-scale investment associated with molecular medicine and drug discovery.

Already in the field of medical genetics a significant investment has been made in the large scale sequence analysis necessary to develop a large number of markers based on Single Nucleotide Polymorphisms or SNP's.[13] These are based on the alternative occurrence of a two or more nucleotides at a particular position of the DNA sequence and can be readily adapted to the generation of DNA genotyping 'chips'. Though the development of this technology requires a high initial level of investment the range of potential applications of the resulting genotyping chips, based on many thousands of SNPs, is enormous, e.g they offer the prospect of rapidly screening for an extremely wide range of human genetic disorders. There are already programmes underway in both the public and private sectors to develop this technology in the most commercially valuable crop species such as maize. The availability of SNP-based genotyping chips is likely to have enormous impact on both the inventory and exploitation of genetic diversity, offering the prospect of fast-track molecular breeding with dramatically improved efficiency.[14, 15]

New high throughput technologies will also enable us to obtain a much clearer understanding of the molecular basis of phenotype. i.e. they will enable us to characterise the molecular basis of such traits as yield, quality and disease resistance. Already significant progress has been made in the development of technologies to simultaneously monitor the expression patterns of many thousands of genes using expression microarrays.[16, 17] These are based on the immobilisation of many thousands of oligonucleotides or cloned cDNAs in gridded arrays. High throughput methods are also becoming available for the characterisation of most of the proteins in a given tissue sample, using a combination of 2D gel analysis for separation and techniques such as matrix-assisted laser desorption ionisation - time-of-flight (MALDI-TOF) mass spectrometry followed by comparison with a reference database.[18-20]

Though technologically more demanding because of the range of chemical entities that must be analysed, technologies are also becoming available for high-throughput characterisation of the full range of metabolites that are found within living cells.[21] These techniques utilise an array of spectroscopy-based techniques e.g. pyrolysis mass spectrometry, Fourier-transform infrared spectroscopy and dispersive Raman spectroscopy. Their integration with DNA, RNA and protein analysis methods will provide a comprehensive framework for the characterisation of living tissues and the quantification of molecular biodiversity.

The development and utilisation of these new high-throughput technologies, however, requires an increasingly sophisticated data management and analysis infrastructure as well as new software tools.[22] Increasingly, with the rapid advance in molecular technologies, the Bioinformatics component has become the rate-limiting factor. The scale of the data avalanche can be seen from the continued exponential growth in the available DNA and protein sequence in public repositories. The latest, March 2000, release of the EMBL DNA database contains some 23 Gbtyes of data (http://www.ebi.ac.uk) and we are only beginning to contemplate the problems of archiving and exploiting microarray data.

The rapid progress of the full repertoire of 'omics' technologies, provided they can be utilised in a 'cost effective' manner, offers the prospect of the development of a sophisticated scientific framework to efficiently exploit the genetic diversity that is available in crops plants and their relatives in the agriculture of the New Millenium. This exploitation can be undertaken within the gene pool of crop species through well-established conventional routes.[23] Alternatively, an array of molecular techniques are now available for gene transfer to plants,[24] effectively removing any taxonomic restriction on the genepool that is available for a given crop species. Such transgenic approaches may also be utilised to engineer novel pathways to obtain new products from plants.[25]

These new 'omics' technologies will also enable us to efficiently identify 'novel' compounds of value to both the Agrochemical and Pharmaceutical industries. It is of crucial importance that the data that results from high throughput analyses of the genome, transcriptome, proteome and metabolome of crop plants and their relatives is archived in an appropriately structured and indexed manner. This will enable it to be fully integrated and data-mined over an extensive period as our ability to formulate increasingly sophisticated queries on these data sets develops.

References

1   E. O. Wilson, *The diversity of life*, Penguin, London, 1992.
2   G. T. Prance, *Biodiversity and Conservation*, 1995, **4**, 490.
3   Anon., Report of the United Nations Conference on Environment and Development, Rio de Janeiro, 1992.

4   A. H. D. Brown and M. T. Clegg, *Current Topics in Biological and Medical Research*, 1983, **11**, 285.

5   J. R. Harlan, *Scientific American*, 1972, **235**, 89.
6   J. R. Harlan, *Science*, 1975, **188**, 618.
7   A. J. Hamilton, G. W. Lycett and D. Grierson, *Nature*, 1990, **346**, 284.

8   Food and Agriculture Organisation, *The State of the World's Plant Genetic Resources for Food and Agriculture*, Rome, 1998.

9   A. Karp and K. J. Edwards, *Molecular techniques in the analysis of the extent and distribution of genetic diversity*, International Plant Genetic Resources Institute Workshop: *Molecular Markers for Plant Genetic Resources*, Rome, 1995

10   S. Kresovich, J. R. McFerson and A. L. Westman, *Using molecular markers in genebanks*, International Plant Genetic Resources Institute Workshop: *Molecular Markers for Plant Genetic Resources*, Rome, 1995

11   M. S. Dixon, D. A. Jones, J. S. Keddie, C. M. Thomas, K. Harrison and J. D. Jones, *Cell*, 1996, **84**, 451.

12   R. W. Michelmore, *Current Opinions in Biotechnology*, 1995, **6**, 145.

13   J. E. Reynolds, S. R. Head, T. C. MacIntosh, L. P. Vrolijk and M. T. Boyce-Jacino, in *DNA Markers: Protocols, Applications and Overviews*, G. Caetano-Anolles and P. Gresshoff, Eds, Wiley-Liss, 1997, pp199-211.

14   R. Lande, in *Plant Breeding in the 1990s*, H. T. Stalker and J. P. Murphy, Eds., CAB International, 1991, pp 437-451.

15   B. J. Mazur and S. Tingey, *Current Opinions in Biotechnology*, 1995, **6**, 175.

16   P. O. Brown and D. Botstein, *Nature Genetics*, 1999, **21** suppl., 33.

17   C. Debouck and P. N. Goodfellow, *Nature Genetics*, 1999, **21** suppl., 51.

18   W. P. Blackstock and M. P. Weir, *Trends in Biotechnology*, 1999, **17**, 121.

19   A. Dove, *Nature Biotechnology*, 1999, **17**, 233.

20   B. J. Walsh, M. P. Malloy and K. L. Williams, *Electrophoresis*, 1999, **19**:,1883.

21   M. K. Winson, N. Kaderbhai, A. Jones, B. K. Alsberg, A. M. Woodward A. D. Shaw, R. Goodacre, J. J. Rowland, D. B. Kell, M. Todd and B. A. M. Rudd, in *Challenges and issues in high throughput screening: a pharmaceutical and agrochemical industry perspective*, G. K. Dixon, J. S. Major and M. J. Rice, Eds, SCI, London, 1998, pp 151-152.

22   D. E. Bassett, M. B. Eisen and M. S. Boguski, *Nature Genetics*, 1999, **21** suppl., 51.

23   N. W. Simmonds, *Biological Reviews*, 1993, **68**, 539.

24   R. Walden and R. Wingender, *Trends in Biotechnology*, 1995, **13**, 324.

25   V. C. Knauf, *Current Opinions in Biotechnology*, 1995, **6**, 165.

NATURAL PRODUCTS DRUG DISCOVERY AT THE NATIONAL CANCER INSTITUTE. PAST ACHIEVEMENTS AND NEW DIRECTIONS FOR THE NEW MILLENNIUM

Gordon M. Cragg, Michael R. Boyd, Yali F. Hallock, David J. Newman, Edward A. Sausville, and Mary K. Wolpert.

Developmental Therapeutics Program, Division of Cancer Treatment and Diagnosis, National Cancer Institute, Frederick Cancer Research and Development Center, Fairview Center, P.O.Box B, Frederick, MD 21702-1201

1 ANTINEOPLASTIC AGENTS DERIVED FROM NATURAL SOURCES

An analysis of the number and sources of anticancer agents, reported mainly in the Annual Reports of Medicinal Chemistry from 1984 to 1995 covering the years 1983 to 1994, indicates that over 60% of the approved drugs developed are of natural origin.[1] Of the 92 anticancer drugs commercially available prior to 1983 in the United States and approved worldwide between 1983 and 1994, approximately 62 can be related to natural origin. Two additional antitumor agents derived from natural sources were approved for marketing in 1995 and 1996, with three more from structures first identified from a natural product and five more from synthetic sources (Table 1).

1.1 Plant Sources

Plants have a long history of use in the treatment of cancer,[2] though many of the claims for the efficacy of such treatment should be viewed with some skepticism because cancer, as a specific disease entity, is likely to be poorly defined in terms of folklore and traditional medicine.[3] Of the plant-derived anticancer drugs in clinical use, the best known are the so-called vinca alkaloids, vinblastine and vincristine (Figure 1), isolated from the Madagascar periwinkle, *Catharanthus roseus*. *C. roseus* was used by various cultures for the treatment of diabetes, and vinblastine and vincristine were first discovered during an investigation of the plant as a source of potential oral hypoglycemic agents. Therefore, their discovery may be indirectly attributed to the observation of an unrelated medicinal use of the source plant.[3] Vinorelbine, a semisynthetic derivative of vinblastine has been approved for the treatment of non-small cell lung cancer, while several other semi-synthetic derivatives, including vindesine and vinfosiltine, are in clinical trials for the treatment of a variety of cancers.[4]

The two clinically-active agents, etoposide and teniposide (Figure 1), are semisynthetic derivatives of the natural product epipodophyllotoxin, an isomer of podophyllotoxin which was isolated as the active antitumor agent from the roots of various species of the genus *Podophyllum*. These plants possess a long history of medicinal use by early American and Asian cultures, including the treatment of skin cancers and warts, so these agents may be considered as being more closely linked to a plant originally used for the treatment of "cancer".[3] Several water-soluble derivatives of etoposide are currently in clinical trials.[4]

Table 1.   **Agents from Natural Sources Approved for Marketing in the 1990s***

Compound		Year
Bicalutamide		1995
Capecitabine		1998
Bisantrene		1990
Cytarabine ocfosfate		1993
Docetaxel	(Taxotere)	1995
Formestane		1993
Fludarabine phosphate		1991
Gemcitabine		1995
Idarubicin		1990
Irinotecan	(Camptosar)	1994
Interferon, gamma-1a		1992
Miltefosine		1993
Paclitaxel	(Taxol)	1993
Pegaspargase		1994
Pentostatin		1992
Porfimer sodium		1993
Raltitrexed		1996
Topotecan	(Hycamptin)	1996
Valrubicin		1998
Zinostatin stimalamer		1994

* Agents are either pure natural products, semi-synthetic modifications or the pharmacophore is from a natural product.

More recent additions to the armamentarium of naturally-derived chemotherapeutic agents are the taxanes and camptothecins. Paclitaxel initially was isolated from the bark of *Taxus brevifolia*, collected in Washington State as part of a random collection program by the U.S. Department of Agriculture for the National Cancer Institute (NCI).[5] The use of various parts of *T. brevifolia* and other *Taxus* species (e.g., *canadensis, baccata*) by several Native American tribes for the treatment of some non-cancerous conditions has been reported,[2] while the leaves of *T. baccata* are used in the traditional Asiatic Indian (Ayurvedic) medicine system,[6] with one reported use in the treatment of "cancer".[2] Paclitaxel, along with several key precursors (the baccatins), occurs in the leaves of various *Taxus* species, and the ready semi-synthetic conversion of the relatively abundant baccatins to paclitaxel, as well as active paclitaxel analogs, such as docetaxel,[7] has provided a major, renewable natural source of this important class of drugs (Figure 1). Likewise, the clinically-active agents, topotecan (hycamptamine), irinotecan (CPT-11) and 9-aminocamptothecin (Figure 1), are semi-synthetically derived from camptothecin, isolated from the Chinese ornamental tree, *Camptotheca acuminata*.[8] Camptothecin (as its sodium salt) was advanced to clinical trials by NCI in the 1970s, but was dropped because of severe bladder toxicity.

Other examples of plant-derived agents currently in investigational use, are homoharringtonine, isolated from the Chinese tree, *Cephalotaxus harringtonia* var. *drupacea* (Sieb and Zucc.), and elliptinium, a derivative of ellipticine, isolated from species of several genera of the *Apocynaceae* family, including *Bleekeria vitensis*, a Fijian medicinal plant with reputed anticancer properties.[3] Homoharringtonine has shown efficacy against various leukemias, while elliptinium is marketed in France for the treatment of breast cancer.[9] The flavone, flavopiridol, currently in Phase I clinical trials, is scheduled to be advanced to Phase II trials against a broad range of tumors.[10] While flavopiridol is totally synthetic, the basis for its novel structure is a natural product isolated from *Dysoxylum binectariferum*.

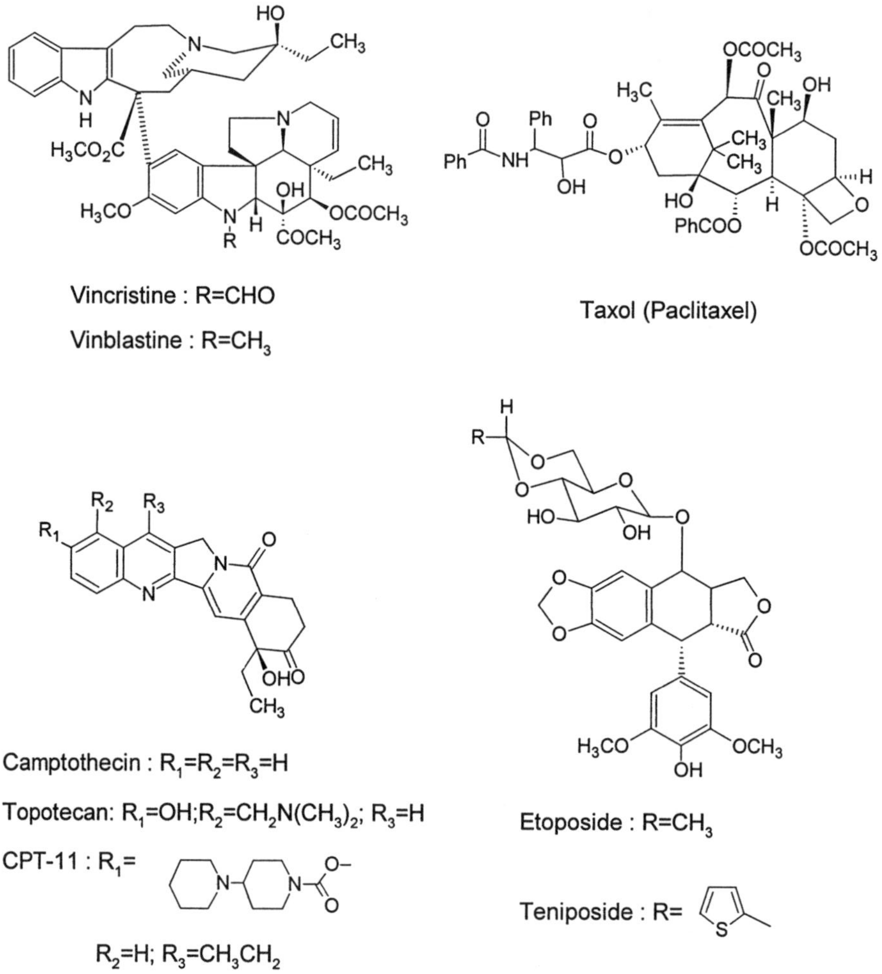

Vincristine : R=CHO

Vinblastine : R=CH$_3$

Taxol (Paclitaxel)

Camptothecin : R$_1$=R$_2$=R$_3$=H

Topotecan: R$_1$=OH;R$_2$=CH$_2$N(CH$_3$)$_2$; R$_3$=H

CPT-11 : R$_1$=

R$_2$=H; R$_3$=CH$_3$CH$_2$

Etoposide : R=CH$_3$

Teniposide : R=

Figure 1. *Plant Derived Anticancer Agents*

Ipomeanol, a pneumotoxic furan derivative produced by sweet potatoes (*Ipomoea batatas*) infected with the fungus, *Fusarium solani*, has been in clinical trials for the treatment of lung cancer.[9] A number of other plant-derived agents were entered into clinical trials and were terminated due to lack of efficacy or unacceptable toxicity. Some examples are acronycine, bruceantin, maytansine and thalicarpine.[9]

1.2 Marine Sources

Prior to the development of reliable scuba diving techniques some 40 years ago, the collection of marine organisms was limited to those obtainable by "skin diving". Subsequently, depths from approximately 10 feet to 120 feet became routinely attainable, and the marine environment has been increasingly explored as a source of novel bioactive agents. Deep water collections can be made by dredging or trawling, but these methods suffer from disadvantages, such as environmental damage and non-selective sampling. These disadvantages can be partially overcome by use of manned submersibles or remotely operated vehicles (ROVs); however the high cost of these forms of collecting precludes their extensive use in routine collection operations.

The first notable discovery of biologically-active compounds from marine sources was the serendipitous isolation of the C-nucleosides, spongouridine and spongothymidine, from the Caribbean sponge, *Cryptotheca crypta*, in the early 1950s. These compounds were found to possess antiviral activity, and synthetic analog studies eventually led to the development of cytosine arabinoside (Ara-C) as a clinically useful anticancer agent approximately 15 years later.[11] The systematic investigation of marine environments as sources of novel biologically active agents only began in earnest in the mid-1970s. During the decade from 1977-1987, about 2,500 new metabolites were reported from a variety of marine organisms. These studies have clearly demonstrated that the marine environment is a rich source of bioactive compounds, many of which belong to totally novel chemical classes not found in terrestrial sources.[12]

As yet, no compound isolated from a marine source has advanced to commercial use as a chemotherapeutic agent, though several are in various phases of clinical development as potential anticancer agents. The most prominent of these is bryostatin 1 (Figure 4), isolated from the bryozoan, *Bugula neritina*.[11] This agent exerts a range of biological effects, thought to occur through modulation of protein kinase C. Phase II trials are either in progress or are planned against a variety of tumors, including ovarian carcinoma and non-Hodgkin's lymphoma.[10]

The first marine-derived compound to enter clinical trials was didemnin B, isolated from the tunicate, *Trididemnum solidum*.[11] Unfortunately, it has failed to show reproducible activity against a range of tumors in Phase II clinical trials, while always demonstrating significant toxicity. Ecteinascidin 743 (Figure 2), a metabolite produced by another tunicate *Ecteinascidia turbinata*, has significant *in vivo* activity against the murine B16 melanoma and human MX-1 breast carcinoma models, and currently is undergoing clinical evaluation in Europe and the United States.[12] The sea hare, *Dolabella auricularia,* an herbivorous mollusc from the Indian Ocean, is the source of more than 15 cytotoxic cyclic and linear peptides, the dolastatins. The most active of these is the linear tetrapeptide, dolastatin 10 (Figure 4), which has been chemically synthesized and is currently in Phase I clinical trials.[13] It is interesting to note that analogs of the dolastatins have been isolated from marine cyanobacteria (*Symploca hydnoides*). *Dolabella auricularia* is known to be a generalist herbivore, and it probable that the dolastatins are of cyanobacterial origin.[14]

Sponges are traditionally a rich source of bioactive compounds in a variety of pharmacological screens,[12] and in the cancer area, halichondrin B (Figure 2), a macrocyclic polyether initially isolated from the sponge, *Halichondria okadai* in 1985, is currently in preclinical development by the NCI. Halichondrin B and related compounds have been isolated from several sponge genera, and the present source, a *Lissodendoryx* species, is being successfully grown by in-sea aquaculture in New Zealand territorial waters.[15] Another sponge-derived agent of considerable interest is discodermolide (Figure 2), isolated from *Discodermia dissoluta*; which has been shown to act by a similar mechanism of action to paclitaxel.[16] Discodermolide is currently in preclinical development. Another active agent, eleutherobin (Figure 2), isolated from the soft coral, *Eleutherobia aurea*, also acts in a similar manner to paclitaxel.[17] In view of the clinical success of paclitaxel and related analogs, such as docetaxel, there is considerable interest in other classes of compounds sharing the same basic mechanism of action.

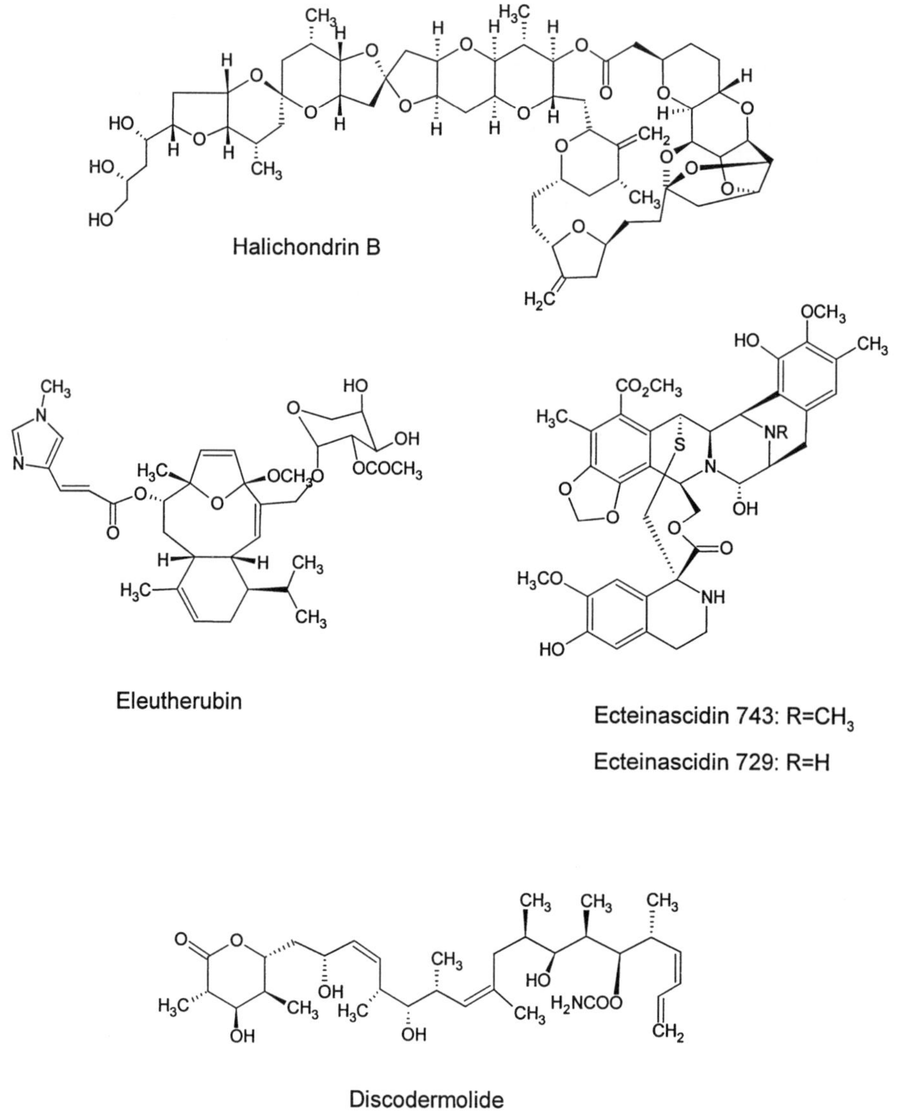

Halichondrin B

Eleutherubin

Ecteinascidin 743: R=CH₃

Ecteinascidin 729: R=H

Discodermolide

Figure 2. *Marine Organism Derived Anticancer Agents*

The synthesis of the potent *in vitro* active compound, cephalostatin 1, isolated from a rare marine worm, *Cephalodiscus gilchristi*, has recently been reported,[18] and scale-up synthesis is currently being undertaken in order to obtain sufficient quantities for preclinical development. The mechanism whereby this compound exerts its cytotoxic action has yet to be determined.

1.3 Microbial Sources

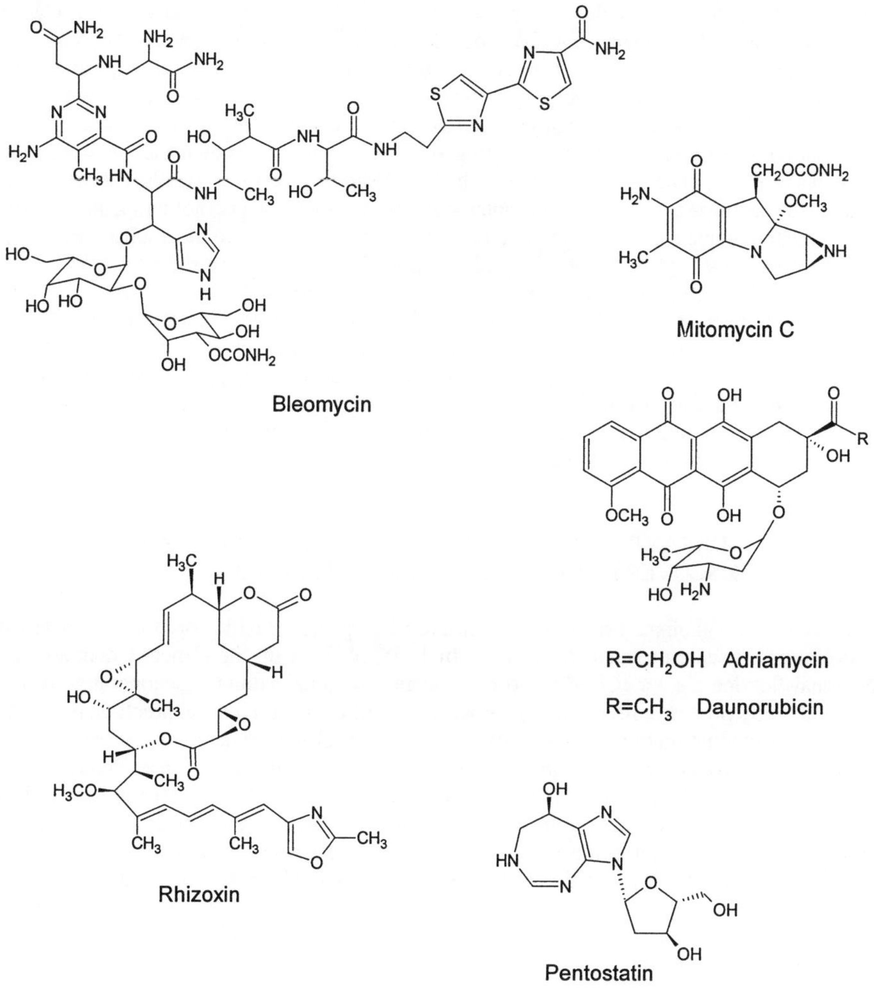

Figure 3. *Microbial Derived Anticancer Agents*

Antitumor antibiotics are amongst the most important of the cancer chemotherapeutic agents, which include members of the anthracycline, bleomycin, actinomycin, mitomycin and aureolic acid families.[19] Clinically useful agents from these families are the daunomycin-related agents, daunomycin itself, doxorubicin, idarubicin and epirubicin; the glycopeptidic bleomycins A_2 and B_2 (blenoxane); the peptolides exemplified by dactinomycin; the mitosanes such as mitomycin C; and the glycosylated anthracenone, mithramycin (Figure 3). All were isolated from various *Streptomyces* species. Other clinically active agents isolated from *Streptomyces* include streptozocin and deoxycoformycin.

Microbial metabolites in past or present clinical trials include acivicin, aclacinomycin, deoxyspergualin, echinomycin, elsametrocin, fostriecin, menogaril, porfiromycin, quinocarmycin and rhizoxin, as well as the glycinate of aphidicolin (Figure 3). Microbial products predominate amongst the agents under development by the Division of Cancer Treatment and Diagnosis (DCTD) of the NCI.[10] These include UCN-01 (7-hydroxystaurosporine), isolated from a *Streptomyces* species, and FR901228, a novel bicyclic depsipeptide isolated from a *Chromobacterium violaceum* strain, as well as derivatives of quinocarmycin (DX-52-1), spicamycin (KRN5500), CC-1065 (bizelesin), rapamycin, and rebeccamycin (Figure 4). Recent exciting discoveries are the epothilones isolated from myxobacteria.[20] This class of compounds has been shown to act by a similar mechanism of action to paclitaxel and could complement the taxanes as chemotherapeutic agents. Cyanobacteria have also been shown to produce a range of bioactive metabolites, including the cryptophycins, cytotoxic depsipeptides originally isolated from the genus *Nostoc*. A synthetic analog, cryptophycin-52, is currently in clinical trials.[21]

The large number of microbial agents reflects the major role played by the pharmaceutical industry in this area of drug discovery and development. Generally, industry has focused on the *Actinomycetales,* but expansion of research efforts, often supported by government funding, to the study of organisms from diverse environments, such as shallow and deep marine ecosystems and deep terrestrial subsurface layers, has demonstrated their potential as a source of novel bioactive metabolites.[22]

2   CURRENT   STATUS OF THE NCI NATURAL PRODUCT DRUG DISCOVERY AND DEVELOPMENT PROGRAM

The NCI was established in 1937, its mission being "to provide for, foster and aid in coordinating research related to cancer." In 1955, NCI set up the Cancer Chemotherapy National Service Center (CCNSC) to coordinate a national voluntary cooperative cancer chemotherapy program, involving the procurement of drugs, screening, preclinical studies, and clinical evaluation of new agents. By 1958, the initial service nature of the organization had evolved into a drug research and development program with input from academic sources and substantial participation of the pharmaceutical industry. The responsibility for drug discovery and preclinical development at NCI now rests with the Developmental Therapeutics Program (DTP), a major component of the Division of Cancer Treatment and Diagnosis (DCTD). Thus, NCI has, for the past forty years, provided a resource for the preclinical screening of compounds and materials submitted by grantees, contractors, pharmaceutical and chemical companies, and other scientists and institutions, public and private, worldwide, and has played a major role in the discovery and development of many of the available commercial and investigational anticancer agents. During this period, more than 400,000 chemicals, both synthetic and natural, have been screened for antitumor activity.

Initially, most of the materials screened were pure compounds of synthetic origin, but the program also recognized that natural products were an excellent source of complex chemicals with a wide variety of biological activities. From 1960 to 1982 over 180,000 microbial-derived, some 16,000 marine organism-derived, and over 114,000 plant-derived extracts were screened for antitumor activity, mainly by the NCI, and, as illustrated above, a number of clinically effective chemotherapeutic agents has been developed. Most of the drugs currently available for cancer therapy, however, are effective predominantly against rapidly proliferating tumors, such as leukemias and lymphomas, and (with some notable exceptions, such as the camptothecins, doxorubicin and the taxanes), show little useful activity against the slow-growing solid tumors usually associated with adults, such as lung, colon, prostatic, pancreatic, and brain tumors. In the early 1980s, the NCI program was discontinued because it was perceived that few novel active leads were being isolated from natural sources. Of particular concern was the failure to yield agents possessing activity against the solid tumor disease-types. This apparent failure might, however, be attributed more to the nature of the primary screens being used at the time, rather than to a deficiency of nature. Continued use of the primary P388 mouse leukemia screen appeared to be detecting only previously identified active compounds or chemical structure types having little or no activity against solid tumors.[23]

2.1 Drug Discovery

During 1985-1990 the NCI developed a new *in vitro* screen based upon a diverse panel of human tumor cell lines.[24] The screen comprised sixty human cancer cell lines derived from nine cancer types, organized into subpanels representing leukemia, lung, colon, central nervous system, melanoma, ovarian, renal, prostate and breast. In early 1999, a prescreen comprising three cell lines [MCF-7 (breast), NCI H460 (lung), SK268 (CNS)] which detect >95% of the materials found to exhibit activity in the 60 cell line screen was introduced. Materials exhibiting a predetermined level of activity at a single high dose in the prescreen are automatically advanced to the 60 cell line screen for 5-dose testing. The profile of activity shown in the 60 cell line screen provides useful information in terms of possible mechanism of action through comparison with the profiles of a set of standard agents using the COMPARE program.[25]

With the development of this new screening strategy, the NCI once again turned to nature as a potential source of novel anticancer agents, and a new natural products acquisition program was implemented in 1986. Contracts for the cultivation and extraction of fungi and cyanobacteria, and for the collection of marine invertebrates and terrestrial plants, were initiated in 1986, and with the exception of fungi and cyanobacteria, these programs continue to operate. Marine organism collections originally focused in the Caribbean and Australasia, but have now expanded to the Central and Southern Pacific and to the Indian Ocean (off East and Southern Africa) through a contract with the Coral Reef Research Foundation, which is now based in Palau in Micronesia. Terrestrial plant collections have been carried out in over 25 countries in tropical and subtropical regions worldwide through contracts with the Missouri Botanical Garden (Africa and Madagascar), the New York Botanical Garden (Central and South America), and the University of Illinois at Chicago (Southeast Asia), and have been expanded to the continental United States through a contract with the Morton Arboretum.

In carrying out these collections, the NCI contractors work closely with qualified organizations in each of the source countries. To date, botanists and marine biologists from source country organizations have collaborated in field collection activities and taxonomic identifications, and their knowledge of local species and conditions has been indispensable to the success of the NCI collection operations. Source country organizations provide facilities for the preparation, packaging, and shipment of the samples to the NCI's Natural Products Repository (NPR) in Frederick, Maryland. The collaboration between the source country organizations and the NCI collection contractors has, in turn, provided support for expanded research activities by source country biologists, and the deposition of a voucher specimen of each species collected in the national herbarium or repository is expanding source country holdings of their biota. When requested, NCI contractors also provide training opportunities for local personnel through conducting workshops and presentation of lectures. In addition, through its Letter of Collection (LOC) and agreements based upon it, the NCI invites scientists nominated by Source Country Organizations to visit its facilities, or equivalent facilities in other approved U.S. organizations for 3-12 months to participate in collaborative natural products research, while representatives of most of the source countries have visited the NCI and contractor facilities for shorter periods to discuss collaboration.[26]

Dried plant samples (0.3-1 kg dry weight) and frozen marine organism samples (~ 1 kg wet weight) are shipped to the NPR in Frederick where they are stored at -20°C prior to extraction with a 1:1 mixture of methanol: dichloromethane and water to give organic solvent and aqueous extracts. All the extracts are assigned unique NCI barcode numbers and returned to the NPR for storage at -20°C until requested for screening or further investigation. After testing in the *in vitro* human cancer cell line screen, active extracts are subjected to bioassay-guided fractionation to isolate and characterize the pure, active constituents. Agents showing significant activity in the primary *in vitro* screens are selected for secondary testing in several *in vivo* systems.[27] Those agents exhibiting significant *in vivo* activity are advanced into preclinical and clinical development.

2.2 Preclinical Development

Those agents showing significant *in vivo* activity are presented to the NCI Division of Cancer Treatment and Diagnosis (DCTD) Decision Network Committee (DNC), and, if approved by the DNC, the agent is entered into preclinical and clinical development. The Decision Network Process divides the preclinical drug development process into stages, as described below.

- An adequate supply of natural product is procured to permit preclinical and clinical development.
- Formulation studies are performed to develop a suitable vehicle to solubilize the drug for administration to patients, generally by intravenous injection or infusion in the case of cancer. The low solubility of many natural products in water poses considerable problems, but these can be overcome by use of co-solvents or emulsifying agents (surfactants) such as Cremophore EL (polyoxyethylated castor oil).
- Pharmacological evaluation determines the best route and schedule of administration to achieve optimal activity of the drug in animal models ,the half-lives and bioavailability of the drug in blood and plasma, the rates of clearance and the routes of excretion, and the identity and rates of formation of possible metabolites.
- In the final preclinical step, toxicological studies are performed to determine the type and degree of major toxicities in rodent and dog models. These studies help to establish the safe starting doses for administration to human patients in clinical trials.

An alternative mechanism for preclinical development which is available to academic investigators is the Rapid Access to Intervention Development (RAID) program. This program is discussed in Section 3.2.

2.3 Clinical Development

Phase I studies are conducted to determine the maximum tolerated dose (MTD) of a drug in humans and to observe the sites and reversibilities of any toxic effects. Once the MTD has been determined and the clinicians are satisfied that no insurmountable problems exist with toxicities, the drug advances to Phase II clinical trials. These trials generally are conducted to test the efficacy of the drug against a range of different cancer disease types. In addition, the new drug may be tried in combination with other effective agents to determine if the efficacy of the combined regimen exceeds that of the individual drugs used alone.

Of the 50 anticancer agents in active preclinical or Phase I development (excluding biologics) as of July, 1999, 24 are either natural products or derived from natural products, with the source organisms being 12 microbial, 3 marine, 6 plant and 3 animal in origin, together with 7 nucleoside derivatives. Some of the more promising anticancer agents currently in development by the NCI are shown in Figure 4.

3 COLLABORATION IN DRUG DISCOVERY AND DEVELOPMENT

As noted above, much of the NCI drug discovery and development effort has been and continues to be, carried out through collaborations with research organizations and the pharmaceutical industry worldwide.

Many of the naturally derived anticancer agents were developed through such collaborative efforts. Thus, the discovery and preclinical development of etoposide and teniposide, semisynthetic derivatives of the natural product epipodophyllotoxin, were performed by Sandoz investigators, and the NCI played a substantial role in the clinical development. Though paclitaxel (taxol®) was discovered by Wall and Wani with NCI contract support, the key to solving the supply problem was the semi-synthetic conversion of baccatin III derivatives to paclitaxel (and taxane analogs) pioneered by the French group led by Poitier, followed by the development of alternative conversion methods by the Holton group, supported by the NCI, and Bristol-Myers Squibb.[5] The semi-synthetic analog, taxotere (docetaxel), produced through a collaborative agreement between the Centre National de la Recherche Scientifique (CNRS) and Rhone-Poulenc Rorer, after undergoing extensive clinical evaluation in Europe and North America under auspices of organizations, such as the European Organization for Research and Treatment of Cancer (EORTC), and the Canadian and U.S. National Cancer Institutes,[10] is now in clinical use in Europe and North America. Indeed, there is close collaboration between the EORTC, the United Kingdom Cancer Research Campaign (CRC) and the NCI in the preclinical and clinical development of many anticancer agents, including agents such as bryostatin 1, dolastatin 10, aphidicolin glycinate, rhizoxin, pancratistatin and phyllanthoside.

Drugs, such as bleomycin, aclacinomycin and deoxyspergualin, were discovered by the Umezawa group at the Institute of Microbial Chemistry in Japan and developed in collaboration with the NCI; a number of the agents currently in advanced preclinical development at the NCI, such as UCN-01, and quinocarmycin and spicamycin analogs (Fig. 4), are the result of collaborations between Japanese companies, such as Kyowa Hakko Kogyo, Fujisawa Pharmaceutical Co. Ltd, and Kirin Brewery Ltd, and the NCI.[10]

The DTP of the NCI thus complements the efforts of the pharmaceutical industry and other research organizations through undertaking the development of positive leads, which industry might consider too uncertain to sponsor, and conducting the "high risk" research necessary to determine their potential utility.

With the approach of the new millennium, the DTP is committed to even stronger interaction and collaboration with the external community. Particular emphasis is being placed on serving as matchmaker between chemists and biologists in facilitating and expediting the drug discovery process, as well as assisting the academic community in the preclinical development of promising new agents.

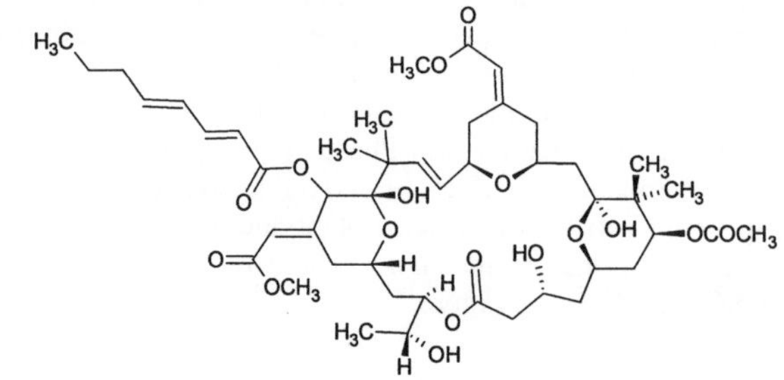

Bryostatin-1

NSC 339555

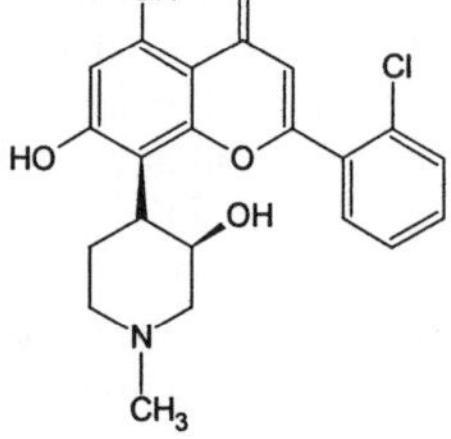

Flavopiridol

NSC 649890

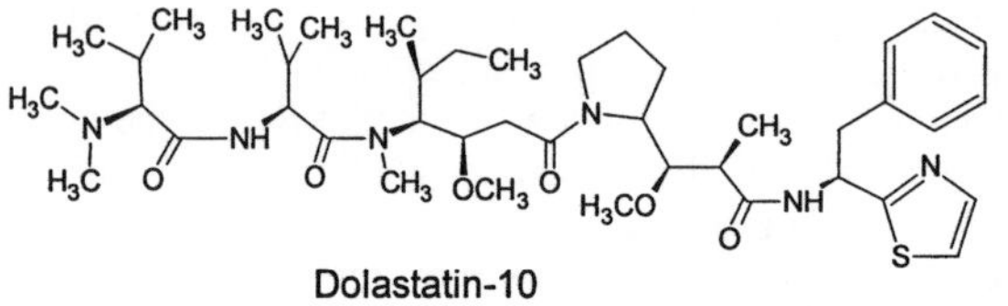

Dolastatin-10

NSC 376128

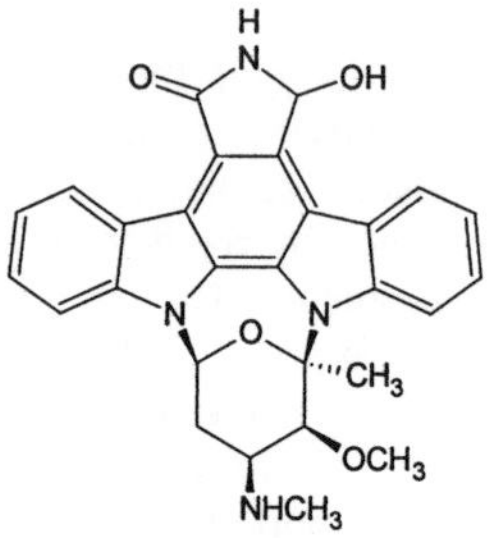

UCN-01

NSC 638850

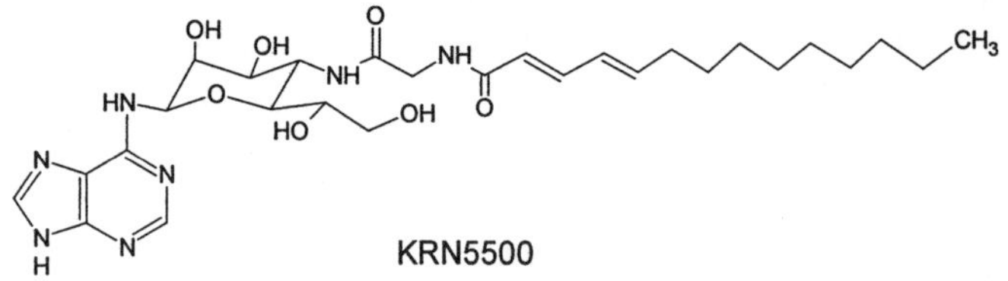

KRN5500

NSC 650426

Figure 4. *Some Anticancer Agents Currently Under Development by the NCI*

In promoting drug discovery and development, the DTP/NCI has formulated various mechanisms for establishing collaborations with research groups worldwide.

3.1 Screening Agreement Between Compound Providers and the NCI DCTD.

In the case of organizations wishing to have pure compounds tested in the NCI drug screening program, such as pharmaceutical and chemical companies or university research groups, the DTP/NCI has formulated a screening agreement which includes terms stipulating confidentiality, patent rights, routine and non-proprietary screening and testing versus non-routine and proprietary screening and testing, and levels of collaboration in the drug development process. Individual scientists and research organizations wishing to submit pure compounds for testing generally consider entering into this agreement with the NCI DCTD.

Should a compound show promising anticancer activity in the routine screening operations, the NCI may propose the establishment of a more formal collaboration, such as a Cooperative Research and Development Agreement (CRADA) or a Clinical Trial Agreement (CTA) (see NCI Technology Development and Commercialization Branch Website: http://tdcb.nci.nih.gov).

3.2 Collaboration in Preclinical Drug Development: Rapid Access to Intervention Development (RAID)

RAID is a program designed to facilitate translation to the clinic of novel, scientifically meritorious therapeutic interventions originating in the academic community. The RAID process makes available to the academic research community, on a competitive basis, NCI resources for preclinical development of drugs (see Section 2.2), and functions as a collaboration between the NCI and the originating laboratory. While the RAID process is similar to the Decision Network Process discussed in Section 2.2, the products of the RAID program are returned directly to the originating laboratory for proof-of-principle clinical trials. It is assumed that most of the products in the RAID program will be studied clinically under investigator-held INDs (Investigational New Drug approvals granted by the FDA) within the originating (or a collaborating) institution. NCI may consider assuming responsibility for clinical trials sponsorship if unanticipated circumstances develop precluding clinical development by the originating institution.

The RAID process cannot be used by private industry (which can interact with NCI through the DN process), though it can be used to develop a product licensed to a small business (defined as having 500 or fewer employees). The existence of research collaborations between the academic investigators and large companies does not affect the eligibility for support from RAID for an individual product, provided the product is not licensed to the company.

Of the 25 agents currently being developed, 15 are biologics, 5 are synthetics, and 5 are of natural origin. Full details may be obtained from the DTP Website (see Section 3.7).

3.3 National Cooperative Drug Discovery Group (NCDDG) Program

In the late 1970s and early 1980s, many significant discoveries were being made in such fields as biochemistry, molecular biology, embryology and carcinogenesis, that had the potential for the development of new strategies and agents for cancer treatment; most investigators, however, were working only in their own areas of expertise without the benefit of close liaison with experts in the many disciplines required to discover and develop new therapies and strategies. In response to the need to coordinate these research efforts, the NCI initiated the NCDDG Program in the early 1980s with the goal of bringing together scientists from academia, industry and government, in the form of consortia, in a focused effort aimed at the discovery of new drugs for cancer treatment.[28] The inclusion of an industrial component in almost all consortia has had strong positive effects in helping to orient the academic component(s) towards drug development, and maintaining a focus on the final outcomes of drug discovery in terms of clinical trials and marketable products, as well as contributing high quality scientists and resources to the Program. Involvement of NCI Staff has enabled the NCI to contribute its considerable resources and expertise in cancer drug development, including extensive computerized databases and repositories of compounds and extracts tested over more than 35 years, primary and secondary screening systems, and all the resources necessary for preclinical development of agents meeting selection criteria of the NCI Decision Network or RAID processes.

The consortia, headed by a Principal Investigator, submit proposals based on independent ideas, rather than in response to specific topics proposed by the NCI, thereby permitting the widest scope and the greatest degree of innovative science, and encouraging diversity in the discovery of new drugs and therapeutic approaches.

The National Cooperative Natural Product Drug Discovery Group (NCNPDDG) Program is one of four such programs, the other three being directed at studies of Mechanisms of Action, Specific Diseases (e.g., lung and colon cancer), and Preclinical Model Development. Since 1989, twelve NCNPDDGs have been awarded encompassing the study of all natural sources, including plants, marine bacteria and invertebrates, microalgae, cyanophytes and dinoflagellates, and using a variety of assays, such as molecular targets, mechanism-based assays, cell lines and *in vivo* systems. Most of these projects involve collaboration with organizations in source (mainly developing) countries, and the principal investigators are required to provide documented evidence that acceptable policies for collaboration with, and potential compensation of, source countries are in place before final approval of funding is granted. The program is currently being re-competed, and new awards are anticipated in the fall of 2000.

One product developed with substantial involvement of the NCDDG program, topotecan (Section 1.1), has been approved for commercial use, while two others, cryptophycin (Section 1.3) and the combination of cordycepin and deoxycoformycin, are in clinical trials.

3.4 International Cooperative Biodiversity Group (ICBG) Program

The ICBG Program resembles the NCDDG Program in structure, in that consortia are formed comprising academic, industrial and U.S. government organizations, but organizations from developing countries are also required components. This Program is jointly sponsored by the National Science Foundation (NSF), the U.S. Department of Agriculture, and components of the National Institutes of Health (NIH), including the NCI, the National Institute of Allergy and Infectious Diseases (NIAID), the National Heart, Lung and Blood Institute (NHLBI), and the National Institute of Mental Health (NIMH). The goals of the Program are research into drug discovery from natural sources, linked to the identification, inventory and conservation of biological diversity, a primary concern of the NSF, and economic development in developing countries.[28] All these goals are linked to the provision of suitable training and infrastructure building. The program is administered through the NIH Fogarty International Center.

Five awards, four involving projects in Argentina, Chile, Costa Rica, Mexico, Peru and Suriname, and one involving the West African countries of Cameroon and Nigeria, were awarded in 1993 and 1994. In 1998, the program was re-competed, and projects are now ongoing in Argentina, Chile, Mexico, Panama, Peru, Suriname, Cameroon, Madagascar, Nigeria, Laos and Vietnam. A significant challenge in the development of the ICBGs is the establishment of principles related to intellectual property rights and the protection of the rights of the participating source (developing) countries, including communities and indigenous peoples. While it is possible to develop guidelines for use in negotiating contracts and agreements, no single set of contractual terms can apply to all participants, and awardees have developed unique mechanisms and agreements to suit the particular circumstances of the organizations and countries involved.[29]

As integrated conservation and development projects, the long-term evaluation of this Program will depend on how successful the projects are in demonstrating the economic value of biodiversity in providing new pharmaceuticals and sustainable natural products-based industries for the participating developing countries.

3.5 Source Country Collaboration

3.5.1 Drug Discovery. As discussed earlier, the collections of plants and marine organisms have been carried out in over 25 countries through contracts with qualified botanical and marine biological organizations working in close collaboration with qualified source country organizations. The recognition of the value of the natural resources (plant, marine and microbial) being investigated by the NCI, and the significant contributions being made by source country scientists in aiding the performance of the NCI collection programs, have led the NCI to formulate its Letter of Collection (LOC) specifying policies aimed at facilitating collaboration with, and compensation of, countries participating in the NCI drug discovery program (22).

With the increased awareness of genetically-rich source countries to the great value of their natural resources and the confirmation of source country sovereign rights over these resources by the U.N. Convention of Biological Diversity, organizations involved in drug discovery and development are increasingly adopting policies of equitable collaboration and compensation in interacting with these countries.[30] Particularly in the area of plant-related studies, source country scientists and governments are committed to performing more of the operations in-country, as opposed to the export of raw materials. The NCI has recognized this fact for several years, and has negotiated Memoranda of Understanding (MOU) with a number of source country organizations suitably qualified to perform in-country processing. In considering the continuation of its plant-derived drug discovery program, the NCI has de-emphasized its contract collection projects in favor of expanding closer collaboration with qualified source country scientists and organizations. In establishing these collaborations, the NCI undertakes to abide by the same policies of collaboration and compensation, as specified in the LOC. A number of other organizations and companies have implemented similar policies.[30]

Through this mechanism collaborations have been established with organizations in Bangladesh, Brazil (3 organizations), China, Costa Rica, Iceland, Korea, Mexico, New Zealand, Pakistan, Panama, Russia, South Africa (2 organizations), and Zimbabwe.

3.5.2 Drug Development. In 1988, an organic extract of the leaves and twigs of the tree, *Calophyllum lanigerum*, collected in Sarawak, Malaysia in 1987, through the NCI contract with the University of Illinois at Chicago (UIC) in collaboration with the Sarawak Forestry Department, showed significant anti-HIV activity. Bioassay-guided fractionation of the extract yielded (+)-calanolide A (Figure 5) as the main *in vitro* active agent.[31] Attempted recollections in 1991 failed to locate the original tree, and collections of other specimens of the same species gave only trace amounts of calanolide A. In 1992, a detailed survey of *C. lanigerum* and related species was undertaken by UIC and botanists of the Sarawak Forestry Department. As part of the survey, latex samples of *Calophyllum teysmanii* were collected and yielded extracts showing significant anti-HIV activity. The active constituent was found to be (-)-calanolide B (Figure 5) which was isolated in yields of 20 to 30%. While (-)-calanolide B is slightly less active than (+)-calanolide A, it has the advantage of being readily available from the latex which is tapped in a sustainable manner by making small slash wounds in the bark of mature trees without causing any harm to the trees. A decision was made by the NCI DNC to proceed with the preclinical development of both the calanolides, and, in June of 1994, an agreement based on the NCI Letter of Collection was signed between the Sarawak State Government and the NCI. Under the agreement a scientist from the University of Malaysia Sarawak was invited to visit the NCI laboratories in Frederick to participate in the further study of the compound.

The NCI obtained patents on both calanolides, and, in 1995, an exclusive license for their development was awarded to Medichem Research, Inc., a small pharmaceutical company based near Chicago. Medichem Research had developed a synthesis of (+)-calanolide A[32] under a Small Business Innovative Research grant from the NCI (SBIR: http://www.nih.gov/grants/funding/sbir.html). The licensing agreement specified that Medichem Research negotiate an agreement with the Sarawak State Government. Medichem Research, in collaboration with the NCI, has advanced (+)-calanolide A through preclinical development and has been granted an INDA for clinical studies by the U. S. Food and Drug Administration (FDA). The Sarawak State Government and Medichem Research formed a joint venture company, Sarawak Medichem Pharmaceuticals Incorporated (SMP), in late 1996, and SMP has sponsored Phase I clinical studies with healthy volunteers. It has been shown that doses exceeding the expected levels required for efficacy against the virus are well tolerated. Phase II trials with patients infected with HIV-1 are currently ongoing.

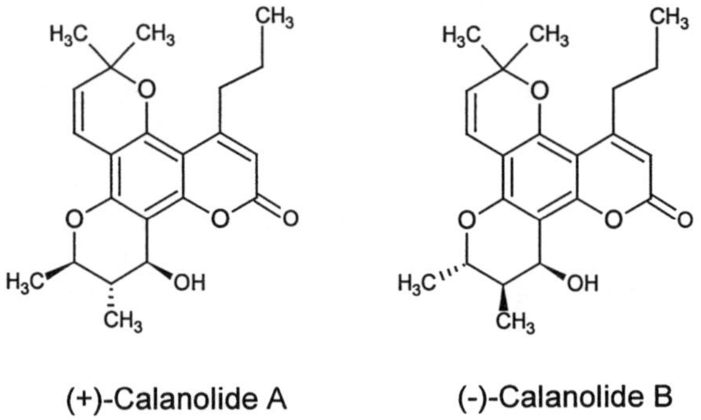

(+)-Calanolide A   (-)-Calanolide B

Figure 5. *Structures of calanolides.*

Meanwhile, by late 1995, the Sarawak State Forestry Department, UIC, and the NCI had collaborated in the collection of over 50 kg of latex of *C. teysmanii*, and kilogram quantities of (-)-calanolide B have been isolated for further development towards clinical trials. The development of the calanolides is being facilitated through the signing of a Cooperative Research and Development Agreement (CRADA) between Medichem Research and the NCI in which NCI is contributing research knowledge and expertise.

The development of the calanolides is an excellent example of collaboration between a source country (Sarawak, Malaysia), a company (Medichem Research, Inc.) and the NCI in the development of promising drug candidates, and illustrates the effectiveness and strong commitment of the NCI to policies promoting the rights of source countries to fair and equitable collaboration and compensation in the drug discovery and development process. The development of the calanolides has been reviewed as a "Benefit-Sharing Case Study" for the Executive Secretary of the Convention on Biological Diversity by staff of the Royal Botanic Gardens, Kew.[33]

3.6 Distribution of Extracts from the NCI Natural Products Repository

In carrying out the collection and extraction of thousands of plant and marine organism samples worldwide, the NCI has established a Natural Products Repository (NPR), which is a unique and valuable resource for the discovery of potential new drugs and other bioactive agents. The rapid progress made in the elucidation of mechanisms underlying human diseases has resulted in a proliferation of molecular targets available for potential drug treatment. The adaptation of these targets to high throughput screening processes has greatly expanded the potential for drug discovery. In recognition of this potential, the NCI has developed policies for the distribution of extracts from the NPR to qualified organizations for testing against all human diseases, subject to the signing of a legally-binding Material Transfer Agreement (MTA) (DTP WWW Homepage, Section 3.7).

To be considered for access to the NPR, organizations are required to submit short proposals outlining the nature of their screening systems and demonstrating the capability to process active extracts and develop any isolated active agents towards clinical trials and commercial production. Approved organizations have to enter into an MTA with DCTD, with one of the key terms being the requirement for the recipient organization to negotiate suitable terms of collaboration and compensation with the source country(ies) of any extract(s) shown to exhibit significant activity in the organization's screens (DTP WWW Homepage, Section 3.7).

3.7 DTP WWW Homepage

The NCI DTP offers access to a considerable body of data and background information through its WWW Homepage:

http://dtp.nci.nih.gov/

Publicly available data include results from the human tumor cell line screen and AIDS antiviral drug screen, the expression of molecular targets in cell lines, and 2D and 3D structural information. Background information is available on the drug screen and the behavior of "standard agents", NCI investigational drugs, analysis of screening data by COMPARE,[25] the AIDS antiviral drug screen, and the 3D database. It must be noted that data and information are only available on so-called "open compounds" which are not subject to the terms of confidential submission.

In providing screening data on extracts, the extracts are identified by code numbers only; details of the origin of the extracts, such as source organism taxonomy and location of collection, may only be obtained by individuals or organizations prepared to sign agreements binding them to terms of confidentiality and requirements regarding collaboration with, and compensation of, source countries. Such requirements are in line with the NCI commitments to the source countries through its LOC and the MTA.

4 CURRENT DEVELOPMENTS AND NEW DIRECTIONS IN NATURAL PRODUCT DRUG DISCOVERY AND DEVELOPMENT

4.1 Development of New Molecular Targets for Cancer

Over the past 20 years, there has been an explosion in the understanding of how cancer cells work. Through the Cancer Genome Anatomy Project (CGAP: http://www.ncbi.nlm.nih.gov/cgap/), it is the goal of the NCI to identify as many as possible of the human genes associated with cancer.[34] Through gene sequence analysis, numerous mutational sites in cancer cells have been, and are being, identified, some of which are unique to specific types of cancer. This knowledge permits the prediction of the structure of the encoded proteins associated with the malignant process, and the discovery of possible molecular targets affecting important aspects of cancer cell function. Anticancer drugs which have emerged from molecular-target approaches are being evaluated in the clinic, and include inhibitors of angiogenesis,[35] farnesyl tranferase, signal transduction and metalloprotease, protein kinase (PK) antagonists, and modulators of gene expression (antisense oligonucleotides).[36]

On the recommendation of a board of external advisors, the NCI has approved promoting the discovery of novel molecular targets for cancer through provision of grant support. Investigators will be expected to identify novel molecular targets, to validate them as a basis for cancer drug discovery, and to develop assays for the targets. Approaches to achieving these goals will be left to the creativity of the investigators, but may include genetic, structural biological or molecular biology approaches, or the identification of the functions of novel cellular targets through exploring their binding patterns to natural products or other ligands. The validation of a new target will be aimed at its development for use in the discovery of new agents for the treatment and prevention of cancer, and is likely to require considerable ingenuity and skill on the part of the investigator. While the ultimate validation can only come from a successful clinical trial, having information on the importance of the target prior to this stage should provide important feedback in cases where clinical therapies are ineffective.

The goal of the research will be the conversion of targets into screens for potential small-molecule effectors of their function, or the determination of their molecular structures as a prelude to drug discovery studies. It is anticipated that the NCI may assist in these steps, as well as facilitating access to compound and natural product extract libraries, and the preclinical development of any promising leads discovered through the process.

The ultimate goals envisaged are the creation of an integrated, cohesive drug discovery program and early clinical trials system that is founded on mechanistic-based approaches, and to make emerging knowledge of cancer biology the basis for drug discovery, development and testing.

4.2 Exploration of New Environments

The potential of the marine environment as a source of novel drugs has already been discussed. The NCI contract collection program has been expanded to the waters off East and Southern Africa, and expansion to under-explored regions, such as the Red Sea is being considered. These collections are performed in close collaboration with organizations based in the countries controlling the relevant waters.

Exciting untapped resources are the deep-sea vents occurring along ocean ridges, such as the East Pacific Rise and the Galapagos Rift. Exploration of these regions is being performed by several organizations, including the Center for Deep-Sea Ecology and Biotechnology of the Institute of Marine and Coastal Sciences, Rutgers University, using deep-sea submersibles such as Alvin, and their rich biological resources of macro and micro organisms are being catalogued.[37, 38] Samples are being evaluated by the NCI in collaboration with chemists at Research Triangle Institute.

Despite the more intensive investigation of terrestrial flora, it is estimated that only 5-15% of the approximately 250,000 species of higher plants have been systematically investigated, chemically and pharmacologically.[39] The potential of large areas of tropical rainforests remains virtually untapped and may be studied through collaborative programs with source country organizations, such as those established by the NCI (Section 3.5).

Another vast untapped resource is that of the insect world, and organizations, such as the Instituto Nacional de Biodiversidad (INBio) in Costa Rica, are investigating the potential of this resource, in collaboration with some pharmaceutical companies and the NCI.

The continuing threat to biodiversity through the destruction of terrestrial and marine ecosystems lends urgency to the need to expand the exploration of these resources as a source of novel bioactive agents.

4.3 The Unexplored Potential of Microbial Diversity

Until recently, microbiologists were greatly limited in their study of natural microbial ecosystems due to an inability to cultivate most naturally occurring microorganisms. In a report recently released by the American Academy of Microbiology entitled "The Microbial World: Foundation of the Biosphere", it is estimated that "less than 1% of bacterial species and less than 5% on fungal species are currently known", and recent evidence indicates that millions of microbial species remain undiscovered.[40] The continuing potential for novel drug discovery from microbial sources was recently highlighted by a report on the isolation of an insulin mimetic having antidiabetic activity in mice from a strain of the fungus *Pseudomassaria* sp. recovered from leaves of an undetermined plant collected in the Democratic Republic of the Congo.[41] Further evidence of this potential has been provided by the recent report of the antimalarial activity of fosmidomycin, isolated from a *Streptomyces* strain; total suppression of the *in vitro* growth of multi-drug resistant *Plasmodium falciparum* strains was observed, as well as cures of mice infected with the rodent malarial parasite, *P. vinckel*.[42]

The recent development of procedures for cultivating and identifying microorganisms will aid microbiologists in their assessment of the earth's full range of microbial diversity. In addition, procedures based on the extraction of nucleic acids from environmental samples will permit the identification of microorganisms through the isolation and sequencing of ribosomal RNA or rDNA (genes encoding for rRNA); samples from soils are currently being investigated, and the methods may be applied to other habitats, such as the microflora of insects and marine animals.[43] Valuable products and information are certain to result from the cloning and understanding of the novel genes which will be discovered through these processes.

Extreme habitats harbor a host of extremophilic microbes (extremophiles), such as acidophiles (acidic sulfurous hot springs), alkalophiles (alkaline lakes), halophiles (salt lakes), baro- and thermophiles (deep-sea vents),[44] and psychrophiles (arctic and antarctic waters, alpine lakes).[45] While investigations thus far have focused on the isolation of thermophilic and hyperthermophilic enzymes,[46] there are reports of useful enzymes being isolated from other extreme habitats (Diversa webpage: http://www.biocat.com/mediakit/extrefls.html), and these organisms are also likely to yield interesting, novel bioactive chemotypes.

As Dr. Rita Colwell, Director of the United States National Science Foundation, commenting on the importance of exploration and conservation of microbial diversity, has stated: "Hiding within the as-yet undiscovered microorganisms are cures for diseases, means to clean polluted environments, new food sources, and better ways to manufacture products used daily in modern society".[47]

4.4 Combinatorial Biosynthesis

Advances in the understanding of bacterial aromatic polyketide biosynthesis have lead to the identification of multifunctional polyketide synthase enzymes (PKSs) responsible for the construction of polyketide backbones of defined chain lengths, the degree and regio-specificity of ketoreduction, and the regiospecificity of cyclizations and aromatizations, together with the genes encoding for the enzymes.[48] Since polyketides constitute a large number of structurally-diverse natural products exhibiting a broad range of biological activities (e.g., tetracyclines, doxorubicin, and avermectin), the potential for generating novel molecules with enhanced known bioactivities, or even novel bioactivities, appears to be high.[48, 50]

The NCI is promoting this area of research through the award of grants to consortia composed of multidisciplinary groups devoted to the application of combinatorial biosynthetic and/or combinatorial chemical techniques to the generation of molecular diversity for testing in high throughput screens related to cancer.

4.5 Total Synthesis of Natural Products

The total synthesis of complex natural products has long posed challenges to the top synthetic chemistry groups worldwide, and has led to the discovery of many novel reactions, and to developments in chiral catalytic reactions.[51] More recently, the efforts of some groups have been focused on the synthesis and modification of drugs that are difficult to isolate in sufficient quantities for development. In the process of total synthesis, it is often possible to determine the essential features of the molecule necessary for activity (the pharmacophore), and, in some instances, this has led to the synthesis of simpler analogs having similar or better activity. Notable examples in the anticancer drug area are the synthesis of synthetic analogs of the marine organism metabolites, bryostatin 1[52] and ecteinascidin 743 (Section 1.2).[53]

The synthesis of the epothilones (Section 1.3) by several groups has permitted the preparation of a large number of designed analogs and detailed structure-activity studies which are reviewed in reference 54. These studies have identified desirable modifications which might eventually lead to more suitable candidates for drug development, but thus far none of the analogs has surpassed epothilone B in its potency against tumor cells.

The similarity in the mechanisms of action of paclitaxel (Section 1.1), the epothilones (Section 1.3), discodermolide and eleutherobin (section 1.2) has led to proposals that these structurally dissimilar substances possess common pharmacophores which could lead to the design and synthesis of analogs having substantially different structures and superior activities.[55]

4.6 Combinatorial Chemistry and Natural Products

In the study of the structure-activity relationships of the epothilones, solid-phase synthesis of combinatorial libraries has been used to probe regions of the molecule important to retention of improvement of activity.[54] The combinatorial approach, using an active natural product as the central scaffold, can also be applied to the generation of large numbers of analogs for structure-activity studies, the so-called parallel synthetic approach.[56]

The split-and-pool solid-phase synthetic approach has also been used to assemble a library of over 2 million natural product-like compounds from 18 chiral tetracyclic scaffolds, 30 terminal alkynes, 62 primary amines, and 62 carboxylic acids, using a six-step reaction sequence.[57, 58] This library will be used to probe complex biological processes, including protein-protein interactions, for which no ligands have, as yet, been identified. This approach of probing complex biological processes by altering the function of proteins through binding with small molecules has been called chemical genetics.[59]

4.7 Targeting Natural Products

A recurring liability of natural products, at least in the area of cancer chemotherapy, is that although many are generally very potent, they have limited solubility in aqueous solvents and exhibit narrow therapeutic indices. These factors have resulted in the demise of a number of promising leads, such as bruceantin and maytansine.

An alternative approach to utilizing such agents is to investigate their potential as "warheads" attached to monoclonal antibodies specifically targeted to epitopes on tumors of interest.[60] While this is not a new area of research to the NCI, the DTP is well established to refine and expand this approach to cancer therapy. The DTP has a wide range of potent, natural product chemotypes to explore as potential "warheads", and also has the capability to produce clinical grade monoclonal antibodies (Mabs) through its Biological Resources Branch. One of the approaches being investigated is the selection of a less potent member of a chemical class (e.g., ansamycin antibiotics) as the "warhead" in order to avoid undue toxic effects in the event of cleavage of the "warhead-Mab" bond prior to delivery of the agent to the desired target tumor cells. The potential for severe toxicity in such instances when using "warheads" of the potency of ricin or calicheamicin is substantial.

Another strategy of interest is the use of antibodies as vectors for enzymes capable of activating a nontoxic drug precursor (prodrug) to a potent cytotoxic moiety.[61] After injection and localization of an antibody-enzyme conjugate at the tumor, a nontoxic prodrug is administered, and while remaining innocuous to the normal tissues, it is converted to the cytotoxin by the enzyme localized at the tumor. This approach, called "antibody-directed enzyme prodrug therapy" (ADEPT) provides further potential for the application of potent natural products to cancer treatment.

5 CONCLUSION

As illustrated in the foregoing discussion, nature is an abundant source of novel chemotypes and pharmacophores, as demonstrated by the many antitumor agents in various stages of current development (Section 2.3). However, it has been estimated that only 5 to 15% of the approximately 250,000 species of higher plants have been systematically investigated for the presence of bioactive compounds,[39] while the potential of the marine environment has barely been tapped.[11] The *Actinomycetales* have been extensively investigated and have been, and remain, a major source of novel microbial metabolites;[62] however, less than 1% of bacterial and less than 5% of fungal species are currently known, and the potential of novel microbial sources, particularly those found in extreme environments, seems unbounded.[37, 38] To these natural sources can be added the potential to investigate the rational design of novel structure types within certain classes of microbial metabolites through genetic engineering, as has been elegantly demonstrated with bacterial polyketides.[48] Molecular diversity can be further increased by the synthesis of large libraries of analogs through combinatorial and parallel synthetic approaches, while total synthesis can be applied to the preparation of simpler analogs possessing equivalent or improved bioactivity.

The proven natural product drug discovery track record, coupled with the continuing threat to biodiversity through the destruction of terrestrial and marine ecosystems, provides a compelling argument in favor of expanded exploration of nature as a source of novel anticancer agents. The application of genetics and molecular biology to the discovery of new molecular targets and the development of high throughput screens should expand and expedite the identification of promising natural product lead compounds which can be modified for optimal activity by combinatorial chemical and biochemical techniques.

References

1.   G. M. Cragg, D. J. Newman, and K. M.. Snader, *J. Nat. Prod.*, 1997, **60**: 52-60.
2.   J. L. Hartwell, *'Plants Used Against Cancer'*, Quarterman, Lawrence, MA, 1982.
3.   G. M.. Cragg, M. R. Boyd, J. H. Cardellina II, D. J. Newman, K. M. Snader, and T. G. McCloud, in *Ethnobotany and the Search for New Drugs*, Ciba Foundation Symposium 185, eds. D. J. Chadwick and J. Marsh, Wiley & Sons, Chichester, U.K., 1994, pp.178-196.
4.   H-K Wang, *IDrugs*, 1998, **1**, 92.
5.   G. M. Cragg, S. A. Schepartz, M. Suffness, and M. R. Grever, *J. Nat. Prod.*, 1993, **56**, 1657-1668.
6.   L. D. Kapoor, *CRC Handbook of Ayurvedic Medicinal Plants*, CRC Press, Boca Raton, Florida, 1990.
7.   J. E. Cortes and R.Pazdur, *J. Clin. Oncol.*, 1995, **13**, 2643-2655.
8.   *'Camptothecins: New Anticancer Agents'*, eds. M. Potmeisel and H. Pinedo, CRC Press, Boca Raton, Florida, 1995.
9.   G. M. Cragg, M. R. Boyd, J. H. Cardellina II, M. R. Grever, S. A. Schepartz, K. M. Snader, and M. Suffness, in *'Human Medicinal Agents from Plants'*, Amer. Chem. Soc. Symposium Series 534, eds. A. D. Kinghorn and M. F. Balandrin, Amer. Chem. Soc., Washington, DC, 1993, pp. 80-95.
10.   M. C. Christian, J. M. Pluda, T. C. Ho, S. G. Arbuck, A. J. Murgo, and E. A. Sausville, *Sem. Oncol.*, 1997, **24**, 219-240.
11.   O. McConnell, R. E. Longley, and F. E. Koehn, in *'The Discovery of Natural Products with Therapeutic Potential'*, ed. V. P. Gullo, Butterworth-Heinemann, Boston, 1994, pp 109-174.
12.   B. K. Carté, *BioScience*, 1996, **46**, 271-286.
13.   J. Poncet, *Current Pharmaceutical Design*, 1999, **5**, 139-162.
14.   G. G. Harrigan, H. Luesch, W. Y. Yoshida, R. E. Moore, D. G. Nagle, and V. J. Paul, *J. Nat. Prod.*, 1999, **62**, 655-658.
15   G. M. Cragg, D. J. Newman, and R. B. Weiss, *Sem. Oncol.* , 1997, **24**, 156-163.
16.   E. Ter Haar, R. J. Kowalski, E. Hamel, C. M. Lin, R. E. Longley, S. P. Gunasekera, H. S. Rosenkranz, and B. W. Day, *Biochemistry*, 1996, **35**, 243-250.
17.   E. Hamel, D. L. Sackett, D. Vourloumis, and K. C. Nicolaou, *Biochemistry*, 1999, **38**, 5490-5498.
18.   G. R. Pettit, V. Gaddamidi, D. L. Herald, S. B. Singh, G. M. Cragg, J. M. Schmidt, F. E. Boettner, M. Williams, and Y. Sagawa, *J. Nat. Prod.*, 1986, **49**, 995.
19.   *'Cancer Chemotherapeutic Agents'*, ACS Professional Reference Book, ed. W. O. Foye, Amer Chem Soc, Washington, DC, 1995.
20.   Bollag DM, McQueney PA, Zhu J, et al., *Cancer Res.*, 1995, **55**, 2325-2333.
21.   D. Panda, K DeLuca, D. Williams, M. A. Jordan, and L. Wilson, *Proc. Natl. Acad. Sci. USA*, 1998, **95**, 9313-9318.
22.   Newman DJ: *Mother nature's pharmacy. A source of novel chemical structures.* SIM News, 1994, **44**, 277-283.
23.   J. M. Venditti, R. A. Wesley, and J. Plowman, *Adv. Pharmacol. Chemotherapeutics*, 1984, **20**, 1
24.   M. R. Boyd, and K. D. Paull, *Drug Dev. Res.*, 1995, **34**, 91-109.
25.   K. D. Paull, R. H. Shoemaker, L. Hodes, A. Monks, D. A. Scudiero, L. Rubinstein, J.

Plowman, and M. R. Boyd, *J. Natl. Cancer Inst.*, 1989, **81**, 1088.

26. T. D. Mays, K. D. Mazan, G. M. Cragg, and M. R. Boyd, in *'Global Genetic Resources: Access, Ownership, and Intellectual Property Rights'*, eds. K. E. Hoagland and A.Y. Rossman, Association of Systematics Collections, Washington, D.C., 1997, pp 279-298.

27. M. G. Hollingshead, M. C. Alley, R. F. Camalier, B. J. Abbott, J. G. Mayo, L. Malspeis, and M. R. Grever, *Life Sciences*, 1995, **57**, 131.

28. M. Sufffness, G. M. Cragg, M. R. Grever, F. Grifo, G. Johnson, J. A. R. Mead, S. A. Schepartz, J. M. Venditti, and M. Wolpert, *Internat. J. Pharmacognosy*, 1995, **33**, Supplement, 5-16.

29 J. Rosenthal, *OECD Proceedings: Investing in Biological Diversity.* The Cairns Conference, Australia, 25-28 March, 1996, OECD Publications, Paris, 1997, pp. 253-273.

30. J. T. Baker, R. P. Borris, B. Carte, G. A. Cordell, D. D. Soejarto, G. M. Cragg, M. P. Gupta, M. M. Iwu, D. R. Madulid, and V. E. Tyler, *J. Nat. Prod.*, 1995, **58**, 1325-1357.

31. Y. Kashman, K. R. Gustafson, R. W. Fuller, J. H. Cardellina, II, J. B. McMahon, M. J. Currens, R. W. Buckheit, S. H. Hughes, G. M. Cragg, and M. R. Boyd, *J. Med. Chem., 1992*, **35**, 2735-2743.

32. M. Flavin, J.D. Rizzo, A. Khilevich, A. Kucherenko, A. K. Sheinkman, V. Vilaychack, L. Lin, W. Chen, E. M. Greenwood, T. Pengsuparp, J. Pezzuto, S. H. Hughes, T. M. Flavin, M. Cibulski, W. A. Boulanger, R. L. Shone, and Z-Q. Xu, *J. Med. Chem.*, 1996, **39**, 1303-1313.

33. K. Ten Kate and A. Wells, *Benefit-Sharing Case Study. The access and benefit-sharing policies of the United States National Cancer Institute: a comparative account of the discovery and development of the drugs Calanolide and Topotecan.* Submission to the Executive Secretary of the Convention on Biological Diversity by the Royal Botanic Gardens, Kew, 1998.

34. R. Klausner, *The Cancer Letter*, 1999, Jan. 15, **25**, 3.

35. D. H. Paper, *Planta Medica*, 1998, **64**, 686-695.

36. A. Budillon and F. Caponigro, *Anticancer Drugs*, 1999, **10**, 249-256.

37. R. A. Lutz, T. M. Shank, D. J. Fornari, *Nature*, 1994, **371**, 663-664.

38. R. A. Lutz, and M. J. Kennish, *Reviews of Geophysics*, 1993, **31**, 211-242.

39. Balandrin MF, Kinghorn AD, Farnsworth NR: Plant-derived natural products in drug discovery and development. An overview. In: *Human Medicinal Agents from Plants. Am Chem Soc Symposium Series 534*, edited by AD Kinghorn, MF Balandrin, Washington, DC, Amer Chem Soc, 1993, pp 2-12.

40. Young P: Major microbial diversity initiative recommended. *ASM News*, 1997, **63**, 417-421.

41. B. Zhang, G. Salituro, D. Szalkowski, Z. Li, Y. Zhang, I. Royo, D. Vilella, M. T. Diez, F. Pelaez, C. Ruby, R. L. Kendall, X. Mao, P. Griffin, J. Calaycay, J. R. Zierath, J. V. Heck, R. G. Smith and D. E. Moller, *Science*, 1999, **284**, 974-977.

42. H. Jomas, J. Wiesner, S. Sanderbrand, B. Altincicek, C. Wiedemeyer, M. Hintz, I. Turbachova, M. Eberl, J. Zeidler, H. K. Lichtenthaler, D. Soldati, and E. Beck, *Science*, 1999, **285**, 1573-1576.

43. J. Handelsman, M. R. Rondon, S. F. Brady, J. Clardy and R. M. Goodman, *Chem. & Biol.*, 1998, **5**, R245-249.

44. A. Persidis, *Nature Biotechnology*, 1998, **16**, 593-594.

45. R. Psenner and B. Sattler, *Science*, 1998, **280**, 2073-2074.

46. M. W. Adams and R. M. Kelly, *Trends in Biotechnology*, 1998, **16**, 329-332.
47. R. R. Colwell, *J. Ind. Microbiol. & Biotechnology*, 1997, **18**, 302-307.
48. C. R. Hutchinson, *Proc. Natl. Acad. Sci. USA*, 1999, **96**, 3336-3338.
49. R. S. Gokhale, S. Y. Tsuji, D. E. Cane, and C. Khosla, *Science*, 1999, **284**, 482-485.
50. R. Rawls, *C&EN*, 1998, Mar. 9, 29-32.
51. R. F. Service, *Science*, 1999, **285**, 184-187.
52. P. A. Wender, J. DeBrabander, P. G. Harran, J-M Jiminez, M. F. T. Koehler, B. Lippa, C-M Park, C. Siedenbiedel, and G. R. Pettit, *Proc. Natl. Acad. Sci. USA*, 1998, **95**, 6624-6629.
53. E. J. Martinez, T. Owa, S. L. Schreiber, and E. J. Corey, *Proc. Natl. Acad. Sci. USA*, 1999, **96**, 3496-3501.
54. K. C. Nicolaou, F. Roschangar and D. Vourloumis, *Angew. Chem. Int. Ed.*, 1998, **37**, 2014-2045.
55. S. Borman, *C&EN*, 1999, April 26, 35-36.
56. K. C. Nicolaou, S. Kim, J. Pfefferkorn, J. Xu, T. Oshima, S. Hosokawa, D. Vourloumis, and T. Li, *Angew. Chem. Int. Ed.*, 1998, **37**, 1418-1421.
57. S. L. Schreiber, D. S. Tan, M. A. Foley, and M. D. Shair, *J. Amer. Chem. Soc.*, 1998, **120**, 8565.
58. A. M. Rouhi, *C&EN*, 1999, July 26, 44-46.
59. S. L. Schreiber, *Bioorg. & Medicinal Chem.*, 1998, **6**,1127-1152.
60. E. A. Sausville, *Encyclopedia of Cancer*, 1997, 1703-1714.
61. R. G. Melton and R. F. Sherwood, *J. Natl. Cancer Inst.*, 1996, **88**, 153-165.
62. A. C. Horan, in *The Discovery of Natural Products with Therapeutic Potential*, ed. V. P. Gullo, Butterworth-Heinemann, Boston, 1994, pp3-30

SECONDARY METABOLITES AS A VITAL SOURCE OF ANIMAL HEALTH PRODUCTS

John C Ruddock
Pfizer Central Research, Ramsgate Road, Sandwich, Kent CT13 9NJ, UK

1 INTRODUCTION

When putting this review together, I had to choose between giving a superficial overview of a large subject area or picking and choosing a few key topics to cover in a bit more detail. Even so, a full detailed treatment is not possible; for example, to cover the history of the discovery, indications, SAR, activities, synthesis *etc.*, of the wide range of livestock antibacterials would occupy the subject matter of several books let alone a short review. Following a brief market overview, I have selected a number of market sectors for which the key scientific and commercial developments, that have made an impact on the market-place, are identified. At the end I have given a personal perspective on the challenges that will mould the future contribution of natural products as animal health drugs.

2 MARKET OVERVIEW[*]

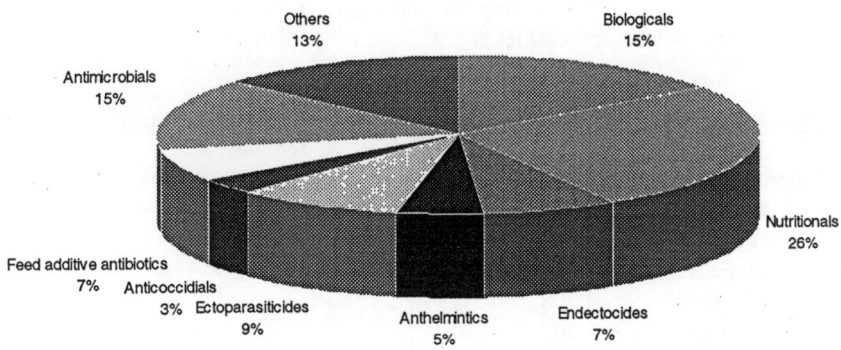

Figure 1 *1998 sales by market sector*

In 1998, the total animal health market was worth just over \$15 billion. Nutritionals, i.e. vitamins, amino-acid supplements, enzymes *etc.* and vaccines account for \$6.3 billion worth of sales, so pharmaceutical type products, the subject of this review, comprise about 60% of the whole (Figure 1).

Livestock products constitute the major franchise in the pharmaceutical sector. As we shall see, antimicrobial, feed additive antibacterial, anticoccidial and endectocide market sectors are dominated by natural product drugs.

---

[*] Pie charts and tables have been compiled from data provided by Wood Mackenzie and are reproduced with their kind permission.

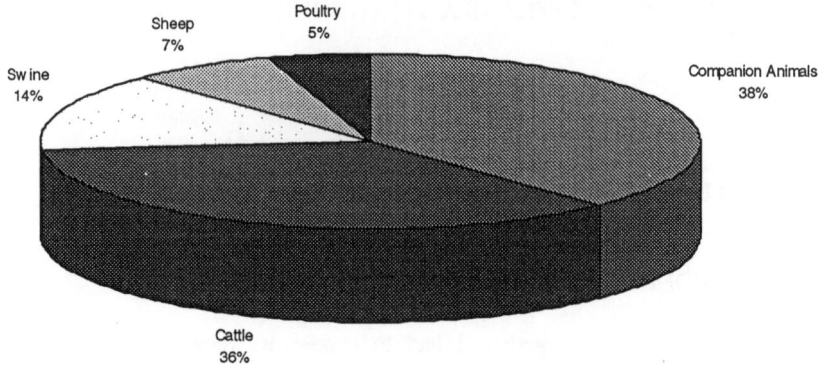

Figure 2   *1998 sales by species*

Analysing the sales data by species (Figure 2) reveals that the cattle and companion animal sectors account for nearly 75% of pharmaceutical sales. Of these two, companion animal sales are growing fast. The limited commercial opportunities in the poultry and sheep markets will strongly influence future product development for these sectors.

The list of top ten products (Table 1) also clearly demonstrates strong sales of cattle and companion animal products. Doramectin and Ivermectin are leading cattle endectocides, Fipronil, Imidacloprid and Lufenuron have made a big impact as flea agents. Before leaving this slide, it is worth noting that the tenth best selling product worldwide is a twenty-year old drug and only sells $120m - there is a substantial opportunity for new chemical entities.

Product	Sales ($m)
Ivermectin	700
Fipronil	301
Oxytetracycline	260
Imidacloprid	214
Chlortetracycline	210
Lufenuron	208
Recombinant BST	207
Enrofloxacin	170
Doramectin	166
Levamisole	120

Table 1   *Top ten products by sales (non-feed)*

Moving on to discuss specific market sectors, I thought I would chart the history of the key milestones in the development of animal health natural product drugs. Whilst the intention is that this review should be an objective discussion of important academic and industrial discoveries, as this is very much a personal perspective, it has been difficult not to include a tinge of commercial bias. Therefore, I chose my starting point as 1849.

Exactly 150 years ago, it was the active principle from *Artemisia absinthia* that inspired two pioneering German chemists to found one of today's major pharmaceutical companies.

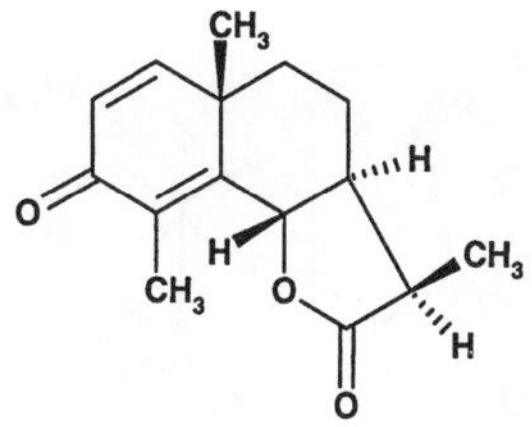

Figure 3 *Structure of santonin.*

Whilst santonin was not primarily an Animal Health drug, it is a natural product with anthelmintic properties and, as we shall see later, today's antiparasitic market is dominated by a single class of natural products. Although since then, plant metabolites have played a commercial role, the major impact has been made by microbial natural products.

3 ANTIBACTERIALS

The great step forward was the development of submerged fermentation processes for antibiotics during the 1940's, also pioneered by Pfizer. The search began for new classes of antibacterials with improved spectrum and potency for use in therapy, both for human and animal health use.

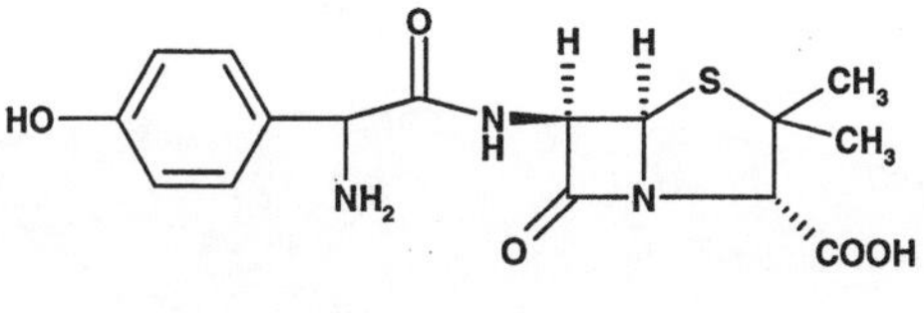

Figure 4 *Structure of amoxycillin.*

In parallel with soil screening programmes, many companies sought improvements through semi-synthesis. Semi-synthetic penicillins had a considerably broader spectrum than the original penicillins – amoxycillin (Figure 4) is an important example. However, they are susceptible to degradation by the β-lactamases produced by important pathogens such as *Staphylococcus, Escherischia coli*, and *Klebsiella*. The discovery of β-lactamase inhibitors such as clavulanic acid, led to the commercialisation of combined products. Synulox™, launched in 1988, is a leading antimicrobial brand in Western Europe.

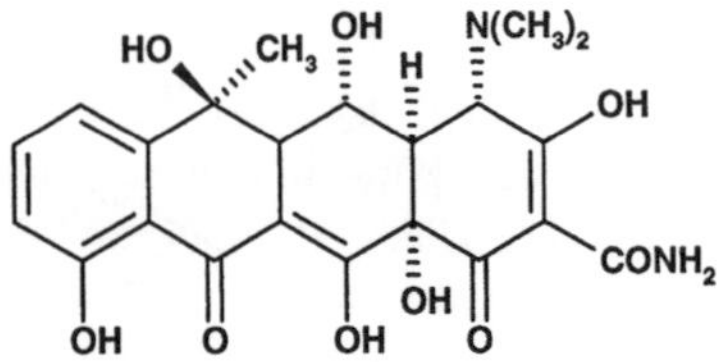

Figure 5 *Structure of oxytetracycline.*

The tetracyclines were discovered towards the end of the 1940's (structure of oxytetracycline shown in Figure 5). They have a broader spectrum of activity than the early penicillins. In addition effects on bacteria are different. The penicillins are bactericidal whereas the tetracyclines are bacteriostatic, reflecting differing modes of action. Tetracyclines disrupt protein synthesis by binding to the bacterial ribosome whilst the β-lactams inhibit bacterial cell wall biosynthesis. During the 60's, 70's and early 80's, tetracycline-based products made the biggest commercial impact in the animal health industry.

With sales of over $2 billion, therapeutic antimicrobials are one of the largest pharmaceutical market sectors. The chart in Figure 6 shows that the natural product classes, tetracyclines, β-lactams, and macrolides account for over 60% of the sales. The more recently introduced quinolones are capturing a substantial market share with enrofloxacin already commanding a place in the top 10 products.

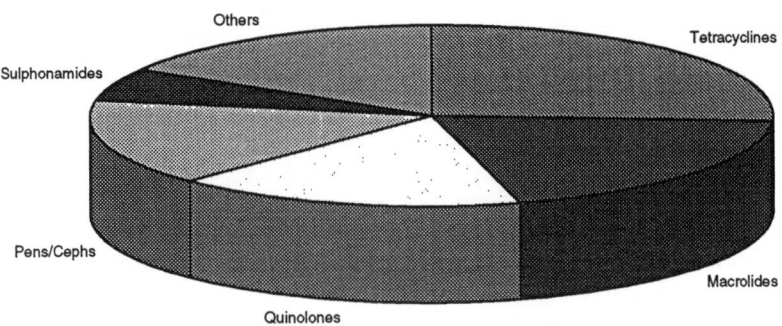

Figure 6 *1998 sales of antibacterials*

4 FEED ADDITIVES

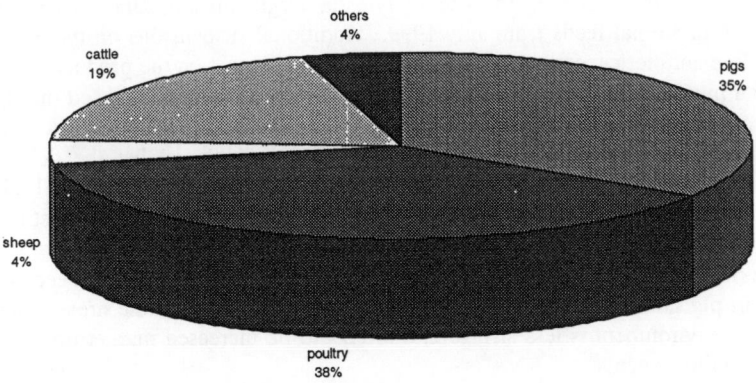

Figure 7 *Sales of feed additives by species*

Moving on from therapeutic antibacterials to feed additive antibacterials, the chart shows that swine and poultry are the largest commercial sectors.

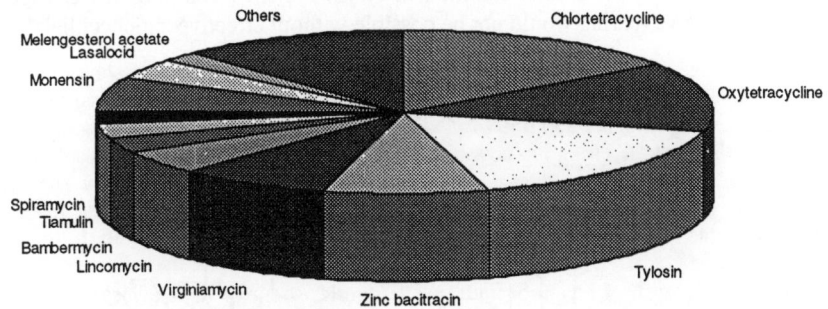

Figure 8 *Sales of feed additives by product*

1998 sales for the sector were just over $1 billion with the major products being chlortetracycline, tetracycline and tylosin. Although the latter drug is used in combination with monensin in feedlot cattle, its major role is in the pig industry. It is used in combination with sulphamethazine in starter rations to prevent atrophic rhinitis and dysentery. In finisher rations its prime role is as a growth promotant. A new claim for porcine ileitis has supported product growth.

More than any other product type, this sector has been influenced by economic and political factors. For a long time there have been concerns that the continuous feeding of sub-therapeutic levels of antibacterials to livestock would generate resistant micro-organisms and resistance would spread to human pathogens. The 1969 Swann report resulted in the withdrawal of β-lactams from livestock feeds in the UK. More recently, European Council directives have restricted the products available in the EU. In 1995, Denmark banned the use of avoparcin, a glycopeptide related to vancomycin, in beef cattle. The rest of the EU countries soon followed suit, though still allowing its use in swine and poultry. However, following an European Union review, a total ban on the use of avoparcin in livestock feeds was implemented in 1997. More recently, Council Directive 2821/98 suspended the use of tylosin, virginiamycin, zinc bacitracin, and spiramycin in animal feeds from July 1999. Additional suspensions on the use of the quinoxaline antibiotics, carbadox and olaquindox will impact on the productivity of the European pig industry in particular. These suspensions are being contested through the courts; although the outcome is unpredictable, it is unlikely that the decision will be based solely on concrete scientific evidence. One certainty is that the debate will continue and additional product suspensions are a possibility. Only time will tell how this will affect the Western European pig industry, but any reduction in competitiveness would result in falling pig numbers in this geographical area. To satisfy a steady, albeit low, worldwide growth in demand for pig meat products, there would need to be a growth in pig numbers in the rest of the world. Ironically, in these areas where the regulatory environment is less stringent, there could be increased market opportunities for feed additive agents.

5 ANTICOCCIDIALS

My final topic on in-feed agents is a major problem in the broiler industry. Coccidiosis is a protozoan disease of poultry caused by various *Eimeria spp.* It is highly prevalent in the intensive rearing conditions common in North America and Western Europe. In fact, such intensive rearing would not be possible without effective anticoccidials.

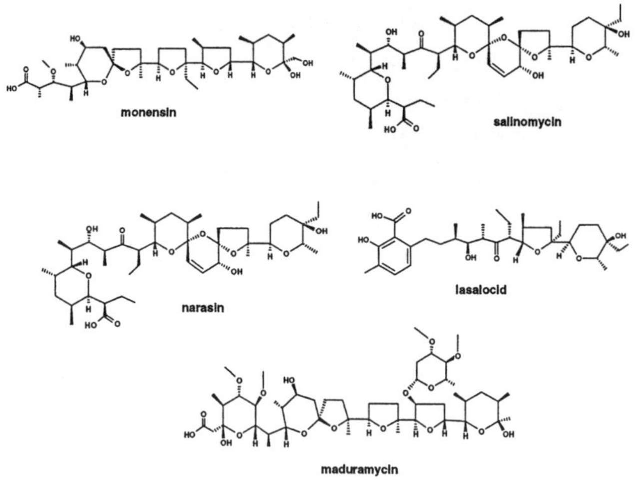

Figure 9 *Structures of commercial polyether anticoccidials*

70% of the market comprises the five polyether ionophores whose structures are shown in Figure 9, with monensin and salinomycin accounting for nearly 50% of the total market. In addition to preventing coccidiosis outbreaks, the polyethers improve feed conversion ratios, an added benefit to the producer. Unfortunately, drug resistance has developed quickly - it is sobering to realise that during its first year, 1971, monensin captured 80% of the US market, primarily because older agents were failing to work. Countering resistance by increasing dose is not an option because of the extremely low therapeutic indices for these compounds. To date, effective prophylactic control is achieved through the use of combination products, shuttle and rotation programmes.

The broiler industry is extremely cost conscious - a producer will only pay about one cent to medicate the feed for the whole of the 42-day broiler lifespan. Increasing integration in the US broiler industry has led to purchasing power being concentrated in a handful of producers. They have forced prices down so effectively, that despite steady increases in poultry numbers, the market value of this sector continues to fall. These are severe challenges for any prospective new chemical entity - only widespread breakdown of current treatments is likely to change this picture.

Chronologically, natural product therapeutic antibacterials entered the market place in the 1940's, shortly followed by the feed additive antibacterials. The first anticoccidial agents appeared at the beginning of the 1970's. As we have seen in all these therapeutic areas, natural product products quickly grew to a high proportion of the market sector.

6 ANTIPARASITICS

I would now like to move forward a decade to describe how the discovery of a single compound class transformed the antiparasitic market. Before reviewing the market sector, I give a brief summary of the major economically important parasites.

Boophilus microplus is a typical single host tick. Attachment to an animal's hide followed by intake of a blood meal can cause severe hide damage and general debilitation of the animal. In many geographical areas, ticks also act as specific disease vectors e.g. theileriosis and babesiosis. If left uncontrolled, tick infestations increase. Fully engorged adults will drop off the host and lay eggs on the pasture. These hatch into larvae which migrate back onto the animal to commence feeding. Following a number of moults, progressing through nymphal stages, development continues through to adults - the life-cycle continues. Effective chemotherapeutic control has to clean up the animal and also have sufficient persistent efficacy to break the life-cycle. Ticks cause production losses in the major markets of South America, Australasia, Africa, Mexico and the Southern States of the USA.

There are many species of parasitic worm. Although it is rare for helminth infections to cause mortality, the effects on nutrient absorption lead to reductions in growth rate, reductions in milk quality and yield - all of these reducing the economic return to the producer. As there are a range of common livestock parasites such as *Trichostrongylus colubriformis, Ostertagia circumcincta, Haemonchus contortus* and *Cooperia* spp. in sheep, *Trichostrongylus spp., Ostertagia ostertagi, Dictyocaulus viviparus, Cooperia* spp., amongst others, in cattle, an effective anthelmintic has to have broad spectrum activity. Drug distribution is also critical as different nematode species inhabit different body compartments e.g. abomasum, small intestines, lungs. As with ticks, the nematode life cycle involves a number of parasitic and free living stages. Infected animals will pass large numbers of eggs in their faeces contaminating the pasture. With favourable environmental conditions, these eggs hatch and develop through a couple of larval stages to an infective L3 larval form. After ingestion by the host, these mature into adults inside the host - egg laying will start and so the lifecycle continues.

The most prevalent ectoparasite of companion animals is the flea. Unfortunately, fleas do not restrict their eating habits to our four-legged friends as most owners will testify. Bites, themselves, are rarely felt, it is the allergic dermatitic reactions to the flea salivary secretions that cause the clinical problems. The degree of irritation varies from individual to individual; in the most severe cases, it can result in secondary infections caused by scratching the irritated skin area.

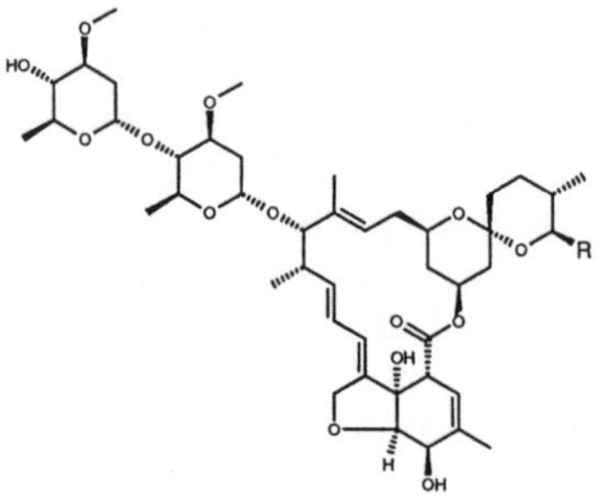

ivermectin   -   mixture   R = iPr, R = secBut

Figure 10   *Structure of ivermectin*

Combatting this wide range of parasites has been a challenge for parasitologists for many years. The antiparasitic marketplace was revolutionised in the 1980's with the discovery of a new class of macrocyclic lactones with potent ectoparasiticidal and anthelmintic activity. The first reported members of the class, the insecticidal milbemycins were discovered by Sankyo scientists in the early 1970's. This was shortly followed by the discovery of disaccharide derivatives, the avermectins, through a collaboration of scientists from Merck and the Kitasato Institute. This latter discovery was noteworthy in that an *in vivo* anthelmintic test was used as the primary screen. The avermectins had both potent broad spectrum anthelmintic and ectoparasiticidal activities. The first product, an injectable formulation of ivermectin (Figure 10), was launched in 1981 for use in cattle. The 1980's saw a period of explosive growth in the market as a result of its ease of use and broad spectrum control. Formulations for use in horses, pigs and sheep soon followed. The development of a pour-on formulation resulted in an even easier mode of administration and this product commands 70% of Merial's sales.

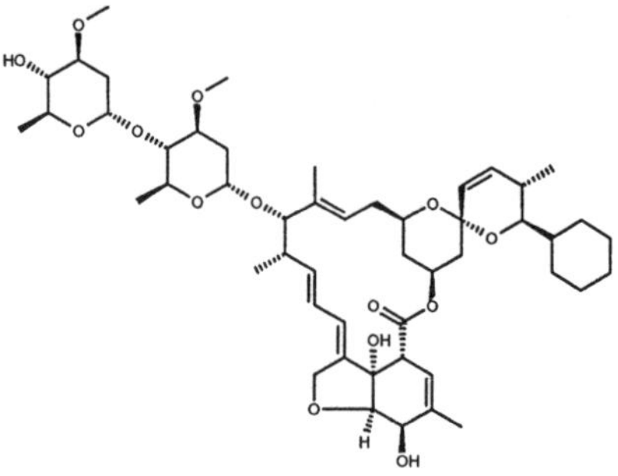

Figure 11   *Structure of doramectin*

The early 90's saw 3 other companies launch competitor products with Pfizer's doramectin (Figure 11) establishing the highest market share. This gives better persistent anthelmintic activity than the early ivermectin products.

It was during the 1970's and early 80's that advances in technology transformed natural product discovery programmes. The advent of HPLC, and later coupling to UV diode-array detectors and then mass spectrometers, improved the efficiency of dereplication procedures. The ability to rapidly separate complex mixtures reduced the time from lead identification to natural product structure. The discovery of doramectin is a fine example of the use of such technology.

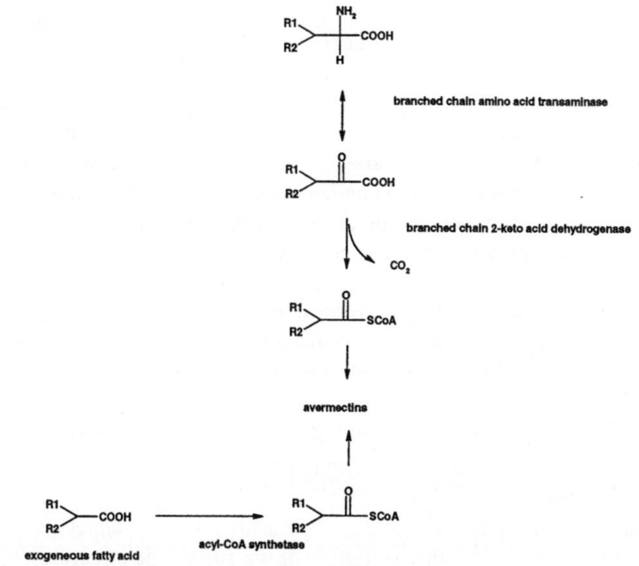

Scheme 1 *Biosynthesis of acids involved in avermectin biosynthesis*

Cane, Schulman and co-workers had demonstrated that the carbon backbone of the avermectins was built up from acetate and propionate units.[1] The biosynthetic scheme shown in scheme 1 was proposed. It was conjectured that an *S. avermitilis* mutant defective in the branched chain 2-keto acid dehydrogenase would be incapable of biosynthesising the coenzyme-A derivatives of isobutyric, isovaleric and 2-methyl-butyric acids from the corresponding branched chain 2-keto acid substrates. Hafner and co-workers isolated a mutant that was defective in this enzymic activity and established that it was incapable of biosynthesising the natural avermectins in fermentations where branched chain and straight chain fatty acids and their potential metabolic precursors were omitted from the medium.[2] In contrast, when the medium was supplemented with (*S*)(+)-2-methyl butyric acid, the little 'a' analogues of the natural avermectins were biosynthesised. Supplementing the medium with other exogeneous fatty acids yielded avermectins with differing C-25 substituents. HPLC with diode-array detection was critical in identifying the novel avermectins in the crude extracts obtained from the fermentations. Using cyclohexane carboxylic acid yielded doramectin.

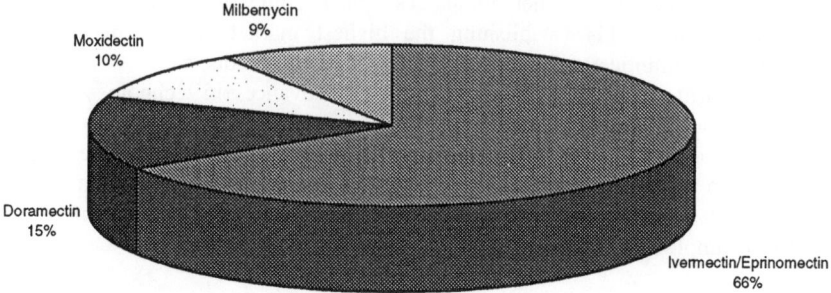

Figure 12 *Sales of endectocides by product*

Moving back to the market in general. The livestock endectocide marketplace is changing rapidly. The launch of generic products, particularly in South America where there are >15 avermectin products available, is forcing prices down. Merial's former dominance of the marketplace is being challenged not only by these generics but also by three major new products. Eprinomectin, despite being the only endectocide licensed for use in dairy cattle, has only had limited commercial success. Before moving on from endectocides, it is worth noting the impact of the avermectins in the companion animal market. Ivermectin tablets for heartworm prophylaxis in dogs were launched in 1987. Since then both milbemycin and moxidectin have been licensed in a number of markets. In 1997, Novartis launched a combination milbemycin oxime/lufenuron product for flea and heartworm control. We have already seen that fipronil, lufenuron and imidacloprid have grown the flea control market to >$700m. The early signs from the milbemycin oxime/lufenuron combination product are that a convenient to use, single agent product could revolutionise this market sector.

For the final topic in my discussion of antiparasitic natural products, we have to turn the clock back to the 1920's for the beginning of the story. The structures of the natural pyrethrins, first isolated from chrysanthemum flower heads, were elucidated in the mid-1920's. Although having good insecticidal activity, they decomposed rapidly in sunlight thus limiting their usefulness in the field. Many groups embarked on the synthesis of photostable analogues. It is almost unfair to have to condense the enormous volume of pyrethroid research into a few sentences. Key advances were made by two groups, one based at Sumitomo and the other being Michael Elliott's group at Rothamsted.

The Japanese group were working initially on making chrysanthemic esters of 3-phenoxybenzyl alcohol leading to phenothrin. Replacing the chrysanthemate with a phenylacetic acid led to fenvalerate, commercialised in 1976. Elliott's group took a different approach, they were exploring dichlorovinyl analogues of chrysanthemic acid esterified with phenoxybenzyl alcohols. Their first photostable pyrethroid, permethrin, ushered in the era of so called 'second generation pyrethroids'. Systematic exploration of SAR led to cypermethrin and deltamethrin.

Table 2 shows the relative toxicities of leading pyrethroids when applied topically to houseflies, deltamethrin being one of the most potent pyrethroids commercialised. More recent product introductions have been flumethrin and cyhalothrin. 1998 global sales of pyrethroids were $190m, they are particularly attractive products for tick control giving rapid clean-up and persistent efficacy.

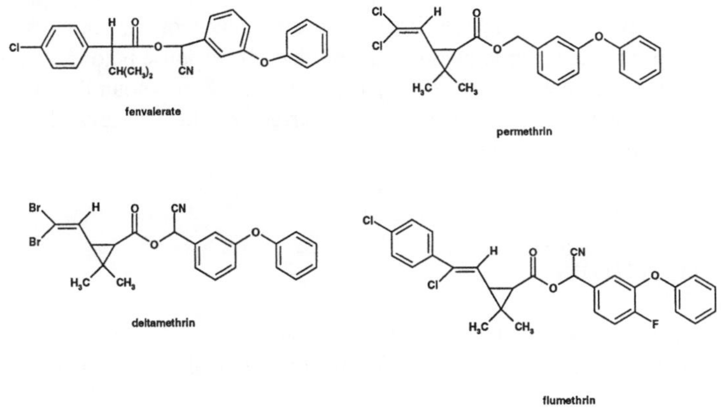

Figure 13   *Pyrethroid structures*

The global ectoparasiticide sales for 1998 were $1.35 billion (Figure 14). As we have noted earlier, the new flea control agents, fipronil, imidacloprid and lufenuron, command a substantial slice of this. Despite concerns about environmental and worker toxicity, organophosphates are still major products in the livestock sector.

Table 2   *Relative toxicity of pyrethroids to houseflies* [#]

Compound	Relative toxicity
Natural pyrethrin 1	1
Allethrin	2
Phenothrin	15
Fenvalerate	20
Permethrin	35
Cypermethrin	80
Deltamethrin	1150

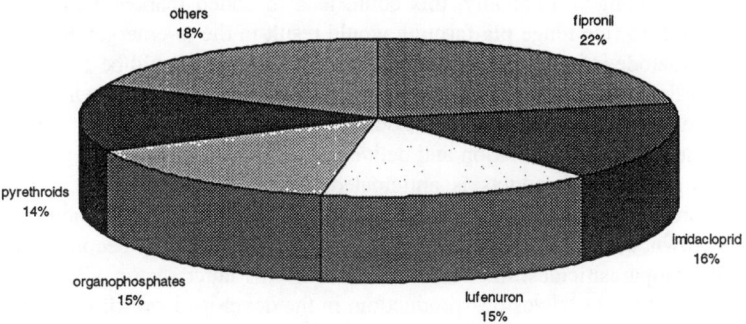

Figure 14   *Sales of ectoparasiticides by product*

[#] Reprinted from *Parasitology Today*, 1988, **4**, page S2, with kind permission of Elsevier Science.

The key drivers are cost and efficacy. The emergence of parasite resistance has blighted most products. In some markets, the producer's response is to increase the dose or to increase the frequency of dosing, thus ultimately compounding the problem further. This completes my overview of the current market scene and how natural products have played a major role.

7 THE FUTURE

In this last section, I would like to give you some personal thoughts on the future of natural products as animal health drugs. There are three main questions I would like to address:

- What is the future of the present products?
- What is the potential for new products?
- What will be the role of natural products in the search for new products?

7.1   Future of Present Products

Public opinion fuelled by media attention is becoming more effective at promoting issues high onto the political agenda. The recent concerns over GM crops are testament to this. The impact will be felt in the Animal Health industry. We have already noted the increasing regulatory pressure on antibacterial feed additives in the European Union. In addition, the FDA wishes to apply stricter controls over such agents. This trend is likely to continue. Although, probably many years away, my prediction is that in both the European Union and the US, the continuous use of antibacterials in feeds, purely to enhance productivity yields, will be banned completely. The debate will not be restricted to drug usage, but will extend to other aspects of animal management. For example, support is growing for the lobby that opposes the use of intensive rearing practices in the swine and poultry industries. Whilst it is unlikely that consumers' meat demands can be met by complete reversion to non-intensive farming methods, some producers will change. Ironically, this could lead to added market opportunities. A large-scale shift to free range pig farming would result in the re-emergence of ascariasis and other nematode infections for which antiparasitic therapy would be required. Again, we noted earlier, that generic ivermectin products are forcing prices down in the South American cattle endectocide market. Looking forward to the next 5 - 10 years, the patent expiries of both formulation and delivery patents will increase the pressure. The entry of more generic products is anticipated, the profits for those companies that pioneered products will diminish and although volume sales could increase, as the market is grown, cost sales would lag behind or even fall. As prices become comparable to those of ectoparasiticides, the decline of sales in this latter sector will decline more rapidly. A third factor - livestock production, in the developed world, is depressed; we are all familiar with the media images of lambs being "given away". Although the human population is rising continually, cattle and swine numbers are predicted to remain flat - there is a trend away from red meat to white meat - poultry numbers are growing rapidly. There is a different picture in the developing world with the *per capita* consumption of meat products increasing - the swine and poultry industries being the profitable sectors. The net result will be opportunities for existing agents in developing countries, but little or no growth in developed markets.

7.2   Potential for New Products

The factor that most influences the search for new products is parasite resistance. In both the anticoccidial and ectoparasiticide areas, resistance emerges rapidly to new agents. Although this is managed presently by rotating products or increasing dosing frequency, one might anticipate a future breakdown in all present therapies. In anticipation of this, the search for new chemical entities continues.

I have already alluded to the static livestock populations. Over the past 5 years, there has been a significant increase in the value of the companion animal market - 8.6% annual increase. If choices have to be made, then focussing research on this sector would seem to be more profitable.

7.3   Future of Natural Products

What contribution will natural products play in future product portfolios? There are a number of factors that influence the answer to this question. The pace of drug discovery increases. Natural product programmes need to compete with other sources of drug leads. Combinatorial chemistry has given access to large numbers, tens of thousands, of potential lead structures. Increased miniaturisation and sophistication of high throughput screens has resulted in the ability to screen hundreds of thousands, or even millions of samples in weeks rather than months. Project chemists may have finished their lead optimisation or candidate seeking programmes before the structures of any natural product leads have been resolved - scale-up to provide substrate for semi-synthesis taking longer still. Keeping pace with other lead-seeking approaches will tax the ingenuity of natural product scientists.

Most of the structures we have seen today have been stereochemically complex. The only cost effective manufacturing process is by fermentation. In addition, we have seen that products for the livestock market have to have minimal cost to command commercial acceptance - this effectively limits total synthesis to simple structures, made in a few steps. Such favourable cost of goods is a point in favour of natural product drugs, though even this becomes weakened if more than one or two additional semi-synthetic steps are required to yield final product.

One of the curses of natural product research is the number of times we rediscover the wheel. Although todays' highly efficient dereplication procedures minimise the time and effort wasted on this, the key to novel compound discovery is the sourcing of a diverse range of microorganisms. The early days of soil screening programmes saw collectors scouring the globe for soil, plant and other samples for microorganism isolation. Todays' realisation that, like mineral deposits, the flora, fauna *etc.* are part of a country's economic wealth means such collections are more regulated. Contractual agreements between the originating countries and industrial partners can require lengthy negotiations and normally include financial compensation to the originating country either as license fees or milestone payments. These add to the image that natural product research is difficult and complicated, providing yet more ammunition for those who are sceptical of the benefits of such programmes. One of the criticisms, often heard, is that there is nothing new in natural products. Just two well known examples illustrate how technological advances are opening up new opportunities. We have already seen how screening diverse organisms leads to novel structures. It has been conjectured that less than 1% of the microorganisms in a soil sample can be isolated by classical methods. In addition there are a large number of microorganisms which grow so slowly that they are useless for the practical production of secondary metabolites. A number of groups are applying genetic techniques to unfold the hidden opportunities. DNA, either extracted from the soil or from a slow growing organism, can be cloned into a heterologous host. These libraries of clones can then be screened for secondary metabolite production.

The increasing understanding of the genetics of biosynthesis is opening new routes to novel compounds. An example from the Leadlay/Staunton group - the transfer of the loading domain from *S. avermitilis*, the avermectin producer, to *S. erythrea*, the erythromycin producer, resulted in the production of erythromycins with novel C-13 substituents.[3] Targeted modification to other positions of the erythromycin backbone have also been achieved. Time will tell how far the technology can be pushed - the goal must be to design a polyketide natural product, assemble the modules or domains required to make it in an appropriate host to yield gm/litre titres of a novel compound. The take home message is that continuous investment in new technologies will widen the opportunities for finding new natural products.

In summary, there are challenges ahead for natural product drug discovery, natural products have played a major role in the Animal Health market; predicting the future will be a judgement call for those involved in the industry.

References

1. D.E. Cane, T.C. Liang, L. Kaplan, M.K. Nallin, M.D. Schulman, O.D. Hensens, A.W. Douglas, and G. Albers-Schonberg., *J. Am. Chem. Soc.*, 1983, **105**, 4110
2. E.W. Hafner, B.W. Holley, K.S. Holdom, S.E. Lee, R.G. Wax, D. Beck, H.A.I. McArthur, and W.C. Wernau., *J.Antibiotics*, 1991, **44**, 349
3. M.S. Pacey, J.P. Dirlam, R.W. Geldart, P.F. Leadlay, H.A.I. McArthur, E.L. McCormick, R.A. Monday, T.N. O'Connell, J. Staunton, and T.J. Winchester., *J. Antibiotics*, 1998, **51**, 1029

THE RELATIONSHIP BETWEEN NATURAL PRODUCTS AND SYNTHETIC CHEMISTRY IN THE DISCOVERY PROCESS

Stephen Brewer

Consultant, Bioproducts Technology
'Poldhune' Parc Owles, Carbis Bay, St. Ives, Cornwall TR26 2RE, UK.

Natural-product drugs have been discovered in traditional medicines, by screening for a functional activity and, more recently, from isolated enzyme screens. These natural products are often suitable for use as medicines as they are, without the need for synthetic chemistry, and, in any case, the natural products are often too complex to allow synthetic chemists to explore their structure activity relationships. It is almost a pre-requisite for a complex natural-product drug that it is potent, has the desired metabolism-distribution, and low toxicity. In contrast, chemistry is most powerful when a synthetically accessible lead molecule is found which becomes the basis for a chemical program to convert the lead into a drug with the desired metabolism, distribution and toxicity profile. Analogues are made which progressively approach the desired *in vivo* activity in the disease model based on a learning cycle of *in vivo* and *in vitro* testing. This process needs considerable pharmacological knowledge, the synthesis of 1-2 thousand compounds and 60-120 person years of chemistry and biological effort. Predictive pharmacological models of disease are essential to the synthetic chemist, otherwise a decision as to whether an analogue is closer to, or further away from, being a drug cannot be made. Thus, the extraordinarily rapid development of the cyclo-oxygenase 2 inhibitors was enabled by a long history of anti-inflammatory agents, which stretches back to the natural product, salicin, and by predictive animal models of disease. It is interesting that nearly all the major areas of pharmacology, from cardio-vascular to anti-depressants, trace their origins back to natural products. This process is still occurring. In the new areas of pharmacology such as anti-cancer, immuno-modulation and, now, diabetes, natural products are the pathfinder compounds. There-fore, rather than competing with synthetic chemistry, natural products are best used at the cutting edge of drug discovery where the new pharmacological insights they provide will speed up the process of developing improved medicines using rational design.

1 INTRODUCTION

The pharmaceutical industry is under enormous pressure to decrease the 12-15 years currently needed to discover and develop new medicines. The technology needed to do this is seen to be the combination of genomics and high-throughput mechanism based screening. The starting point for the discovery of new medicines is now at the gene level. This is the final phase of a technology which is perceived as moving the basis of drug discovery from functional activity in whole animals to cells, then proteins and now into genes. In this pursuit of technology, what has been forgotten is that the most useful mechanisms have been identified as a result of the understanding of the mechanism of action of certain pathfinder chemicals° which were originally discovered by their functional activity. This functional activity provided the pharmacologist with a means to develop and test *in vivo* animal models of disease, provided medicinal chemists with a baseline for improvement and allowed biochemists to determine a pathway and a molecular basis for the physiological effect. Thus, the original pathfinder drugs put in place the technology needed for the development the next generation of drugs reversing the process in which the movement is from mechanism to functional activity.

When a functionally active chemical has been found which links a particular enzyme or receptor to a disease state, there are fast and inexpensive ways to find alternative low molecular weight chemicals which interfere with the *in vitro* activity of the target enzyme or receptor protein. These 'hits' must then progress from mechanistic *in vitro* activity to functional *in vivo* activity in animal models of disease. The challenge for the chemist is to move a chemical with *in vitro* activity to one with *in vivo* activity. The chemical must reach a sufficient concentration to cause the selective inhibition of the target protein *in vivo* but physical diffusion barriers to chemical adsorption, active defence mechanisms that metabolise and excrete 'foreign' chemicals and the micro-environment of the protein, all conspire against this goal being achieved. However, with considerable chemistry and biology effort (costing in the region of $20 million), it is possible to achieve this goal with a high level of certainty. The cost of the effort required for this medicinal chemistry programme means that this activity, not the production of mechanistic hits, is currently the bottleneck in drug discovery.[1]

It is possible to hypothesise a link between an enzyme or receptor and a disease state. The pharmacologist establishes a theoretical causal chain of logic between a target protein and the disease. However well researched and constructed the hypothesised link may be, the complexity of the biochemistry often brings these linear chains of reason to no account. In my experience, only a small number of 'pure' mechanism based targets actually become the subject of major pre-clinical medicinal chemistry programs and even when they do, they have a much larger chance of failure than those which are following in the wake of a pathfinder chemical. Alternative reasons for failure; the lack of delivery by the chemists of the right compounds *versus* the pharmacologist's chain of reasoning, are the subject of hot debate and considerable acrimony in the hallways of pharmaceutical research companies!

In this paper I will argue that the real importance of natural products is in their role as functionally active pathfinder chemicals in the drug discovery process. These pathfinders validate the learning cycle between biologists and chemists, which is used to turn small molecular weight synthetic chemical leads into clinical trial candidates. I will argue that by putting them in direct competition with high-throughput mechanistic screens of synthetic chemical libraries we are devaluing natural products. Instead, I will suggest that natural products would be put to better use by pursuing their ability to show activity in low throughput functional screens in research programs directed at the cutting edge of pharmacology. In this way, their value as pathfinders into new areas of pharmacology can be fully exploited.

° I am indebted to Dr. M. Legg (Zeneca Agriculture) for introducing me to the term 'Pathfinder' to describe succinctly the role of these biologically active chemicals in the development of new drugs and pesticides.

2 NATURAL PRODUCTS AND THE ELUCIDATION OF MECHANISM

With the enormous advances in biochemistry seen in the latter half of the twentieth century, it has been possible to pin the mechanism of medicines down to activity against a single enzyme or receptor. Once this was achieved, the way was opened to develop improved or novel drugs. For example, the ability of certain snake venoms to cause a drop in blood pressure was shown to be due to the inhibition of angiotensin converting enzyme (ACE). For patients suffering from high blood pressure, the modulation of this enzyme could be a useful therapy and this pharmacological reasoning, combined with excellent medicinal chemistry, lead to one of the great milestones in medicinal chemistry, the development of the ACE inhibitors. The understanding that the mode of action of penicillin involved inhibition of a very specific protease unique to bacterial cell wall formation stimulated the discovery of other enzymes unique to infectious organisms, which became the targets of anti-infective programs. Thus when a protease specific for AIDS replication was identified, chemists were confident of their ability to develop protease inhibitors and pharmacologists were confident it would be a good target for anti-viral drug development. Finally, the ability of medicinal chemists to develop chemicals to selectively inhibit sub-classes of the same receptor was spearheaded by the development of the anti-histamines; one class (H1) for motion sickness and the other (H2) for stomach ulcers. Thus, when two forms of the target enzyme for the aspirin based NSAIDS (Cyclooxygenase (COX) I & II) were discovered and it was recognised that one of the iso-enzymes (COXII) was absent from the gut, it was possible rapidly to development an 'aspirin' which did not cause bleeding in the stomach. In all of these programs, functionally active natural products acted as pathfinders in understanding the pharmacology and in developing the subsequent medicinal chemistry programs which yielded excellent synthetic chemical drugs.[2]

I have previously speculated that natural products have functional activity not only because they are part of an organism's natural chemical defences and biological signalling functions, but also because they have been selected to survive the biological environment and are thus bio-available molecules.[3] The combination of chemical defence function, divergent evolution of chemical receptors, and general biological availability, make natural products an especially rich source of functionally active leads.

In order to understand the importance of functionally active leads in the discovery process, I would like to describe a model for drug development that I have used in my work as a consultant. It is based on the need to establish a learning cycle between biologists and chemists, which progressively develops a chemical lead into a drug that is a candidate for clinical trials. This process is really the one on which technical innovations should be focused, because it is the rate-limiting step in the discovery process.

3 MEDICINAL CHEMISTRY-BIOLOGY LEARNING CYCLE

The pre-clinical drug development process has as its end product a drug candidate that is suitable for testing in human clinical trials. To achieve this, the candidate must pass pre-clinical toxicology testing, have its adsorption, distribution, metabolism and excretion understood in suitable animal models, and have good scientific evidence that it will provide medicinally beneficial effects. The latter normally requires showing efficacy in animal models of the disease. The early phase discovery process has usually produced a chemical lead that falls far short of these desired properties. The major cost and rate-limiting step is the process that changes an early lead into a chemical that is suitable for testing as a drug candidate. In this process, cycles of chemical modification based on the results of biological testing are used to move the lead chemical closer to the desired biological properties. The medicinal chemist takes a lead role in deciding which chemical modifications will be attempted in order to move the chemical towards a drug candidate. This is a complex judgement based on prior knowledge derived from biological data from testing earlier analogues in the lead series, on the results of other chemists' efforts to improve the properties of the lead, on personal experience from other projects, on general knowledge and inputs from theoretical and practical chemistry and biology. The process is distinct from random screening because it has an actual drug as its endpoint, for which it uses low throughput *in vivo* tests requiring a highly skilled diagnosis to judge the quality of the test chemical.

The quality of the biological data is paramount in the decision as to whether a given chemical is closer to, or further away from, the desired properties. This high level diagnosis is made from data derived from numerous sources. The confidence level is very much determined by the extent to which the chemistry and mechanism of action of the drug has been previously described. Since a disease model is only proven when it has been shown to predict a chemical activity in the human, it is only validated retrospectively. A novel therapy will require untested animal models of disease and so the confidence in the data derived from the model is substantially reduced. Similarly, new chemical entities having unknown metabolites are much more problematic in the prediction of adverse side effects. Therefore a chemist is conservative and more likely to work with classes of compounds previously shown to be safe.

Although the chemistry is far from deterministic, the process is rationally guided by judgement where experience guides the hand of chance to evolve chemicals into the desired property space by a series of incremental improvements. On top of this is the competition from other pharmaceutical companies, with patents to be avoided or circumvented. This tends to over-shadow the technical challenges in producing new chemicals with the required structure and purity in sufficient scale for testing in animals. By synthesising in the region of 1,000-2,000 chemicals and testing a large number in advanced models of disease and toxicity, the process has a one in three chance of producing a chemical with the required safety and efficacy to test in humans. The medicinal chemistry program is complete and the proof of the pharmacology which originally drove the program now in the hands of the clinicians. Only one in three of the programs that reach this stage will succeed in producing a medicine for human use!

4 MECHANISTIC LEAD DISCOVERY AND NATURAL PRODUCTS

Chemists have a number of avenues from which to derive the lead that is to become part of the medicinal chemistry program. The most favoured is that which has been designed by the chemist using a rationale based on the known substrate or hormone which is the substrate of an enzyme or the effector of a receptor. Protease inhibitors are particular favourites for this approach where the substrate is specific for the enzyme. By substituting non-hydrolysable bonds, it is relatively easy to produce a specific enzyme inhibitor. Computational chemistry can also be used to assist in the design of the molecule, but it is often more efficient to change substituents empirically on the lead to investigate how to make the binding more specific and tight. With a protease as the target, selective inhibitors active at nM concentration can be rapidly produced. Combinatorial chemistry can often be used to great effect where *in vitro* tests of efficacy are available.[4] However, the issues of metabolism, distribution and toxicity, still mean that many hundreds of analogues will usually need to be made in order to home in on the correct pharmacological profile. One draw back with this rational approach is that the knowledge of the target's substrate is known to competitors, therefore a rational approach is likely to lead to the same classes of compounds being developed. As a result, there is a high probability of patent infringement, with a number of chemistry groups from different pharmaceutical companies all pursuing the same chemistry. Rational people all arrive at the same end point!

In many cases, the substrate or effector may be of a more complex nature, making chemistry time-consuming and expensive. For example, carbohydrates and lipids are particularly difficult to work with. In this case, random screening becomes a powerful tool for finding new leads. It is also a way of breaking free from the rational approach into areas of chemistry that would not normally be pursued. For screening to be successful, it is necessary to collect a large and diverse collection of chemicals, and to have a rapid method for looking at their activity. Given a library of 1,000,000 or so diverse chemicals, it is relatively easy to find several classes of compounds that are potential leads for a chemistry program and show selective sub-micromolar inhibition of small molecule-protein interactions. It is likely that a component of the chemical collection used for random screening will be natural products. These are normally screened as extracts, in which case purification is needed in order to identify the structure of the chemical. The use of HPLC and mass spectrometry has greatly improved the ability to identify the chemical entity responsible for the hit. The other approach is to screen pre-purified natural products. If natural product extracts are tested in a head-to-head race with a synthetic chemical library, then the natural product structure activity relationships emerge several months after the screens are completed. There is also the high probability that it will be difficult to obtain materials for more advanced testing, and that the chemical structure will not be easily accessible for making analogues. In those cases where a rational approach is also being used, natural products tend to come in third place after synthetic chemistry hits in the race for chemistry support.

When competing with a successful rational design program, any chemical or natural product screening hits will need some special characteristics in order to have chemistry effort diverted to pursuing the hit. One of the most successful ways of making a random screening hit stand out above the rest is to show some activity in a functional test using an animal model or cell-based assay. In order to do this, it is useful to have at least milligramme amounts of material on hand. Re-synthesis of chemical hits takes time and effort, and chemical support for non-validated hits is limiting. Most chemists are working on rate-limiting and resource-intensive medicinal chemistry programs. Therefore the reality is that where competition exists for the application of rational design, hits from a random screen will be in second place for chemistry support, and their only chance for redemption is to show significantly improved functional activity and toxicity against that currently achieved by the rationally designed lead. Natural products do have a particularly useful role in this area because they often have the ability to show functional activity without further chemistry. Therefore, one approach is to isolate and determine the structure of natural products when they have show functional activity in animal models of disease.

An increasing number of targets is emerging from genomics and proteomics.[5] Thus, the priority will be to develop ultra-high throughput screens able to keep up with the numbers of emerging targets. In these screens, the extra time needed to work with natural products makes them even less competitive. Because medicinal chemistry is rate limiting, it is argued that these screening hits will be used to test the validity of the target in a specific disease process. This requires a functionally active chemical to be found directly by screening but this can only be achieved if expensive, and low-throughput, models of disease are available. What is more likely to happen is that leads which do not have detailed pharmacology to support their role in disease processes will not be followed-up. Therefore, I believe that proteomics and ultra-high throughput screens will do little to address the need to reduce the cost and timelines of drug development.

5 THE REQUISITE USE OF NATURAL PRODUCTS

Against these rather depressing trends, what are the prospects for natural products? Natural products have made a massive impact to date with over 50% of prescription drugs being based on them, and they are continuing to contribute to the development of new drugs with 44% of New Chemical Entities registered by the FDA being natural products or natural product derived.[6] Despite a massive technology focus on mechanism, natural products are still at the cutting edge of pharmacology producing functionally active molecules for cancer therapy, (e.g. taxol), immuno-suppressants (cyclosporin) and more recently in the enormously technically-demanding area of small molecule mimics of insulin .[7] This past and recent history begs that we should ensure that they continue to be used as pathfinders in functional screens directed at advancing pharmacology.

Natural products have an important role in random screening, but should only be pursued if they demonstrate functional activity. Only by becoming drugs, or advanced drug leads, can they bypass the expensive synthetic chemistry development route used for chemical leads. This requires the advanced disease models and the commitment to produce sufficient material to allow functional testing. I would reserve my precious natural product resource to screens that demonstrate this capability. In the absence of functional activity, natural products will not compete with rational design or synthetic chemicals emerging from random screenings and their pursuit will be a waste of effort. With so many potential mechanistic targets for each disease, it is also a waste to test a functionally active natural product against only one target. Thus, extracts selected using ethnobotanical information should be used in functional but not mechanism-based screens. Great care needs to be taken to ensure that functional screens do not rediscover the same activity, but the rationale that 'novel chemistry means novel mechanism of action' - a role served admirably by natural products- is the key to finding new leads by this means. Finally, natural products should be used at the cutting edge of pharmacology research where the chemistry is not available, e.g. in the search for small molecular weight mimetics of proteins.

6 CONCLUSION

After considering the current needs of drug discovery and the contribution made by natural products, I believe that natural products should be returned to their proven role as pathfinders for pharmacology and medicinal chemistry. They do this by showing *in vivo* functional activity in models of disease. When they are treated as 'just another chemical in our vast libraries used for mechanism based screening', not only are they non-competitive against synthetic chemicals and rational design-based projects, but they are also being seriously undervalued. As pathfinders, they address the current medicinal chemistry bottleneck in drug discovery, which rather than not finding leads, is the enormous cost of progressing leads through medicinal chemistry programs to turn them into candidates for clinical trials.

References

1 Pamphlet *'Re-inventing Drug Discovery', Pharmaceutical & Medical Products Executive Briefing*, published by Anderson Consulting UK, 1996.
2 J. S. Miller & S. J. Brewer, *'Conservation of Plant Genes'*, Ed. Adams R.P. & Adams J.E., Acedemic Press 1992, pp119-134; W. Sneader, *'Drug Discovery, the Evolution of Modern Medicines'*, John Wiley& Sons, 1985.
3 S. J. Brewer, *'Screening for New Therapeutic Agents'* Abstract of 139th Meeting Society for General Microbiology, 1998.
4 D. J. Tapolczay, R. J. Kobylecki, L. J. Payne & B. Hall, *Chemistry & Industry*, 5 October 1999, 772.
5 C. Ashton, *Chemistry & Industry*, 7 June 1999, 422.
6 G.M. Cragg, D.J. Newman & K. M. Snader, *J. Natural Products*, 1997, **60**, 52.
7 B. Zhang, *Science*,1999, **284**, 974.

NATURAL PRODUCTS *vs.* COMBINATORIALS: A CASE STUDY

Dwight Baker, Ursula Mocek and Cheryl Garr

Integrated Discovery Division, New Chemical Entities, Inc., 18804 North Creek Parkway, Bothell, Washington 98011 USA

Historically pharmaceuticals have largely been derived from natural product sources. With the development of combinatorial approaches to chemical synthesis in the last decade, drug discovery programs have adapted to large-scale screening programs of combinatorial chemical libraries of up to several hundred thousands of wells. Natural product screening programs are perceived by some in the pharmaceutical industry as antiquated, inefficient or even unproductive, despite a steady flow of natural product derived New Chemical Entities (NCE's) into the market. Natural products hold great potential for novel drug discovery. Newer, automated and high-throughput technologies, which make high-throughput combinatorial chemical synthesis possible can be adapted to improve the efficiency of natural product discovery programs.

Libraries of synthetic combinatorial chemicals complement libraries of natural product metabolites, but do not duplicate or replace them. Examples of side-by-side screening of synthetic combinatorial chemicals and microbial fermentation extracts indicate that both sets of chemical diversity can provide unique leads. The quantity of leads generated by each method is irrelevant, if the quality of the lead is not considered. Therefore, scientific research should be focussed not on which type of library is better, but rather how to take advantage of both resources in a cost-effective and timely manner.

To position natural products to meet the current drug discovery paradigm of high to ultra-high throughput random screening, certain well-known technologies should be used. Biological characterization as well as separation chemistries for semi-purification or full purification can be employed prior to screening to reduce the number of compounds in the screening mixture and to create links from the physical entities to databases containing chemical and biological characteristics. Lead candidate compounds from both synthetic and natural product sources can both be used to commence computational approaches to analogue compound generation for optimal lead drug development.

1 INTRODUCTION

Natural products, defined as naturally derived metabolites or by-products of microorganisms, plants or animals, have provided both pre-historic and historic sources of health remedies, drugs and medicines for man and animals.[1,2] The discovery of new drugs has frequently been serendipitous throughout the history of mankind, but in the recent past, the pharmaceutical and agrochemical industries have attempted to make the discovery process more predictable and rigorously statistical. Thus, random screening of large libraries of chemical compounds has become a typical strategy in drug discovery programs.

Natural product discovery by analytical chemistry is a craft that may be described as an art form as well as a precise science. Many of the discoveries of novel drugs have been associated with particular scientists,[3] or scientific teams. The technologies of random synthetic chemical library production and drug discovery programs do not have this same association with particular scientists or laboratories. In this paper, the outputs of discovery programs based on the random screening of natural products are compared with those based on the random screening of chemical libraries derived from combinatorial chemistries.

2 COMPARING NATURAL PRODUCTS WITH COMPOUNDS SYNTHESIZED BY COMBINATORIAL CHEMISTRY

In the last decade, the pharmaceutical industry has whole-heartedly embraced the creation of large libraries of synthetic chemical compounds derived from parallel synthesis or combinatorial chemistry.[4-10] As a result, the industry has attempted to compare the productivity of discovery programs focussed on natural products with those focussed on combinatorial chemicals. This direct comparison is fair because both strategies attempt to provide a broad cross-section of chemical diversity for screening in cellular or biochemical assays. Assays do not discriminate between chemicals on the basis of their source. A compound derived from a natural source is no more, or no less, interesting or unique because it was not synthesized *de novo*. An important aspect of the massive random screening programs for new bioactive compounds, is the need to screen the broadest range of chemical diversity to identify new structure activity relationships. Natural products provide unique chemical diversity, distinct from that found in synthetic or combinatorial chemical libraries currently available. It is important to note however, that chemists working predominantly in the fields of either natural products or combinatorial chemistry, can demonstrate a particular prejudice in favour of the value of each type of chemical diversity,[11] i.e. beauty is in the eye of the beholder.

Combinatorial chemicals are perceived to have certain disadvantages in relation to compounds derived from biological systems. Combinatorial chemicals are viewed as insufficiently complex and as having limited structural rigidity. Since synthetic compounds by definition are not "natural," they cannot *a priori* be considered either biologically relevant or compatible without testing. While the solubility or bioavailability of a synthetic compound may be estimated, it is not known precisely without testing. In the actual testing of combinatorial chemicals, the specific purity of the test sample, even if known, can cause problems in the identification of any bioactivity. Minor impurities, or unreacted reagents, or side products in low concentration in the test sample, can be responsible for bio-activities observed in primary screening.

Combinatorial chemicals have many qualities of value to the pharmaceutical industry. Their production, even in a large scale, is simple because they have already been synthesized. Libraries of large numbers of compound samples can be prepared so that there is a direct relationship between the compound and information held in a database.[12] The diversity of the chemical structures in such a library can be mathematically computed and additional diversity designed.[13, 14] Such libraries are usually composed of samples of less complex mixtures of compounds, if they are not in fact entities approaching 100% purity.

On the other hand, samples in natural product libraries are usually mixtures of compounds extracted from even more complex mixtures found in living tissues or environments. Such samples are derived and not prepared, and therefore the concentrations of major constituents will not be uniform. Thus, generally, there will be no direct relationship between the physical compound and information held in a database of chemical characteristics. Natural product metabolites are frequently cited as being too complex for synthetic production, despite many successes at full synthesis of very complex naturally derived compounds.[15-18] More frequently, the major disadvantage cited for natural products is the length of time required for structural elucidation of natural metabolites. This time factor is decreasing, however, as new technological advances make possible greater numbers of analyses per unit time.

Natural products have significant value. By definition, they are biologically compatible and relevant to cellular systems. Many are structurally rigid, making them inherently stable. They exhibit extremely broad chemical diversity. The number of natural products available is limited only by the ability to extract, purify and identify them. They clearly have a proven track record for providing novel pharmaceuticals, agrochemicals and medicines. For commercialisation purposes, their production is possible by fermentation and cultivation. With the advances made in molecular genetics, the modification of natural products is now possible *via* genetic, rather than strictly chemical, routes.

3 NATURAL PRODUCTS, COMBINATORIALS AND HIGH THROUGHPUT SCREENING

In high throughput drug discovery screening (HTS) programs, there are significant differences in the handling of natural products and combinatorial chemicals which influence their use. For the scientist conducting the screening, it is exciting to run a screen, and immediately relate assay activity data to the structures of single test compounds. This apparent direct link of structure and activity can be deceiving. Many combinatorial chemical libraries are not sufficiently pure (even at purities exceeding 85%) to make a direct link between structure and activity without analytically profiling the test samples. Thus combinatorial chemicals are not much different from mixtures of natural products, which also must be analytically profiled after screening. The real difference is

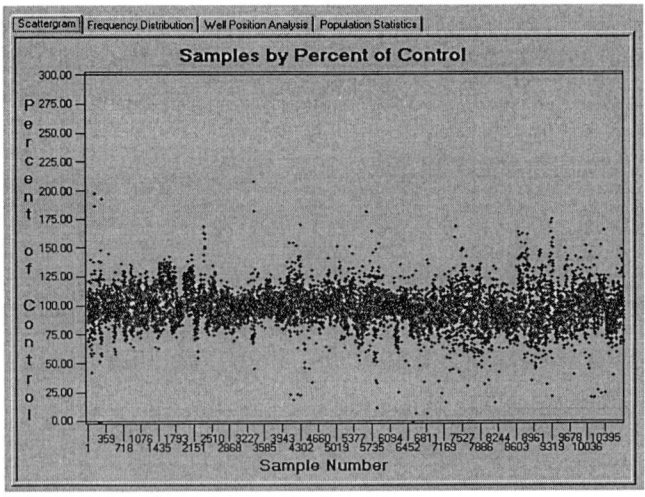

Figure 1 *Scatterplot of activity detected in an HTS assay for combinatorial chemicals*

in the purpose of the analytical profiling. For combinatorial chemicals, analytical profiling checks whether the intended test compound is present in the test sample. For natural products, analytical profiling of the test sample attempts to identify the active molecule.

Another aspect of handling combinatorial chemicals versus natural products is "tidiness." If one looks at a scatterplot of activity from an assay for combinatorial chemicals, the data look relatively tidy (Figure 1). The majority of samples demonstrate no appreciable activity, and it is not difficult to identify hits. A scatterplot for natural product extracts in the same assay will show a broader range of activity (Figure 2). This breadth in assay response is proportional to the greater breadth of chemical diversity inherent in the natural product mixtures. It seems logical that the greater breadth of activity would be desirable, but this is frequently perceived to be an impediment because the HTS interpretative software is biased toward a one-to-one, sample-to-structure, system.

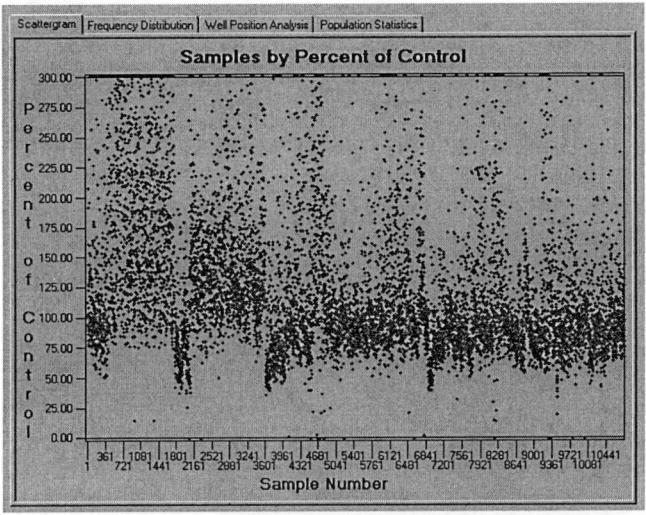

Figure 2   *Scatterplot of activity detected in an HTS assay for natural product extracts*

In a review of the performance of combinatorial chemicals alongside natural product extracts in several biochemical assays, there no comprehensive significant differences (Table 1). Natural products and combinatorial chemicals were screened in three different biochemical assays of current pharmaceutical relevance. In one assay, natural products resulted in more confirmed active samples than combinatorials. In another assay, the reverse was true. In a third assay, both sample types performed about the same. In the majority of assays formatted for high throughput, both combinatorial chemicals and natural products will provide adequate numbers of active samples for drug candidate lead identification programs. There is a perception however, that for biochemical assays for which combinatorial chemicals fail to produce a significant number of active samples, natural products are likely to provide some possibilities.

Table 1 *Combinatorial chemicals vs. natural product extracts: a comparison of primary screening data*

Assay	Combinatorial Chemicals No. of primary hits	No. of confirmed hits	Natural Products No. of primary hits	No. of confirmed hits
"A"	250/11000 (2.3%)	0/200 (0%)	200/10000 (2.0%)	30/200 (15%)
"B"	80/25000 (0.3%)	17/80 (21%)	120/15000 (0.8%)	25/120 (21%)
"C"	50/10000 (0.5%)	10/50 (20%)	200/10000 (2.0%)	4/200 (2%)
"D"	85/25000 (0.3%)	2/85 (2.4%)		
"E"			120/15000 (0.8%)	18/120 (15%)

None of the phenomena described above makes natural products incompatible with HTS. Groups working with natural products will need to apply a little more finesse and less brute force to be successful in the identification of novel drug leads. Likewise, those screening combinatorial libraries should avoid the misconception that they can bypass analytical characterisation of the active compound. As an example of intelligent approaches to some of the problems, we have applied data visualization techniques to address the identification of similarities among bioactive samples identified in HTS. Figure 3 illustrates how bioactivities among fractionated samples might be visually displayed so that samples having the highest potential will be advanced. Ultimately, processes like these which are only semi-automated will become fully automated to speed the discovery process.

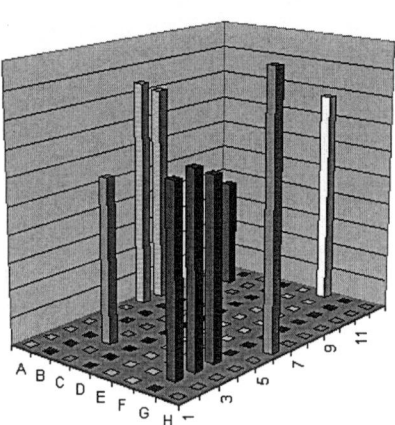

Figure 3 *Plot illustrating display of bioactivity of fractionated samples to facilitate selection for further investigation*

4 THE FUTURE OF DRUG DISCOVERY

We believe that the future of drug discovery programs will not lie in the exclusive use of combinatorial chemicals or natural products. Both of these sources provide a quantity and breadth of chemical diversity, which are best used synergistically. For HTS systems to make most effective use of natural products, some technologies made possible by the widespread use of automation will be beneficial. The creation of semi-purified or fully purified natural product libraries will reduce the number of compounds in sample pools, and make a direct information link from sample to activity possible. More advanced analytical instrument technologies, such as the coupling of liquid chromatography (LC) with nuclear magnetic resonance (NMR) or mass spectrometry (MS), can reduce the time for structural elucidation. Advances in informatics, and the application of sophisticated software for pattern analysis will make possible the automated identification of known natural product compounds in sample libraries within the next decade.[19] Applications of computational design to the development of semi-synthetic analogue libraries around bioactive natural products will increase the speed at which lead candidate development occurs. In conclusion, those pharmaceutical and agrochemical companies that integrate the use of both combinatorial and natural product libraries in their drug discovery programs will be the most successful in building lead candidate pipelines.

References

1. G. M. Cragg, D. J. Newman and K. M. Snader, *J. Nat. Prod.*, 1997, **60**, 52.
2. A. L. Demain, *Nature Biotech.*, 1998, **16**, 3.
3. H. Umezawa, *Ann. Rev. Microbiol.*, 1982, **36**, 75.
4. R. A. Houghten, C. Pinilla, S. E. Blondelle, J. R. Appel, C. T. Dooley and J. H. Cuervo, *Nature (London)*, 1991, **354**, 84.
5. A. Furka, F. Sebestyen, M. Asgedom and G. Dibo, *Int. J. Pept. Protein Res.*, 1991, **37**, 487.
6. B.A. Bunin and J. Ellman, *J. Am. Chem. Soc.*, 1992, **114**, 10997.
7. S. H. Dewitt, J. S. Kielly, C. J. Stankovic, M. C. Schroeder, D. M. R. Cody and M. R. Pavia, *Proc. Natl. Acad. Sci. USA*, 1993, **90**, 6909.
8. W.H. Moos, G. D. Green and M. R. Pavia, *Ann. Rep. Med. Chem.*, 1993, **28**, 315.
9. M.A. Gallop, R. W. Barrett, W. J. Dower, S. P. A. Fordor and E. M. Gordon, *J. Med. Chem.*, 1994, **3**, 1233.
10. E. M. Gordon, R. W. Barrett, W. J. Dower, S. P. A. Fordor and M. A. Gallop, *J. Med. Chem.*, 1994, **37**, 1385.
11. P. Zurer, *Chem. Eng. News*, 1999, **March 29**, 28.
12. C.D. Garr, J. R. Peterson, L. Schvitz, A. R. Oliver, T. L. Underiner, R. D. Cramer, A. M. Ferguson, M. S. Lawless and D. E. Patterson, *J. Biomolec. Screening*, 1996, **1**, 179.
13. A. M. Ferguson, D. E. Patterson, C. D. Garr and T. L. Underiner , *J. Biomolec. Screening*, 1996, **1**, 65.
14. D.E. Patterson, A. M. Ferguson, R. D. Cramer, C. D. Garr, T. L. Underiner and J. R. Peterson, in *High Throughput Screening, The Discovery of Bioactive Substances*, ed. J.P Devlin, Marcel Dekker, New York, 1997, pp. 243-250.
15. R. F. Service, *Science*, 1999, **285**, 184.
16. R. W. Armstrong, J. M. Beau, S. H. Cheon, W. J. Christ, H. Fujioka, W. H. Ham. L. D. Hawkins, H. Jin, S. H. Kang, Y. Kishi, M. J. Martinelli, W. W. McWhorter. Jr., M.

Mizuno, M. Nakata, A. E. Stutz, F. X. Talmas, M. Taniguchi, J. A. Tino, K. Ueda, J. Uenishi, J. B. White and M. Yonaga, *J. Am. Chem. Soc.*, 1989, **111**, 7530.

17. P. A. Wender, N. F. Badham, S. P. Conway, P. E. Floreanig, T. E. Glass, C. Graenicher, J. B. Houze, J. Jaenichen, D. Lee, D. G. Marquess, P. L. McGrane, W. Meng, T. P. Mucciaro, M. Muehlebach, M. G. Natchus, H. Paulsen, D. B. Rawlins, J. Satkofsky, A. J. Shuker, J. C. Sutton, R. E. Taylor and K. Tomooka, *J. Am. Chem. Soc.*, 1997, **119**, 2755.

18. P. A. Wender, N. F. Badham, S. P.Conway, P. E. Floreanig, T. E. Glass, J. B. Houze, N. E. Krauss, D. Lee, D. G. Marquess, P. L. McGrane, W. Meng, M. G. Natchus, A. J. Shuker, J. Sutton and R. E. Taylor, *J. Am. Chem. Soc.*, 1997, **119**, 2757.

19. D. J. Hook, E. J. Pack, J. J. Yacobucci, and J. Guss, *J. Biomolec. Screening*, 1997, **2**, 145.

2 Microbial Natural Products Discovery

MUSHROOMS, MICROBES AND MEDICINES

Tony Buss and Martin A. Hayes

Bioprocessing Unit, GlaxoWellcome Medicines Research Centre, Stevenage, UK

1 INTRODUCTION

1.1 Early Records Of Medical Use And Toxicity

Fungi have been utilised by humans for many different purposes since ancient times. Three thousand years ago the Mayans from Guatamala and Southern Mexico used a fungus ("cuxum"), which was grown on roasted green corn, to treat ulcers and intestinal infections. They even made curious stone carvings of mushrooms, the so-called Mayan mushroom stones, which had human features and were probably associated with the use of hallucinogenic fungi during religious cult ceremonies.

The Egyptian medical texts contained in the Ebers papyrus (1550 BC) lists fermenting yeast, along with goat's dung, as a treatment for burns. Puffballs (*Calvatia gigantea*) were recovered from archaeological sites in Britain dating back approximately 2,000 years and are believed to have been used to stop wounds bleeding (a practice that was deployed during the First World War whenever bandages were in short supply).

The Chinese have long used fungi as components of traditional herbal medicines. The medicinal properties of *Lentinula edodes*, the shiitake mushroom, were first recorded in the 15th Century, during the Ming dynasty. Several hundred years later, Chen Cun Ren's "Chinese Pharmacopoeia" records the use of *Cordyceps sinensis* as a medicine by the Emperors Yong and Qian. *Polyporus umbellatus* has been used as a diuretic for treating edema and even today *P. umbellatus* is produced by the Institute of Chinese Drugs - Academy of Chinese Traditional Medicine for the treatment of lung cancer and leukaemia.

The Greeks and Romans relished truffles, mushrooms and puffballs as food and it was the Greek writer and physician Dioscorides who in the First Century, classified fungi for the first time into edible and poisonous varieties. Of course, during this period the Roman Emperor Claudius Caesar was murdered by his wife Agrippina in AD54, who poisoned him with *Amanita phalloides*, the "angel of death" or death cap mushroom. As little as one cubic centimeter of the mushroom flesh can be fatal to an adult human. It is interesting to note that "Toadstool" is a corruption of the German "Todestuhl" or death's chair! A related fungus, *Amanita muscaria* is responsible for a large number of poisonings, the principle toxin being muscarine (Figure 1) which acts on the central nervous system. It is also known as fly-agaric because when mixed with water and sugar it serves as a fatal attractant to houseflies.

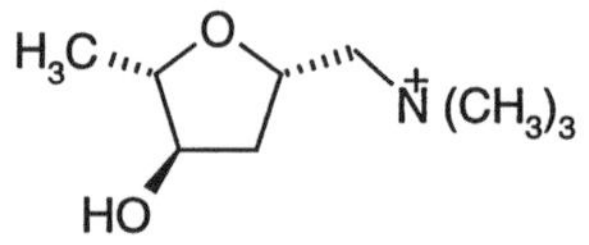

Figure 1 *Structure of muscarine*

Another common toxic fungus is the inky cap, *Coprinus atramentarius*, which contains coprine, a derivative of L-glutamine (Figure 2). It is particularly dangerous if eaten before drinking alcohol (even several days before drinking) because the aminoacid inhibits one of the key enzymes involved in alcohol metabolism, acetaldehyde dehydrogenase.

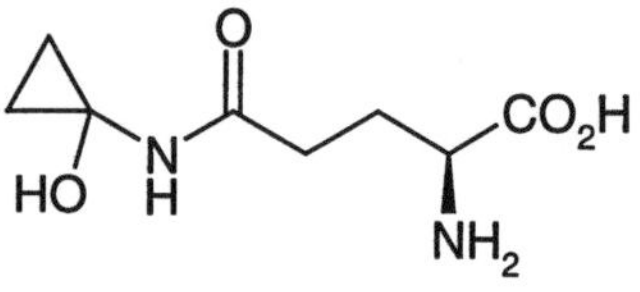

Figure 2 *Structure of coprine*

1.2   Ergot

There are numerous accounts of the use of poultices prepared from mouldy bread for treating burns, cuts and abrasions. At the turn of the century, many farmhouses in Europe had a loaf of mouldy rye bread kept in the kitchen which could be sliced and mixed with water to form a paste. Of course, rye bread was the source of another fungal contaminant, ergot. Ergot is the sclerotium, or reproductive structure, of the fungus *Claviceps purpurea* and inclusion of the sclerotia in milled flour caused devastation in Europe during the Middle Ages when ergotism killed tens of thousands of people. Ergotism is marked by vomiting, convulsions, intense burning sensations, hallucinations and lesions on the hands and feet. Eventually, the limbs develop gangrene and blacken - the term "St Anthony's Fire" was coined for the disease because of these symptoms, together with the name of the religious order established in Southern France to care for those afflicted with ergotism.

The active principles in ergot are the alkaloids ergotamine, ergotaminine, ergometrine and ergometrinine (Figure 3) which stimulate smooth muscle and block the sympathetic nervous system. Before it was realised that ergot was linked to those dreadful poisonings, it was used from the 16th Century as an obstetric "to quicken labour". Even today the ergot alkaloids are used to stimulate uterine contractions and for the treatment of bleeding after childbirth. A recent review highlights current medicinal interest in the ergot alkaloids.[1]

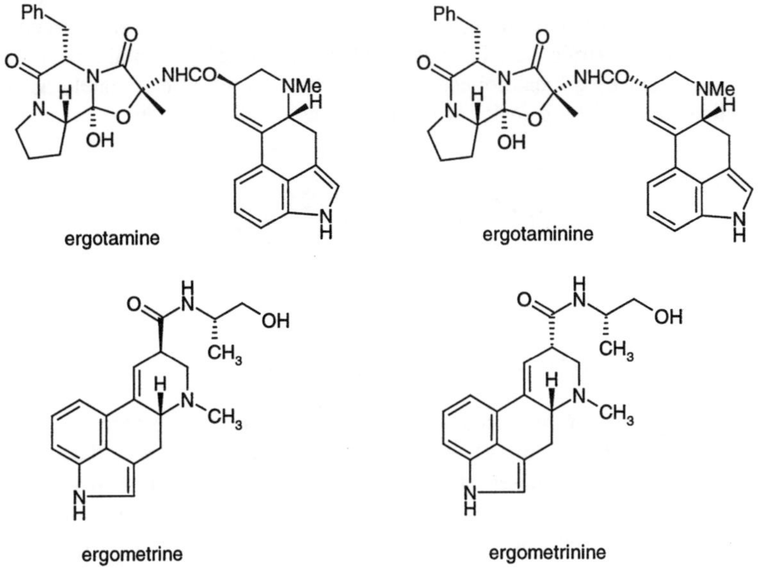

ergotamine

ergotaminine

ergometrine

ergometrinine

Figure 3 *Structures of ergot alkaloids*

1.3 Penicillins

The first scientific report of a fungus exhibiting antimicrobial activity was made by John Burdon-Sanderson (Medical Officer of Health for Paddington), who, in 1870, noted that bacteria did not grow in the presence of *Pencillium* mould. The observation was made at St Mary's Hospital in Paddington and 59 years later, Alexander Fleming (Director of the Inoculation Dept of the same hospital) published the results of his chance finding, namely that a *Pencillium* mould caused lysis of staphylococcal colonies on an agar plate. He went on to show that the culture filtrate, named penicillin, displayed activity against Gram positive bacteria and Gram negative cocci.

However, it was not until 1940 that the true therapeutic efficacy of penicillin was revealed when Florey, Chain, Heatley *et al* at Oxford successfully treated mice with semi-purified penicillin, that had previously been injected with a lethal dose of streptococci. Initially, the pharmaceutical industry was slow to initiate penicillin production because at that time material could only be produced by surface culture of *P.notatum* and large-scale manufacture presented significant technical problems.

Eventually, companies such as Pfizer and Wyeth Laboratories in the USA and Glaxo in the UK began production, the latter by growing pencillin mould in large, flat-sided bottles from which the liquors below the surface growth of mould were drained and filtered. No doubt, Glaxo's experience in dairy technology proved invaluable for operating these early, rudimentary antibiotic production facilities and by 1944 British penicillin output was sufficient to meet war-time needs for the services. Following a visit by Florey and Heatley in 1941, researchers at the US Department of Agriculture's Northern Regional Research Laboratory in Peoria, Illinois began working on improving penicillin production. Within a few months they discovered that corn-steep liquor, a by-product in the manufacture of cornstarch, gave a twelve-fold increase in yield when added to the culture medium. However, even greater improvements in yield were obtained by the use of a new strain, *P. chrysogenum*, which grew in submerged culture. Serendipitously, the strain had been isolated from a mouldy cantaloupe found at a Peoria fruit market by a local collector, Mary Hunt (known affectionately as "mouldy Mary") who had noticed it growing with "a pretty golden look" on the overripe fruit.

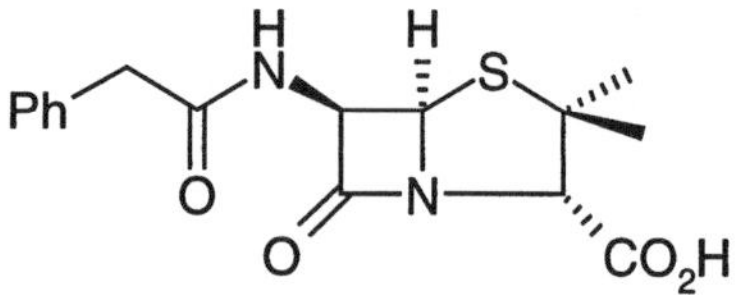

Figure 4 *Structure of benzylpenicillin*

The major penicillin produced by *P. chrysogenum* in submerged culture was benzylpenicillin (PenG) (Figure 4), which was rather unstable under acid conditions and also deactivated by β-lactamases. However, the discovery of 6-aminopenicillanic acid (6-APA) also produced by *P. chrysogenum*, led to the preparation of new semisynthetic derivatives with improved stability *e.g.* methicillin, ampicillin and amoxycillin (Figure 5).

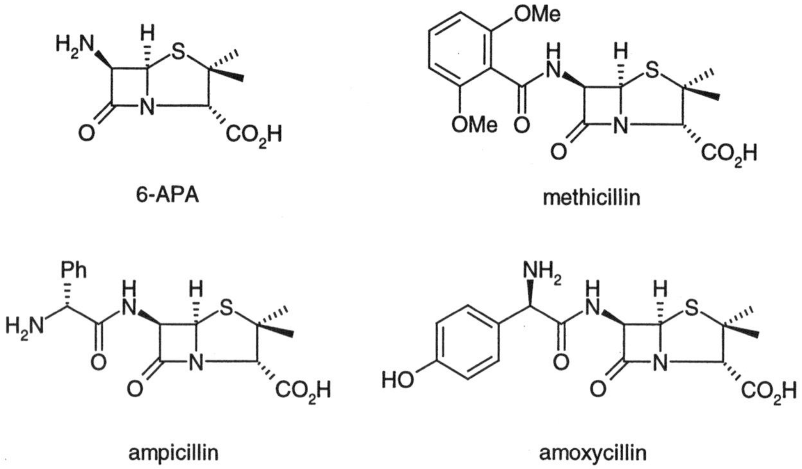

6-APA

methicillin

ampicillin

amoxycillin

Figure 5 *Structures of 6-APA and semi-synthetic penicillins*

1.4   Cephalosporins

In 1948, Giuseppe Brotzu during a search for antibiotic-producing organisms isolated the fungus *Cephalosporium acremonium* from a water sample collected off the coast of Sardinia. After Brotzu showed that the organism inhibited the growth of typhoid bacilli, a friend arranged for Howard Florey (at Oxford) to receive a culture sample from Italy and eventually an antibiotic substance was isolated and named cephalosporin C (Figure 6).

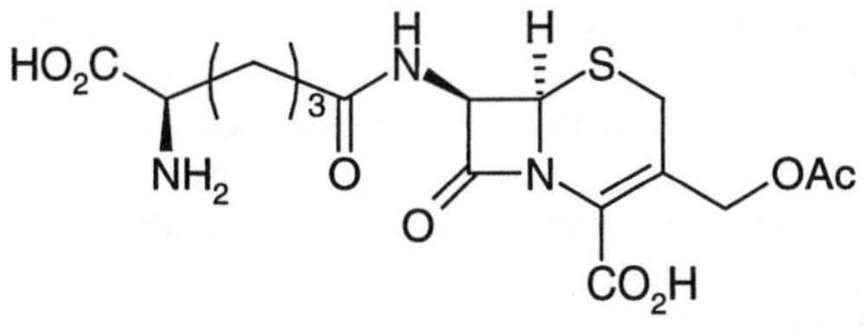

Figure 6 *Structure of cephalosporin C*

The compound (which had a dihydrothiazine ring fused to the β-lactam core) showed resistance to β-lactamases and was less toxic than benzylpenicillin. The discovery that the basic building block, namely 7-aminocephalosporanic acid (7-ACA), could be synthesised, led to the preparation of numerous cephalosporin derivatives eg cephalothin, cephaloglycin (orally active), cefaclor and cefuroxime (Figure 7).

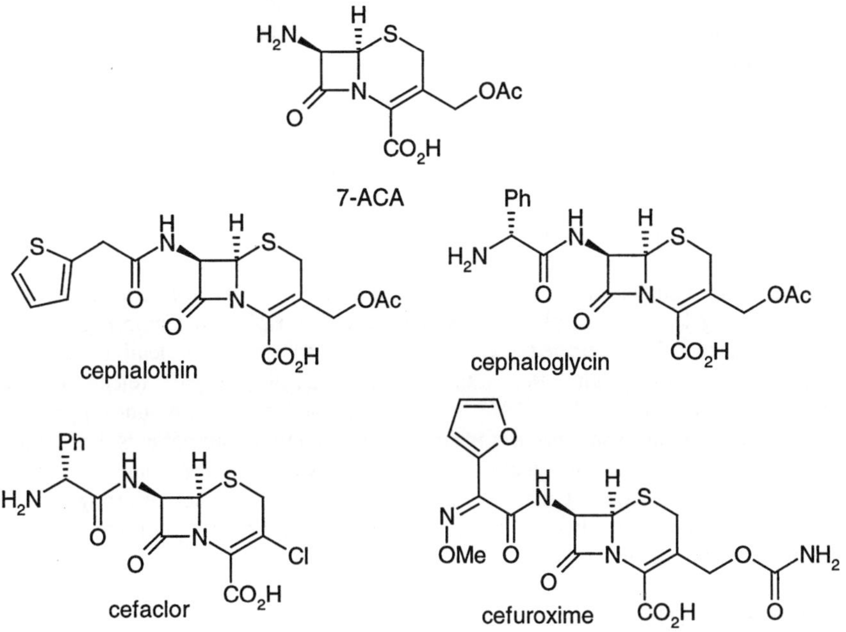

Figure 7 *Structures of 7-ACA and semi-synthetic cephalosporins*

1.5 Cyclosporin A

Following the β-lactam antibiotics, the next milestone for fungal-derived medicines came in the early 1970s. Scientists at Sandoz, in Basle, had embarked on a screening programme for novel antimicrobial agents. After screening over 600 different soil microorganisms, they discovered several closely related peptide antibiotics secreted by two strains of fungi imperfecti, *Tolypocladium inflatum* from a sample obtained in Norway and *Cylindrocarpon lucidum* from a sample collected in Wisconsin.

The most active metabolite was the novel cyclic undecapeptide, cyclosporin A (Figure 8). As an antibiotic, cyclosporin A exhibited only a very narrow spectrum of modest antifungal activity, but in 1972 its true potential was realised.[2] Jean Borel and colleagues at Sandoz discovered that cyclosporin A also neutralised cytotoxic T-cell activity *in vitro* and prevented haemagglutination in mice immunised against sheep erythrocytes.[3] Further studies revealed that cyclosporin A inhibits T-cell proliferation by blocking the synthesis of IL-2.[4]

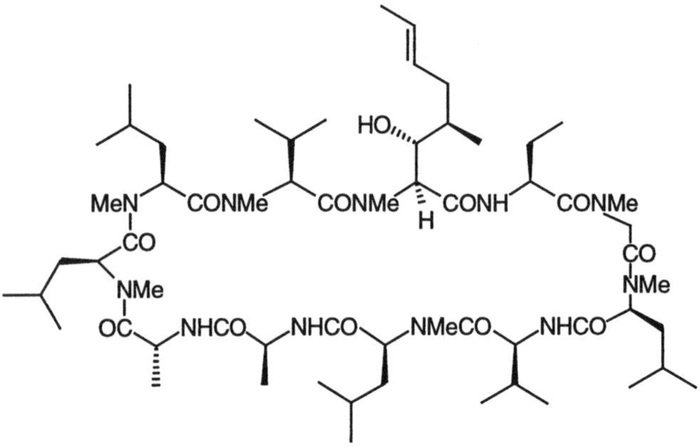

Figure 8 *Structure of cyclosporin A*

Around 1976, samples of cyclosporin A were given to Calne for clinical trials in patients receiving kidney transplants at Addenbrooke's hospital in Cambridge. These and many other trials confirmed the value of cyclosporin A in dealing with the commonest cause of death after transplantation, namely organ rejection. The introduction of cyclosporin A in 1978 was a major advance in immunosuppressive therapy and has been used successfully in several hundred thousand patients. For kidney transplantation, 90% of patients receiving cyclosporin A escape even a single rejection episode of allographs. Cyclosporin A is being evaluated in patients suffering from a variety of autoimmune-mediated diseases including nephrotic syndrome, Grave's disease and rheumatoid arthritis.

1.6   Lovastatin (Mevinolin)

Just a few years after the discovery of cyclosporin, another major fungal-derived drug was identified, namely lovastatin or mevinolin (Figure 9). Introduced by Merck to treat hypercholesterolemia, lovastatin was the forerunner of a family of lipid lowering drugs, the "statins", currently occupying over 80% of this $8 billion market.

High serum cholesterol levels are an important risk factor for coronary heart disease and an effective way of reducing serum cholesterol is to inhibit sterol biosynthesis. The "statins" work by inhibiting 3-hydroxy-3-methylglutaryl CoA reductase (HMG CoA reductase), a major regulatory enzyme in the cholesterol biosynthetic pathway which catalyses the rate-limiting conversion of HMG CoA to mevalonic acid.

The first "statin" was discovered by Endo and coworkers at Sankyo in Tokyo in 1976.[5] After testing over 8,000 microbial extracts, they found a compound, named mevastatin (ML-236B) (Figure 9), from *Penicillium citrinum* which showed specific inhibition of HMG CoA reductase and functioned *in vivo*, lowering serum cholesterol levels. Further development of mevastatin was curtailed because inhibition of cholesterol biosynthesis was not restricted to the liver. The compound enters the lens and adrenals, where it blocks the essential biosynthesis of cholesterol. The same compound was also isolated by a team at the Beecham Research Laboratories in Brockham Park, Surrey.[6] Whilst screening for antifungal activity, they isolated the compound, which they named compactin, from *Pencillium brevicompactum*, but apparently failed to recognise it as a potent inhibitor of HMG CoA reductase.

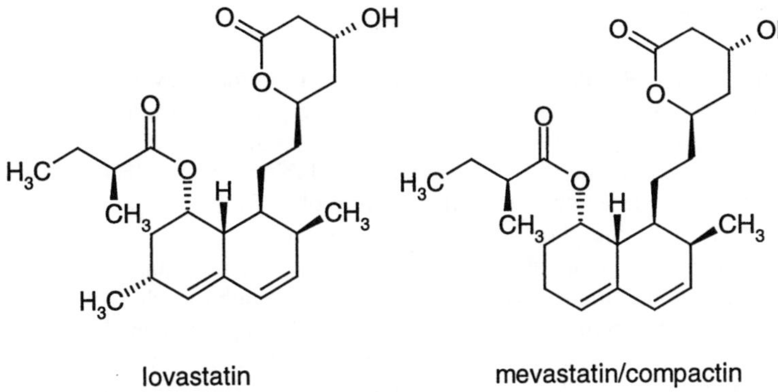

lovastatin   mevastatin/compactin

Figure 9 *Structures of lovastatin and compactin*

After spending 3 years developing systems to search effectively for HMG CoA reductase inhibitors, in 1978 Alberts and others at Merck began screening microbial extracts. At the beginning of only the second week of testing, the group discovered an active extract and by February the following year, lovastatin was isolated in pure form.[7] The producing organism had been isolated from a soil sample collected in Spain and was identified as *Aspergillus terreus*. Lovastatin was given FDA approval in 1987 for patients with high cholesterol levels that could not be reduced by diet.
The drug was later approved for marketing in 42 additional countries. Patients with serum cholesterol levels of 450mg/dl and above had decreases, within weeks of treatment with lovastatin, of 30% or more in serum cholesterol.

This key discovery prompted further efforts to develop improved cholesterol lowering agents. For example, chemical modifications of the methylbutyryl sidechain gave simvastatin (having twice the potency of lovastatin) (Figure 10) which is also marketed by Merck. Pravastatin, marketed by Sankyo and Bristol Myers Squibb, is the 6-hydroxy open hydroxyacid derivative produced by microbial biotransformation of mevastatin (Figure 10).[8]

2   SQUALESTATINS

The product of HMG CoA reductase, mevalonate, is also used for the biosynthesis of non-sterol products such as dolichol and ubiquinone as well as isoprenylation of proteins. There is potential for more selective inhibition of cholesterol biosynthesis by

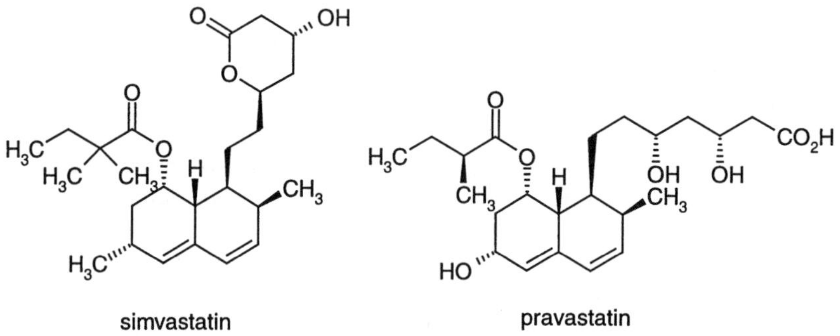

simvastatin

pravastatin

Figure 10 *Structures of simvastatin and pravastatin*

inhibiting steps beyond the branch in the pathway. Therefore, several groups have investigated squalene synthase, the first enzyme in the pathway committed solely to sterol biosynthesis, as a target for new cholesterol-lowering drugs.

At GlaxoWellcome, an assay was developed for inhibition of squalene synthase in rat liver homogenates. Based on a 96 well microtitre plate format, the assay measured incorporation of $[2\text{-}^{14}C]$ farnesyl diphosphate into squalene.

After examining a large number of synthetic compounds, plant and microbial extracts in this array, one fungus was found to produce an inhibitor which was specific to squalene synthase. The fungus was isolated from a soil sample collected from a cliff-top site in the Algarve, Portugal and was later classified as a species of *Phoma*. The fungus proved difficult to handle - producing morphological forms varying from large white pellets to dense black mycelium in supposedly identical cultures. It proved possible to control the problem by homogenising seed cultures by passaging through a syringe. In addition, development of a production medium containing glycerol, soybean oil and cottonseed flour give much improved yields of the active components. The main metabolite was characterised as a novel highly functional dioxabicyclo [3.2.1] octane system which we named squalestatin A1 [9, 10] (subsequently reported by Merck as zaragozic acid A) (Figure 11).[11]

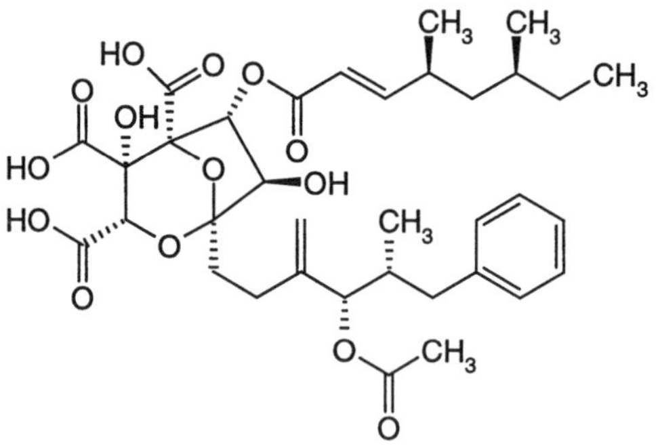

Figure 11 *Structure of squalestatin S1*

The squalestatins proved to be potent inhibitors of both mammalian and fungal squalene synthase. Further testing showed a broad spectrum of anti-fungal activity and inhibition of cholesterol biosynthesis in isolated rat hepatocytes. Dosing marmosets (which possess a similar lipoprotein profile to man) orally at 100mg/kg/day reduced serum cholesterol by up to 75%. Low density lipoprotein levels (characterised by Apo-B) were dramatically reduced whereas high density levels (characterised by Apo-A1) were unchanged.

The squalestatins represent the first novel class of squalene synthase inhibitors to be isolated from a microbial source. Numerous attempts have been made to synthesise inhibitors chemically, but the most active of these is still 10-fold less potent than the squalestatins.

3   SORDARINS

A more recent discovery at GlaxoWellcome shows that there is still the potential for finding novel, fungal-derived antibiotics. During the course of screening for inhibitors of *Candida albicans* protein synthesis, an extract from a strain of *Graphium putredinis* obtained from an excavation site in Leeds was investigated. The extract showed potent activity in the screen but, interestingly, lacked any activity in a mammalian (rabbit reticulocyte) protein synthesis assay. The active principle was isolated and characterised as a novel diterpene glycoside (GR135402) (Figure 12).[12,13] The compound showed good *in vitro* activity against a limited range of pathogenic fungi and *in vivo* activity in a murine model of systemic candidosis (ED$_{50}$ 50mg/kg).

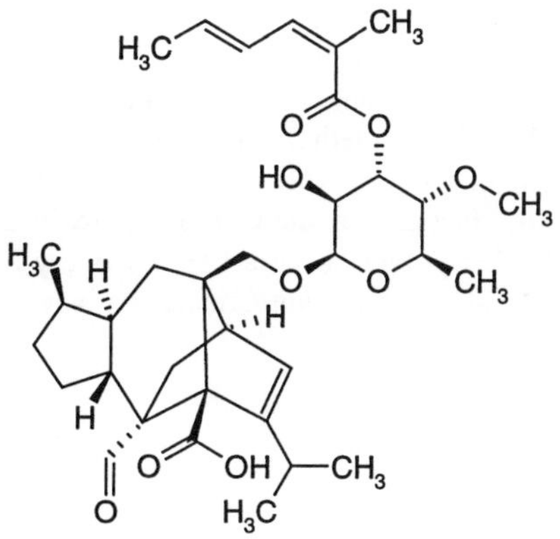

Figure 12 *Structure of GR135402*

Production levels of GR135402 from *G. putredinis* in submerged culture were very low, initially about 1mg/L. Experiments with solid state cultures showed that good levels of GR135402 were produced where the fungus was grown on a mixture of moistened sawdust and bran. Therefore, attempts were made to produce the compound in submerged liquid culture supplemented with wheatbran and eventually the titre was increased to around 15mg/L. The modest level of productivity, coupled with the limited spectrum of activity displayed by GR135402, prompted us to initiate a semi-synthetic chemistry programme based on an alternative starting material.

GR135402 is similar in structure to metabolites, eg zofimarin and sordarin (Figure 13), isolated from other fungi, including *Pencillium, Xylaria* and *Sordaria* species. We obtained a strain, *Sordaria araneosa*, reported in 1969 by Stoll and coworkers at Sandoz to produce sordarin (a closely related diterpene glycoside lacking the 3'-acyl substituent present in GR135402).[14]

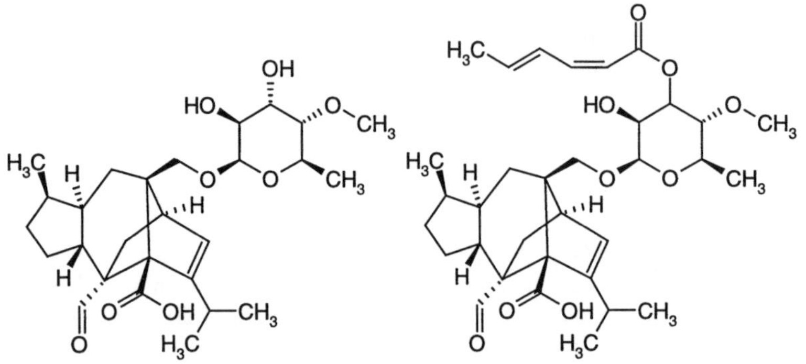

sordarin

zofimarin

Figure 13 *Structures of sordarin and zofimarin*

Excellent fermentation titres, typically 200mg/L, were obtained and sordarin became our preferred starting material. The nature of the molecule meant that opportunities for semi-synthetic derivatisation were limited, particularly modifications to the aglycone (both the acid and aldehyde functions proved essential for antifungal activity). In an attempt to increase titres still further, mutants of *Sordaria araneosa* were generated by treating the ascospores with MNNG (1-methyl-3-nitro-1-nitrosoguanidine). Levels of sordarin up to 0.5g/L were obtained, however, we noticed that one mutant did not produce any sordarin at all. Instead, a new metabolite was obtained, 4'-O-demethylsordarin (Figure 14). This proved to be a key discovery because it served as the starting material for a novel class of semi-synthetic derivatives

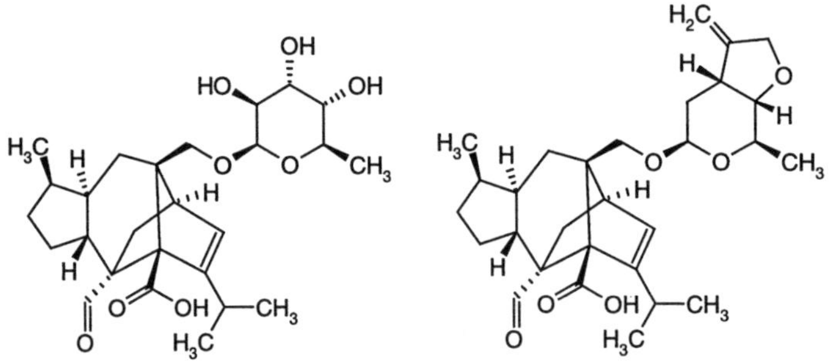

4'-O-demethylsordarin

GM237354

Figure 14 *Structures of 4'-O-demethylsordarin and GM237354*

which progressed to pre-clinical development. One such compound, GM237354 (Figure 14), displays excellent *in vitro* and *in vivo* activity against a range of clinical pathogens, including *C.albicans, Cryptococcus neoformans* and *Pneumocystis carinii*.

4   CONCLUSION

Fungi are the second largest group of eukaryotic organisms and rank second only to the insects in estimated species biodiversity. There are approximately 72,000 recognised species of fungi of an estimated 1 - 1.5 million in total. This means that <5% of fungal species have been described to date.

Given that a quarter of all know biologically active natural products are derived from fungal sources, the world's undescribed fungi can be viewed as a massive resource for potential new medicines waiting to be utilised.

References

1   C. Hart, *Modern Drug Discovery*, 1999, **2**, 20
2   H. F. Stahelin, *Experientia*, 1996, **52**, 5
3   J. F. Borel, C. Feurer, H. U. Gubler and H. Stahelin, *Agents Actions*, 1976, **6**, 468
4   T. Ochiae and K. Isono, *Surgery Today*, 1997, **27**, 88
5   A. Endo, M. Kuroda and Y. Tsujita, *J. Antibiotics*, 1976, **29**, 1346
6   A. G. Brown, T. C. Smale, T. J. King, R. Hasenkamp and R. H. Thompson, *J. Chem. Soc. Perkin Trans. 1*, 1976, 1165
7   A. W. Alberts, J. Chen, G. Kuron, V. Hunt, J. Huff, C. Hoffman, J. Rothrock, M. Lopez, H. Joshua, E. Harris, A. Patchett, R. Monaghan, S. Currie, E. Stapley, G. Albers-Schonberg, O. Hensens, J. Hirshfield, K. Hoogsteen, J. Liesch and J. Springer, *Proc. Natl. Acad. Sci. USA*, 1980, **77**, 3957
8   A. Endo and K. Hasumi, *Nat. Prod. Rep.*, 1993, **10**, 541
9   M.J. Dawson, J.E. Farthing, P.S. Marshall, R.F. Middleton, M.J. O'Neill, A. Shuttleworth, C. Stylli, R.M. Tait, P.M. Taylor, H.G. Wildman, A.D. Buss, D. Langley and M.V. Hayes, *J. Antibiotics*, 1992, **45**, 639
10   P.J. Sidebottom, R.M. Highcock, S.J. Lane, P.A. Procopiou and N.S. Watson, *J. Antibiotics*, 1992, **45**, 648
11   J. D. Bergstrom, M. M. Kurtz, D. J. Rew, A. M. Amend, J. D. Karkas, R. G. Bostedor, V. S. Bansal, C. Dufresne, F. L. VanMiddlesworth, O. D. Hensens, J. M. Liesch, D. L. Zink, K. E. Wilson, J. Onishi, J. A. Milligan, G. Bills, L. Kaplan, M. Nallin Omstead, R. G. Jenkins, L. Huang, M. S. Meinz, L. Quinn, R. W. Burg, Y. L. Kong, S. Mochales, M. Mojena, I. MartinF. Pelaez, M. T. Dietz and A. W. Alberts, *Proc. Natl. Acad. Sci. USA*, 1993, **90**, 80
12   O.S. Kinsman, P.A. Chalk, H.C. Jackson, R.F. Middleton, A. Shuttleworth, B.A.M. Rudd, C.A. Jones, H.M. Noble, H.G. Wildman, M.J. Dawson, C. Stylli, P.J. Sidebottom, B. Lamont, S. Lynn and M.V. Hayes, *J. Antibiotics*, 1998, **51**, 41
13   T.C. Kennedy, G. Webb, R.J.P. Cannell, O.S. Kinsman, R.F. Middleton, P.J. Sidebottom, N.L. Taylor, M.J. Dawson and A.D. Buss, *J. Antibiotics*, 1998, **51**, 1012
14   H.P. Sigg and C. Stoll (Sandoz Ltd): Antibiotic SL2266, UK 1,162,027, August 20, 1969

SIGNAL TRANSDUCTION INHIBITORS FROM MICROORGANISMS

Stephen K Wrigley, David A Kau, Barbara Waters* and Julian E Davies*

TerraGen Discovery (UK) Ltd, 545 Ipswich Road, Slough, Berkshire, SL1 4EQ, UK

*TerraGen Discovery Inc, Suite 300, 2386 East Mall, UBC, Vancouver, BC, Canada, V6T 1Z3

1 INTRODUCTION

The control of cellular proliferation and differentiation in response to external stimuli is achieved by signal transduction pathways, which are regulated in part by the co-ordinated action of protein kinases and phosphatases. In eukaryotic cells the protein kinases involved fall primarily into two classes, those that phosphorylate tyrosine residues and those that are specific for serine and threonine residues. Prokaryotic cells also rely on protein phosphorylation cascades for regulation of cellular activities, but the kinases involved are primarily histidine kinases, which are part of the sensing domain of two-component regulatory systems. These histidine kinases and their associated response regulators are involved in a range of adaptive responses by bacteria.

Natural products have played an important role in the elucidation of signal transduction pathways. The immunosuppressive drugs cyclosporin A, from the fungus *Tolypocladium inflatum* and FK506 (Tacrolimus), from *Streptomyces tsukubaensis*, are widely used for the prevention of organ transplant rejection, while rapamycin, from *Streptomyces hygroscopicus*, is in development for similar applications.[1] All three compounds interfere with T cell activation. Cyclosporin A and FK506 were already in clinical use by the time they were used as research tools to reveal the role of the phosphatase calcineurin in the intracellular signal transduction pathways of the immune response.[1] The indole carbazole staurosporine was discovered as a *Saccharothrix* sp. metabolite in the course of a physico-chemical screening programme for microbial alkaloids and then subsequently found to be a potent inhibitor of protein kinase C (PKC).[2] The tyrosine kinase inhibitor erbstatin was identified as a *Streptomyces* sp. metabolite after detection in an assay for inhibitors of epidermal growth factor receptor (EGF-R) tyrosine kinase.[3] It was later shown to inhibit other tyrosine kinases but not PKC, and to be an inhibitor of angiogenesis.[4] Another compound used extensively as a tyrosine kinase inhibitor is the widespread isoflavonoid, genistein, which has also been shown to be a histidine kinase inhibitor.[5] Two other noteworthy signal transduction inhibitors of microbial origin, wortmannin and geldanamycin, do not directly affect protein kinases. Wortmannin is a terpenoid produced by various fungi, including *Talaromyces wortmannii*, which is a potent inhibitor of phosphatidylinositol 3-kinase.[1, 6] Geldanamycin is a benzoquinone ansamycin produced by *Streptomyces* spp., which indirectly inhibits cellular tyrosine kinases by interacting with the heat shock protein HSP90. A geldanamycin derivative, 17-allylamino-17-demethoxygeldanamycin, is currently in phase 1 clinical trials as an antitumour agent.[1, 7]

Natural products, in particular those produced by microorganisms, have proved to be a rich source of useful signal transduction inhibitors with a variety of mechanisms of action. They continue to be of great interest as a source of new signal transduction inhibitors with potential for therapeutic use in the areas of immunosuppression, oncology and infectious disease.

2   TERRAGEN DISCOVERY INC

2.1   Origins and Technologies

TerraGen was founded in 1996 as TerraGen Diversity Inc in Vancouver, Canada, evolving from the West-East Centre for Microbial Diversity at the University of British Columbia. The original company focus was on developing recombinant genetic approaches to accessing the biosynthetic capabilities of organisms in the environment that are currently uncultivable or not readily cultivable in the laboratory. In Spring 1999, TerraGen expanded significantly through the acquisitions of ChromaXome Corp of San Diego, California, and Xenova Discovery Ltd of Slough, UK, to broaden its technology platform and patent portfolio. The combined company, TerraGen Discovery Inc, is now based in Vancouver and Slough and possesses a wealth of resources and technologies for the discovery, identification and structural manipulation by genetic means of microbial products from both environmental sources and a well-characterised microbial culture collection.

The company's molecular biology-based activities are centred in Vancouver. Interest continues to focus on the direct isolation of high molecular weight DNA from environmental samples such as soil, containing microbial populations which are in large part unrepresented in existing culture collections,[8, 9] or fastidious organisms such as lichen symbionts.[10] These DNA fragments are then cloned and expressed in robust, cultivable surrogate hosts to generate libraries of small molecules for high-throughput screening and drug discovery (NatGen™). A second approach involves the cloning and expression of partial or whole biosynthetic pathways from known organisms in new host organisms and the random recombination and re-arrangement of fragments of these pathways to produce novel compounds or modify lead compounds. TerraGen holds two fundamental issued US patents covering these recombinant genetic approaches to natural product discovery.[11, 12] Both approaches rely on having high-throughput fermentation and screening approaches to identify recombinant strains which have incorporated and expressed new, biosynthetically functional DNA fragments. TerraGen has developed "macrodroplet" technology to encapsulate and ferment recombinant strains in calcium alginate droplets, which can then be screened for the production of biological activity.

TerraGen Discovery (UK) Ltd comprises the company's fermentation, high-throughput screening and natural products chemistry operations. These have been historically based on a culture collection of fungi and actinomycetes which has been assembled with a focus on maximizing taxonomic, ecological and geographic biodiversity with particular emphasis on organisms likely to be involved in many interactions with other organisms.[13] The collection is well-characterised, with approximately 60% of the fungi and 40% of the actinomycetes identified to at least genus level, and has been extensively investigated with respect to its production of biological activity against a range of targets and chemical productivity.[14] Organisms in the collection are fermented using a variety of approaches, including fermentation on solid substrates, to generate compound libraries for screening (NatChem™).

2.2   Signal Transduction Inhibitor Discovery

The remainder of this article will focus on very different strategies developed and used by Xenova Discovery and TerraGen scientists for the discovery of inhibitors of signal transduction processes.   Xenova Discovery Ltd had used a number of different high-throughput screening approaches to discover inhibitors of signaling processes ranging from assays based on isolated enzymes, such as bacterial protein histidine kinase, to cell-based assays for inhibitors of cytokine production and signaling, such as the macrophage activation and CD28 signal transduction screens.   These screens were all productive in terms of the discovery of interesting and novel signal transduction inhibitor leads. On the merger of TerraGen with Xenova Discovery Ltd, these lead compounds proved to be useful tools for the validation of a new assay for the discovery of inhibitors of eukaryotic protein kinases and phosphatases developed by TerraGen. This cell-based assay employs a specific streptomycete as tester organism; inhibition of aerial mycelium formation and sporulation being indicative of the modulation of kinase activity.   The assay had previously been validated with the use of known protein kinase inhibitors such as the tyrphostins.

3   BACTERIAL PROTEIN HISTIDINE KINASE INHIBITORS

3.1   Objective and Screening Assay

The bacterial protein histidine kinase screen was run as one element of a collaboration between Xenova Ltd and Parke-Davis Pharmaceutical Research Division of Warner-Lambert Co. Protein histidine kinases form the sensory element of two component signal transduction systems. These systems are widespread in prokaryotes, show clear homology to one another and play an important role in regulation of gene expression to enable bacteria to adapt to stressful or changing conditions.[15] They are also present in some eukaryotes and plants.[15] Inhibitors of these systems could severely limit the ability of bacteria to colonise and cause disease in a host organism and so they constitute an interesting target for antibacterial therapy.   A screening assay was designed to detect inhibitors of the nitrogen regulator II (NRII) histidine kinase from *Escherichia coli*. The assay was based on autophosphorylation of NRIIc, a fusion protein consisting of maltose binding protein at the amino terminus and NRII at the caboxy terminus.[16] On termination of the reaction the protein was collected by filtration and the amount of radioactivity incorporated from $[^{33}P]$-ATP measured by scintillation counting.[17]

3.2   Screening Results and Compounds Characterised

The histidine kinase assay was used to test 64,000 microbial fermentation (NatChem™) extracts, of which 1.8% was found to be active. Active samples were first prioritized on the basis of their potency and selectivity of action for histidine kinase with respect to an unrelated protein kinase. After chemical fingerprinting by high-performance liquid chromatography with photodiode array and mass spectrometric detection (HPLC-PDA-MS), using the Waters Integrity™ system, and small-scale chemical fractionation studies, a number of extracts where activity was due to the presence of fatty acids were eliminated. Seven organisms progressed to scale-up fermentation and chemical investigations. After assay-guided purification, active compounds from three of these organisms were deprioritised due to low potency. The structures of thirteen active compounds isolated from the other four organisms were determined and, at the time, twelve of these were novel. One example was XR770, a phenalenone derivative produced by *Penicillium* cf. *herquei* 20421, the structure of which is shown in Figure 1. This compound inhibited NRIIc with an IC_{50} value (concentration at which 50% inhibition is observed) of 20 μM. It did not have any significant whole cell antibacterial activity against a panel of test organisms. XR770 has subsequently been published independently as erubalenol B, an inhibitor of cholesteryl ester transfer protein produced by a *Penicillium* sp.[18, 19]

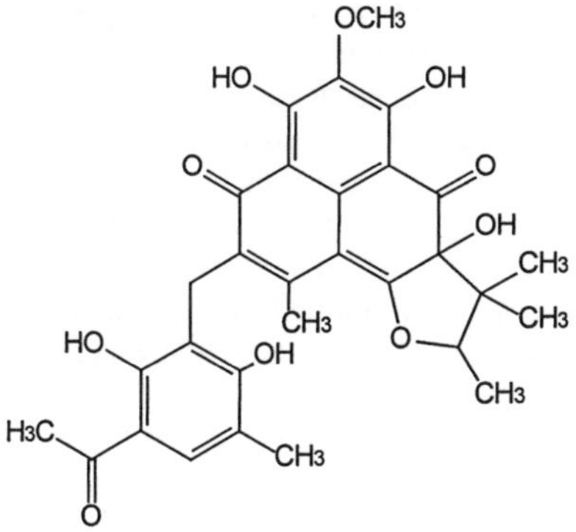

Figure 1 *Structure of XR770, erubalenol B*

The most interesting series of protein histidine kinase inhibitors discovered by this screening programme were produced in fermentations of an actinomycete strain X10/78/978 subsequently identified as *Streptomyces rimosus*.[17] Nine members of this series of pyrrolo [2,1-b][1,3] benzoxazines were isolated and characterized.[17] The structure of the major fermentation product, XR587, is shown in Figure 2. XR587 inhibited NRIIc with an IC_{50} of 20 μM and was also active against a range of Gram positive bacteria, including drug resistant *Staphylococcus aureus* strains, with minimum inhibitory concentration (MIC) values as low as 1 μg/ml.[17] Other compounds of this series were found as minor fermentation products and differed from XR587 with respect to the length of the alkyl side chain, the presence of O-methylation and the degree of halogenation of the pyrrole ring. A brominated analogue of XR587 was generated using fermentation production medium supplemented with sodium bromide. These XR587 analogues showed some variation with respect to their inhibition of NRIIc and antibacterial properties but none was significantly more active than XR587.[17] The fermentation of *S. rimosus* X10/78/978 was rapidly developed to the point where titres of XR587 of over 600 mg/litre were being achieved in 50L fermentations. A number of semi-synthetic and synthetic analogues of XR587 have been made but the compound is no longer being actively investigated. As this work was being completed, XR587 was independently published as streptopyrrole, a metabolite of *Streptomyces armeniacus* with antimicrobial properties.[20]

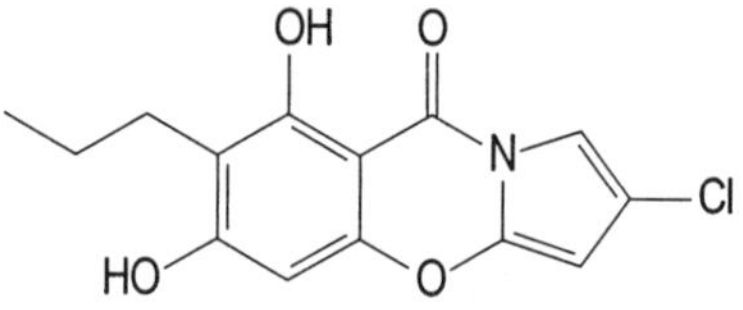

Figure 2 *Structure of XR587, streptopyrrole*

4   CYTOKINE PRODUCTION INHIBITORS

Pro-inflammatory cytokines are important mediators of inflammation and tissue destruction. This section describes two cell-based assays that were used to screen for inhibitors of cytokine production and some of the compounds discovered using these screens. The two screens were important elements of a collaboration between Xenova Ltd and the Suntory Institute of Biomedical Research to find microbial metabolites with potential utility for the treatment of rheumatoid arthritis. Both screens were cell stimulatory assays with similar formats, the principle of which is illustrated in Figure 3. Treatment of cells with a particular stimulus activates a signal transduction pathway, one of the end results of which is production of a cytokine, which is secreted into the assay medium. After a separation step, the cytokine of interest is measured quantitatively in the supernatant by dissociation enhanced lanthanide fluorescence immunoassay (DELFIA) using a europium-labeled tertiary antibody. At the same time, cytotoxic properties of test substances are determined by assessing their effect on proliferation of the separated cells.

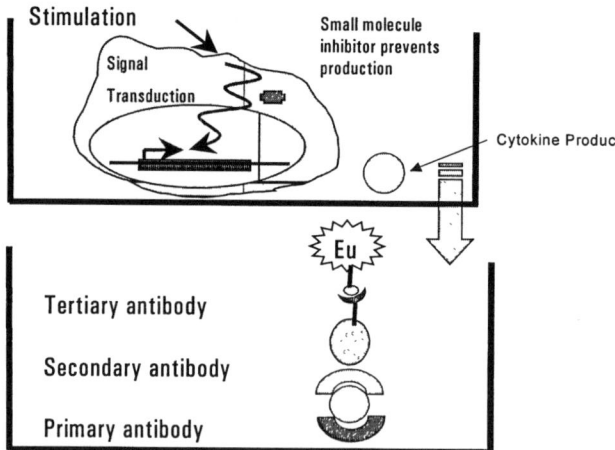

Figure 3   *Principle of cell stimulatory assays for detection of inhibitors of cytokine production*

4.1 Inhibitors of Macrophage Activation

4.1.1 Objectives and Screening Assay. Tumour necrosis factor α (TNFα) and interleukin-1β (IL-1β) are potent pro-inflammatory cytokines produced principally by activated macrophages and which are known to be of great importance in the regulation of immune responses.[21, 22] Both cytokines are important targets for the discovery of potentially useful anti-inflammatory agents for the treatment of conditions such as Crohn's disease, multiple sclerosis and rheumatoid arthritis. A high throughput assay was developed to screen for inhibitors of TNFα production by human histolytic lymphoma U937 cells.[23] After treatment with phorbol myristate acetate (PMA), cells were exposed to test samples followed by lipopolysaccharide (LPS). An eighteen hour-incubation was followed by harvest of the cell culture supernatants for determination of the level of TNFα secreted by DELFIA. Effects of the test sample on cell viability were measured using the tetrazolium salt, XTT (2,3-bis[2-methoxy-4-nitro-5-sulphophenyl]-2H-tetrazolium-5-carboxanilide salt).

4.1.2 Screening Results and Compounds Characterised. The macrophage activation assay was used to test over 95,000 NatChem™ microbial fermentation extracts, of which 1.4% was found to be active. A significant proportion of these hits were due to the presence of a number of relatively common microbial metabolites, which caused potent inhibition of LPS-stimulated TNFα production by U937 cells. A selection of these is shown in Figure 4. Cycloheximide is a well-known *Streptomyces* spp. metabolite which inhibits protein synthesis but did not have an obvious cytotoxic effect on U937 cells over the time period of the assay; brefeldin A is an inhibitor of intracellular protein transport with dramatic effects on the Golgi apparatus,[24] that caused potent activity in this assay and was found in extracts from a number of fungi, predominantly *Penicillium* spp.; geldanamycin, which proved to be a fairly common *Streptomyces* spp. metabolite, and monorden (radicicol), which was produced by a number of fungal strains, cause similar effects on cellular tyrosine kinase activity by inhibiting the ATPase activity of the HSP90 molecular chaperone, involved in the folding of protein kinases.[25] The presence of these and other known active metabolites was routinely detected through chemical fingerprinting by HPLC-PDA-MS, which proved to be an extremely effective tool for chemical dereplication and prioritization of samples for further investigation.

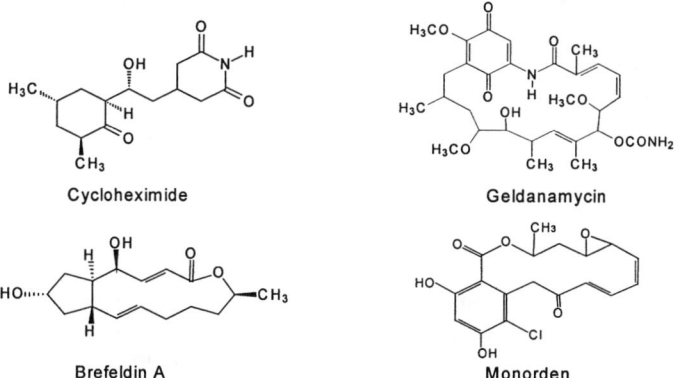

Cycloheximide

Geldanamycin

Brefeldin A

Monorden

Figure 4 *Structures of common microbial metabolites active in the macrophage activation assay*

Among the novel compounds discovered using the macrophage activation screen were two lead series of fungal metabolites, which caused potent inhibition of cytokine production but possessed very different properties. The first of these was a family of resorcylic acid lactones produced in fermentations of the ascomycete, *Cochliobolus lunatus* 20416, the structures of which are shown in Figure 5. The most potent inhibitor of LPS-induced TNFα production from this family was the novel 5Z-7-*oxo*-zeaenol (XR336), with an IC$_{50}$ of 6nM. The isomeric 7-*oxo*-zeaenol (XR318), which had been previously reported,[26] was nearly one hundred-fold less active, as was monorden, another resorcylic acid lactone not produced in fermentations of this particular fungus but commonly detected by this assay. The best-known fungal metabolite of this class, zearalenone, notable for its oestrogenic properties, did not show any activity in the macrophage activation assay. XR336 inhibited LPS-induced TNFα production in whole blood experiments and lowered serum levels of TNFα in mice when administered prior to LPS.[27] It did not show any significant oestrogen or glucocorticoid receptor binding properties. XR336 was also found to inhibit the phosphorylation and activation of mitogen-activated protein kinase (MAPK) induced by LPS, consistent with a mechanism of action at or upstream of MAPK. Work on this lead series was discontinued for several reasons including reports of the same or very similar members of the same class of compounds as cytokine production inhibitors. These include a very close analogue of XR336 named "radicicol analogue A" which was found to act by a mechanism involving induction of mRNA instability.[28]

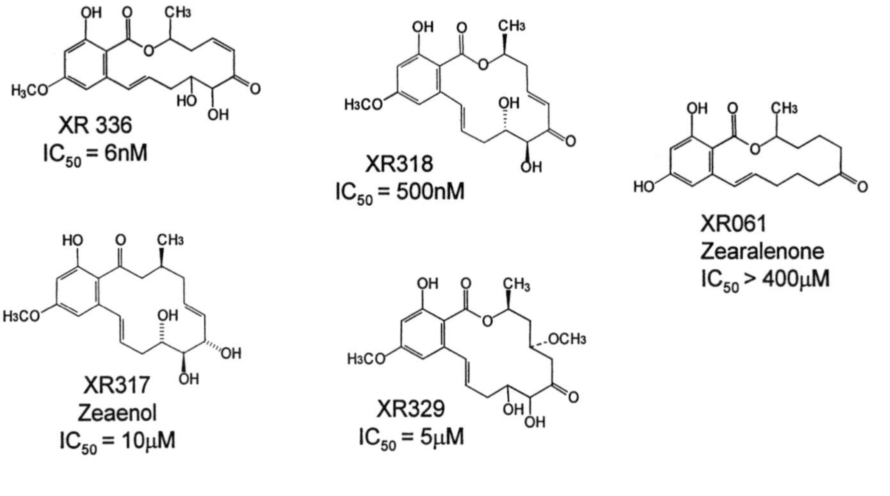

Figure 5 *Structures and IC$_{50}$s in the macrophage activation assay of fungal resorcylic acid lactones*

The second lead series of macrophage activation inhibitors was a family of novel 6-substituted 5,6-dihydro-α-pyrone esters found in fermentations of the fungal strain *Phomopsis* sp. 22502,[29] the structures of which are shown in Figure 6. The first of these to be identified was XR379, a novel (6S)-4,6-dimethyldodeca-2E,4E-dienoyl ester of a prop-2E-enoic side chain-bearing analogue of the known fungal metabolite phomalactone. This inhibited LPS-induced TNFα production by U937 cells with an IC$_{50}$ of 2μM.

Despite this moderate potency, it was deemed worthwhile to investigate the complex mixture of metabolites produced in fermentations of this fungus for the presence of more active members of this metabolite series. The effort paid off with the isolation and characterisation of XR543, (6*S*)-4,6-dimethyldodeca-2*E*,4*E*-dienoyl phomalactone, as a minor fermentation product. XR543 was twenty five-fold more active than XR379 in the macrophage activation assay, with an IC$_{50}$ of 80nM. On secondary evaluation XR543 proved to have some interesting properties. In a more physiologically relevant assay system involving LPS-stimulated peripheral blood monocytes (PBMC), XR543 was found to have a selective inhibitory effect on IL-1β production compared to TNFα and interleukin-6 production.[29] On further investigation this was found to involve a post-translational mechanism of action at the level of IL-1β secretion. This did not appear to involve inhibition of caspase 1 (interleukin-1β converting enzyme-like enzyme).[29] A number of semi-synthetic analogues of XR543 were made to probe the activity of this compound series further. This was facilitated by the discovery of another fungal strain (*Paecilomyces* sp. 3527), which produced phomalactone itself as a major metabolite on fermentation. Not only did this provide an improved process for producing XR543 by esterification of phomalactone with XR652, (6*S*)-4,6-dimethyldodeca-2*E*,4*E*-dienoic acid, produced by *Phomopsis* sp. 22502, but also a convenient template for making a wide variety of derivatives.[30]

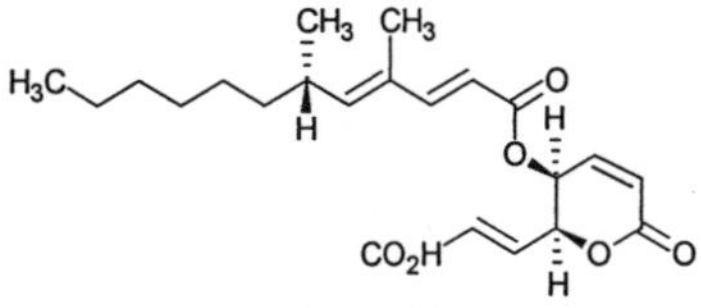

XR379 from *Phomopsis* sp. 22502
IC$_{50}$ = 2μM

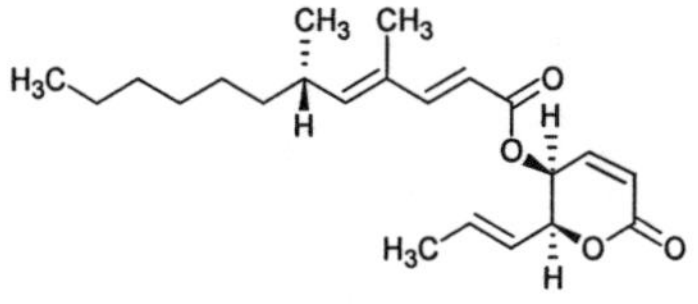

XR543 from *Phomopsis* sp. 22502
IC$_{50}$ = 80nM

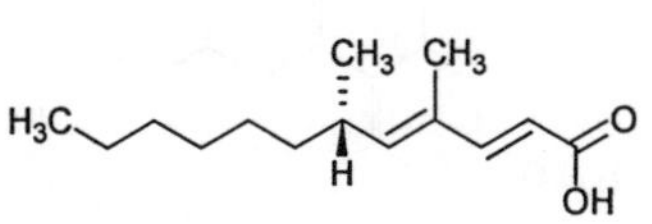

XR652 from *Phomopsis* sp. 22502
IC$_{50}$ = 54μM

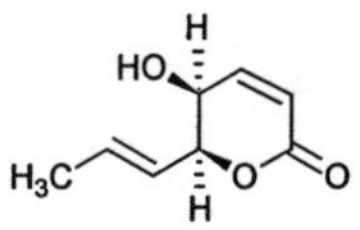

XR665 (phomalactone) from
Paecilomyces sp. 3527
IC$_{50}$ = 2μM

Figure 6 *Structures and IC$_{50}$s of macrophage activation inhibitors from* Phomopsis *sp. 22502 and* Paecilomyces *sp. 3527*

4.2 Inhibitors of CD28 Signal Transduction

4.2.1 *Objectives and Screening Assay.* Optimal T cell activation requires both the recognition of specific antigen-major histocompatibility complexes (MHC) by the T cell receptor and activation by a co-stimulatory molecule such as CD28, a dimeric, immunoglobulin-like molecule expressed on many T cells. CD28 binds two ligands, B7-1 and B7-2, which are expressed on the surface of antigen-presenting cells, both of which induce cytokine production and cell proliferation in T cells.[31] Inhibition of CD28 signalling after an immune response can impede T cell activation and lead to immunosuppression. A high-throughput assay was developed to screen for inhibitors of anti-CD28 induced interleukin-2 (IL-2) production by Jurkat E6-1 cells as potential immunosuppressants.[32] The Jurkat E6-1 cells were pre-treated with PMA and then incubated overnight with test samples before adding anti-CD28 antibody. After incubation for a further twenty hours, the cell culture supernatants were collected and IL-2 concentrations determined by DELFIA. Toxic effects of test substances were assessed by incubation of cells in the presence of the redox potential-sensitive dye, resazurin.[32]

4.2.2   *Screening Results and Compounds Characterised.*   The CD28 signal transduction assay was used to test over 97,000 Nat Chem[TM] extracts, of which 0.96% was found to be reproducibly active. As for the macrophage activation assay, a significant proportion of these hits were due to the presence of several relatively common microbial metabolites with potent, but not apparently toxic, effects, which were routinely detected and identified by HPLC-PDA-MS. A selection of these is shown in Figure 7: K-252c is an indole carbazole related to staurosporine produced in fermentations of some *Streptomyces* spp. which inhibits PKC;[33] strobilurin G is a member of the strobilurin family of β-methoxyacrylate antibiotics produced by various basidiomycetes, which inhibit mitochondrial respiration but only show overt toxic effects in assays such as this over extended time periods;[34] the enniatins are a family of cyclic depsipeptides produced by *Fusarium* spp.;[35] and the zeaenol family of resorcylic acid lactones show similar effects in this assay system as for the macrophage activation assay.

Enniatin B

Strobilurin G

K-252c

5Z-7-oxo-zeanol

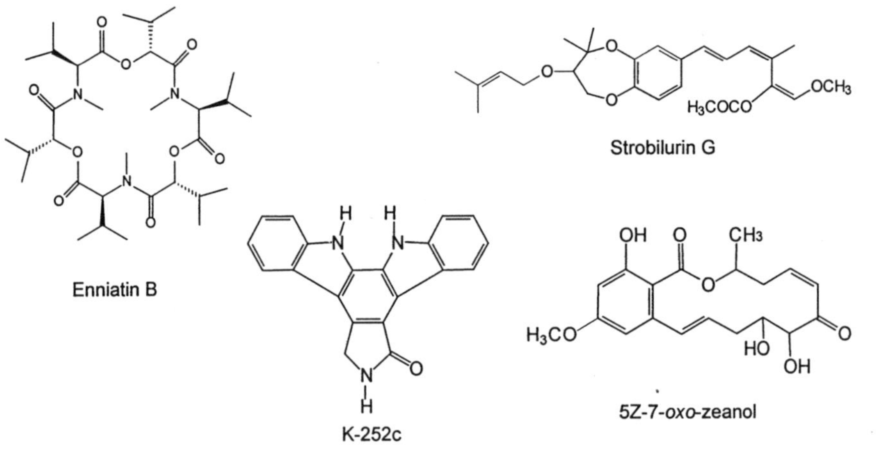

Figure 7   *Structures of common microbial metabolites active in the CD28 signal transduction assay.*

The most interesting series of lead compounds discovered using the CD28 signal transduction assay were found in fermentations of the fungus *Cladosporium* cf. *cladosporioides* 20700. Members of this compound series were found to be novel, reduced benzofluoranthene derivatives related to the known fungal metabolites bulgarein and bulgarhodin,[36] but distinguished structurally from these two highly conjugated, planar compounds by several reduction and one *O*-methylation steps. The structure of the most abundant member of this series found in fermentations of *Cladosporium* cf. *cladosporioides* 20700, XR774, and an oxidised analogue, XR819, are shown in Figure 8. XR774 had the most potent biological effects.   It inhibited anti-CD28 induced IL-2 production by Jurkat E6-1 cells with an IC_{50} of 400 nM and no cytotoxicity or effects on protein synthesis were detected at concentrations up to 14 μM. XR819 showed no effects on cytokine production at concentrations up to 5 μM. XR774 also inhibited anti-CD28 induced IL-2 production and IL-2 mRNA expression in PBMC, indicating a mechanism of action mediated at the transcriptional level. Early steps in the signal transduction process following CD28 ligation by B7-1 or B7-2 are known to involve the activation of protein tyrosine kinases. XR774 was found to inhibit selected tyrosine kinases, including Fyn, Lck, Abl and EGF-R, in *in vitro* assays with IC_{50} values in the range 20-400 nM.[32] No effect was found on inhibition of the serine-threonine kinase, protein kinase A. Kinetic analyses indicated that XR774 was a competitive inhibitor of Fyn kinase with respect to ATP. A number of semi-synthetic derivatives of XR774 have been prepared to investigate the activity of this compound series further.[37] This was expedited by the finding of another strain of *Cladosporium* cf. *cladosporioides*, collected in the same country (Thailand) and at a similar time to the original strain, which produced XR774 titres in 50L fermenters of up to 650 mg/litre.

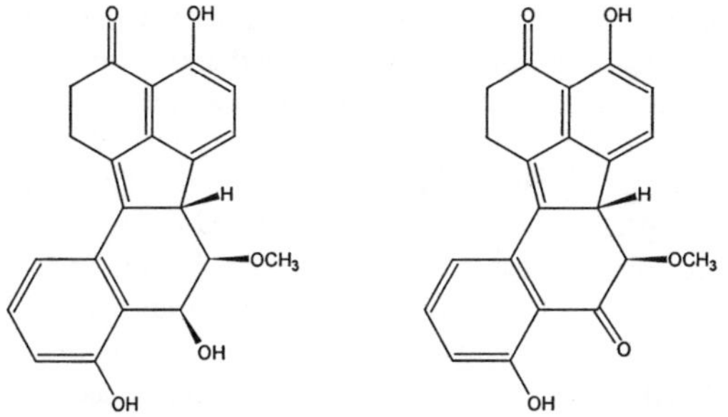

Figure 8 *Structures of XR774 and XR819 from* Cladosporium *cf.* cladosporioides *20700*

5   A PROKARYOTIC WHOLE CELL ASSAY TO IDENTIFY SIGNAL TRANSDUCTION INHIBITORS

5.1 Assay Origins and Development

This alternative approach to the discovery of signal transduction inhibitors was developed following a number of reports that eukaryotic-like kinase and phosphatase activities may complement histidine kinase-based two component regulatory systems in several different types of bacteria. Phosphotyrosine-specific antibodies were used to demonstrate that a variety of proteins in several *Streptomyces* species were phosphorylated on tyrosine residues.[38] Each of six species studied exhibited a unique pattern of protein tyrosine phosphorylation and these patterns were found to vary during growth and according to culture conditions. This suggested that the complex growth cycle of these filamentous, sporulating bacteria and their diverse secondary metabolic pathways might be controlled in part by the action of protein tyrosine kinases and phoshatases. A separate study of the effects of protein kinase inhibitors on *in vitro* protein phosphorylation and cellular differentiation in *Streptomyces griseus* found that staurosporine and K-252a inhibited the formation of aerial mycelia without affecting the growth of substrate mycelia.[39] This provided evidence that *S. griseus* possesses eukaryotic-like protein kinases which are apparently essential for morphogenesis and secondary metabolism. These observations of the effects of known kinase inhibitors on *Streptomyces* morphogenesis led to the proposal of a screening method for the discovery of inhibitors of eukaryotic signal transduction pathway components using a prokaryotic indicator strain.[40] The assay is based on *Streptomyces* strain WEC 85E, which sporulates very readily when grown on particular fermentation media. Solutions of test substances are applied to 6 mm filter paper disks, which are then placed onto freshly seeded agar plates. Active samples of interest are those that inhibit sporulation without affecting the growth of substrate mycelium. This is a rapid, sensitive and inexpensive procedure. It can be used as a simple and inexpensive pre-screen for the selection of extracts already known to be active in a cell-based assay. It is not affected either by the presence of compounds which are toxic to mammalian cells or by protease contaminants which can interfere with *in vitro* enzyme-based assays.

5.2 Assay Validation and Screening

The assay has been validated with a range of known protein tyrosine kinase inhibitors, including genistein and various tyrphostin derivatives, and shown to have the potential to detect compounds with highly selective activities regarding eukaryotic signaling networks.[41] The merger of TerraGen with Xenova Discovery provided an opportunity for further assay validation using signal transduction inhibitors acting with a variety of mechanisms discovered using the assay approaches described in the earlier sections of this article. Compounds were initially tested in 10 µg quantities on paper discs. Notable results are presented in Table 1. Unsurprisingly, the bacterial protein histidine kinase inhibitor XR587 proved to be bactericidal in this assay. The phomalactone derivative XR543, which inhibits IL-1β production in PBMC at the level of secretion, was inactive. The strong activity of the XR543 analogue, XR379, was unexpected and is in the process of being investigated further. The CD28 signal transduction inhibitor and tyrosine kinase inhibitor XR774 was highly active while its oxidized analogue XR819, which is not a CD28 signal transduction inhibitor, was inactive. The resorcylic acid lactones XR336 and XR318, which inhibit the phosphorylation of MAPK, were both weakly active. Geldanamycin did not show any activity. These results provide further proof that this screen based on sporulation of *Streptomyces* strain WEC 85E provides a useful approach for the detection of protein tyrosine kinase inhibitors which is not affected by some of the signal transduction inhibitors acting at other targets which may be found by mammalian cell-based assays.

The *Streptomyces* WEC 85E assay has been used to test fermentation extracts from two thousand actinomycete isolates. Over one hundred strains, representing possibly twenty-seven different species, were found to produce metabolites that affect the onset of sporulation. Purification and characterization of the active metabolites present in some of these extracts is currently in progress.

6   CONCLUSIONS

The natural product discovery experiences described here confirm that microorganisms continue to represent an abundant source of interesting signal transduction inhibitors. A variety of high-throughput screening approaches for the discovery of such compounds were used with varying results. The bacterial protein histidine kinase assay resulted in the

Table 1 *Activities of various signal transduction inhibitors on sporulation of* Streptomyces *strain WEC 85E*

Compound	Result
XR587	Bactericidal
XR543	Inactive
XR379	Highly active, growth inhibition at 100µg
XR774	Highly active, no growth inhibition at 100µg
XR819	Inactive
XR318	Weakly active
XR336	Weakly active
Geldanamycin	Inactive

identification of a number of novel, inhibitory compounds, the most interesting of which belonged to the XR587 (streptopyrrole) lead series. It is debatable, however, whether the antibacterial activity of XR587 was due to its histidine kinase inhibitory properties, given the disparity between its antibacterial MIC and IC_{50} for enzyme inhibition. The mammalian cell-based assays for inhibitors of cytokine production proved to have the ability to detect compounds acting by a wide variety of mechanisms ranging from the early stages of signal transduction through to product secretion. The CD28 signal transduction inhibitor XR774 showed potent inhibitory activity against a number of tyrosine kinases. The resorcylic acid lactone inhibitors of macrophage activation were also active at an early stage of signal transduction, inhibiting the phosphorylation of MAPK. In contrast, the macrophage activation inhibitor XR543 appeared to interfere with IL-1β secretion.

These compounds provided an interesting set of standards to test in the new assay for signal transduction inhibitors using *Streptomyces* strain WEC 85E, the morphogenesis of which had already been shown to be sensitive to a number of protein kinase inhibitors. The results further validate the assay as a tool for the discovery of kinase inhibitors. The strongest sporulation inhibitory effect was found for the tyrosine kinase inhibitor XR774. Of the compounds originally discovered using the macrophage activation assay, no activity was seen for the IL-1β secretion inhibitor XR543, as might be expected. The strong activity observed for the XR543 analogue, XR379 was unexpected and this is being investigated further. The resorcylic acid lactones XR318 and XR336 were weakly active, which may provide extra evidence that these are protein kinase inhibitors. One aspect of the *Streptomyces* WEC 85E assay which remains to be validated is the effect of known protein phosphatase inhibitors. Future priorities also include the characterization of the signal transduction inhibitors detected during the screening programme based on actinomycete strains and adaptation for use with the macrodroplet screening technology. This latter objective will enable the high-throughput screening of recombinant strains for the production of new signal transduction inhibitors not readily detectable by other approaches.

Acknowledgements

The authors are grateful to the many members of Xenova Discovery Ltd and TerraGen staff, past and present, who contributed to the work described here. Particular thanks go to Sally Adams, Lydie Pairet, Martin Hayes, Steve Martin, Martyn Ainsworth, Suzanne Kennedy, Roya Sadhegi and Michael Moore. We would also like to acknowledge the contributions from our collaborators at Parke-Davis Research Division of Warner-Lambert Co., Don Hupe and Eric Olson, and Suntory Ltd, Masashi Matsui, Toshio Tatsuoka and Yasunori Tawaragi. We are also very grateful to Liz Crutch for her help in preparing this manuscript.

References

1   M.E. Cardenas, A. Sanfridson, N.S. Cutler and J. Heitman, *Trends. Biotechnol.*, 1998, **16**, 427.
2   S. Omura, Y. Sasaki, Y. Iwai and H. Takeshima, *J. Antibiotics*, 1995, **48**, 535.
3   H. Umezawa, M. Imoto, T. Sawa, K. Isshiki, N. Matsuda, T. Uchida, H. Iinuma, M. Hamada and T. Takeuchi, *J. Antibiotics*, 1986, **39**, 170.
4   T. Oikawa, H. Ashino, M. Shimamura, M. Hasegawa, I. Morita, S-I. Murota, M. Ishizuka and T. Takeuchi, *J. Antibiotics*, 1993, **46**, 785.
5   J. Huang, M. Nasr, Y. Kim and H.R. Matthews, *J. Biol. Chem.*, 1992, **267**, 15511.
6   B.H. Norman, C. Shih, J.E. Toth, J.E. Ray, J.A. Dodge, D.W. Johnson, P.G. Rutherford, R.M. Schultz, J.F. Worzalla and C.J. Vlahos, *J. Med. Chem.*, 1996, **39**, 1106.
7   T.W. Schulte and L.M. Neckers, *Cancer Chemother. Pharmacol.*, 1998, **42**, 273.
8   P. Hugenholtz, B.M. Goebel and N.R. Pace, *J. Bacteriol.*, 1998, **180**, 4765.
9   Y. Wang, Z.S. Zhang, J.R. Ruan, Y.M. Wang and S.M.Ali, *J. Ind. Microbiol. & Biotechnol.*, 1999, **23**, 178.
10   V.P.W. Miao and J.E. Davies, *Developments in Industrial Microbiology,* R.H Baltz, G.D. Hegeman and P.L. Skatrud, Eds., Soc. Ind. Microbiol., 1997, p57.
11   T.C. Peterson, L.M. Foster and P. Brian. U.S. Patent 5,783,431. Issued July 21, 1998.
12   K.A. Thompson, L.M. Foster, T.C. Peterson, N.M. Nasby and P. Brian. U.S. Patent 5,824,485. Issued October 20, 1998.
13   M.I. Chicarelli-Robinson, S. Gibbons, C. McNicholas, N. Robinson, M. Moore, U. Fauth and S.K. Wrigley, *Phytochemical Diversity : A Source of New Industrial Products,* S. Wrigley, M. Hayes, R. Thomas and E. Chrystal, Eds., RSC. Cambridge, 1997, p30.
14   A.M. Ainsworth, S.K. Wrigley and U. Fauth. *Bio-exploitation of Fungi,* S.B. Pointing and K.D. Hyde, Eds., Fungal Diversity Research Series 2, Fungal Diversity Press, 2000, In press.
15   L.I. Alex and M.I. Simon, *Trends in Genetics,* 1994, **10**, 133.
16   E.S. Kamberor, M.R. Atkinson, P. Chandran and A.J. Ninfa, *J. Biol. Chem,* 1994, **269**, 28294.
17   S.J. Trew, S.K. Wrigley, L. Pairet, J. Sohal, P. Shanu-Wilson, M.A. Hayes, S.M. Martin, R.N. Manohar, M.I. Chicarelli-Robinson, D.A. Kau, C.V. Byrne, E.M.H. Wellington, J.M. Moloney, J. Howard, D. Hupe and E. Olson, *J. Antibiotics*, 2000, **53**, 1.

18 H Tomoda, N. Tabata, R. Masuma, S-Y. Si and S. Omura, *J. Antibiotics*, 1998, **51**, 618.

19 N. Tabata, H. Tomoda and S. Omura, *J. Antibiotics*, 1998, **51**, 624.

20 J. Breinholt, H. Gürtler, A. Kjaer, S.E. Nielsen and C.E. Olson, *Acta. Chem. Scand.*, 1998, **52**, 1040.

21 M. Feldmann, F.M. Brennan and R.N. Maini, *Ann. Rev. Immunol.*, 1996, **14**, 397

22 W.P. Arend and J.M. Dayer, *Arth. Rheum.*, 1995, **38**, 151.

23 T. Mander, S. Hill, A. Hughes, P. Rawlins, C. Clark, G. Gammon, B. Foxwell and M. Moore, *Int. J. Immunopharmacol.*, 1997, **19**, 451.

24 V. Betina, *Folia Microbiol.*, 1992, **37**, 3.

25 S.M. Roe, C. Prodromou, R. O'Brien, J.E. Ladbury, P.W. Piper and L.H. Pearl, *J. Med. Chem.*, 1999, **42**, 260.

26 G.A. Ellestad, F.M. Lovell, N.A. Perkinson, R.T. Hargreaves and W.J. McGahren, *J. Org. Chem.*, 1978, **43**, 2339.

27 P. Rawlins, T. Mander, R. Sadhegi, S. Hill, G. Gammon, B. Foxwell, S. Wrigley and M. Moore, *Int. J. Immunopharmacol.*, **21**, 799.

28 T. Kastelic, J. Schnyder, A. Leutwiler, R. Traber, B. Streit, H. Niggli and A. MacKenzie and D. Cheneval, *Cytokine*, 1996, **8**, 751.

29 S.K. Wrigley, R. Sadhegi, S. Bahl, A.J. Whiting, A.M. Ainsworth, S.M. Martin, W. Katzer, R. Ford, D.A. Kau, N. Robinson, M.A. Hayes, C. Elcock, T. Mander and M. Moore, *J. Antibiotics*, 1999, **52**, 862.

30 M.A. Hayes, D.J. Hardick, J.S. Tang, H. Ryder, A.J. Folkes, T. Tatsuoka and M. Matsui, GB Patent 2336362, published 20 October, 1999.

31 S.G. Ward, *Biochem. J.*, 1996, **318**, 361.

32 R. Sadhegi,, P. Depledge, P. Rawlins, S. Bahl, N. Dhanjal, A. Manic, M. Ainsworth, S. Martin, D. Kau, S. Wrigley, B. Foxwell and M. Moore, *submitted for publication.*

33 S. Nakanishi, Y. Matsuda, K. Iwahashi and H. Kase, *J. Antibiotics*, 1986, **39**, 1066.

34 K.A. Wood, D.A. Kau, S.K. Wrigley, R. Beneyto, D.V. Renno, A.M. Ainsworth, J. Penn, D. Hill, J. Killacky and P. Depledge, *J. Nat. Prod.*, 1996, **59**, 646.

35 H. Tomoda, H. Nishida, X-H. Huang, R. Masuma, Y.K. Kim and S. Omura, *J. Antibiotics*, 1992, **45**, 1207.

36 R.L. Edwards and H.J. Lockett, *J. Chem. Soc. Perkin Trans. I*, 1976, 2149.

37 R.M. Sadhegi Guilani, S.K. Wrigley, S. Bahl, S.M. Martin, D.A. Kau, J.S. Tang, M. Moore and D.J. Hardick, Int. Patent Appl. WO 2000002839, published 20 January, 2000.

38 B. Waters, D. Vujaklija, M.R. Gold and J. Davies, *FEMS Microbiol. Lett.*, 1994, **120**, 187.

39 S.K. Hong, A. Matsumoto, S. Horinouchi and T. Beppu, *Mol. Gen. Genet.*, 1993, **236**, 347.

40 J.E. Davies and B. Waters, U.S. Patent 5,770,392, Issued June 23, 1998.

41 B. Waters and J.E. Davies, manuscript in preparation.

NOVEL INHIBITORS OF LIPOPROTEIN ASSOCIATED PHOSPHOLIPASE A₂ PRODUCED BY *PSEUDOMONAS FLUORESCENS* DSM 11579

Jan Thirkettle

Department of Fermentation and Natural Product Sciences, SmithKline Beecham Pharmaceuticals, New Frontiers Science Park North, Third Avenue, Harlow, Essex, CM19 5AW, U.K.

1 INTRODUCTION

Lipoprotein associated phospholipase A_2 (LpPLA$_2$) is a serine dependent lipase of novel structure which is associated with the presence of LDL in plasma. It hydrolyses oxidised phospholipids to generate lysophosphatidylcholine and oxidised free fatty acids (Scheme 1),[1] both of which are potent chemoattractants for circulating monocytes.[2] Accumulation of lysophosphatidylcholine results in macrophage proliferation[3] and the endolithial dysfunction observed in patients with atherosclerosis.[4,5] Inhibition of LpPLA$_2$ would thus be expected to stop plaque build-up and provide an attractive strategy in the treatment of atherosclerosis.

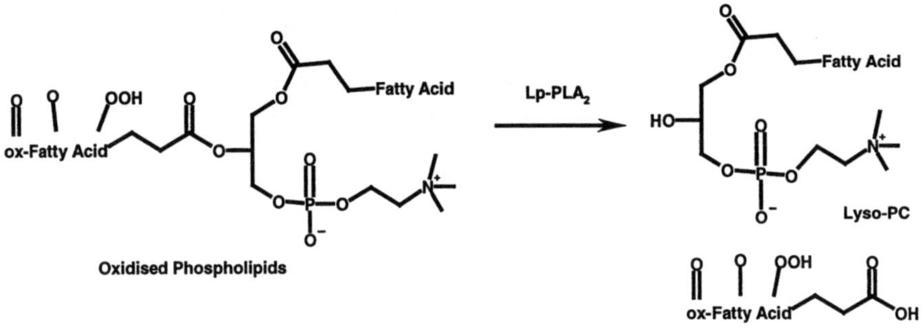

Scheme 1 *Action of LpPLA$_2$*

Early work at SmithKline Beecham identified a number of synthetic β-lactams which were inhibitors of LpPLA$_2$. These compounds, however, only possessed modest activity[6] and so in addition to further synthetic work, a screen of microbial extracts was undertaken.

2 DISCOVERY OF LpPLA₂ INHIBITORS

Fifteen hits were obtained from a primary screen of more than 34,000 microbial extracts using recombinant LpPLA$_2$ as the target. After applying selectivity and dose response criteria four hits were selected for isolation and initial structural elucidation studies.

From the initial structural data and further activity data only one hit culture, from *Pseudomonas fluorescens* DSM 11579 was progressed, and it was seen that this culture produced three related compounds with potent inhibitory activity. DSM 11579 did not produce pseudomonic acid, and equally, a number of known pseudomonic acid producers were tested and found not to produce inhibitors of LpPLA$_2$. Accordingly this culture was progressed for fermentation development and isolation of the active species.

3   FERMENTATION OF DSM 11579

3.1 Fermentation Development

In order to optimise the production of the active species and discover how DSM 11579 responded to variations in production conditions, the culture was put through a rapid fermentation improvement program. This investigation was carried out in a highly efficient array experiment using shake flasks. A large number of nitrogen and carbon sources, and various physical parameters, which could be employed at larger scales were evaluated. The medium and incubation conditions identified by these experiments gave a 20-fold titre increase (to 90-100 mg/litre) in shake flasks, an improvement which was retained upon scaling to 75 L fermenters. As a consequence of the titre increase it was possible to identify a large number of related active species in the broths and these were targeted for isolation.

3.2 Production in 75 Litre Fermenters

In order to provide material for synthetic chemistry and for further analysis, two 50 L fermentations were carried out. After a 20-hour lag phase the titre of the major active species SB-253514 (**3b**) steadily increased and was accompanied by a drop in pH (Fig. 1, Profiles for two 50 L cultures). Following a plateau in accretion of SB-253514 at 78 hours, the culture was pasteurised by heating to 75°C then the broth was cooled.

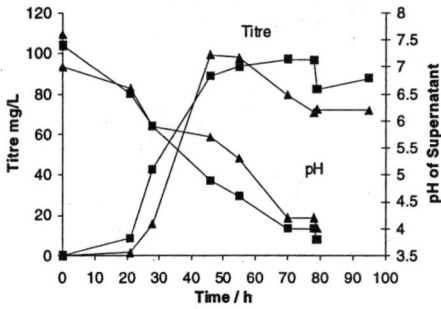

Figure 1 *Profiles for two 50 L Cultures*

4   ISOLATION OF INHIBITORS

Partitioning experiments carried out on small scale indicated that over 80% of the available actives were associated with the cells and insolubles in the fermentation

(Figure 2, Partitioning of actives). On this basis the pasteurised broth was centrifuged to recover the small but high yielding solids fraction. The resulting cell slurry was diluted with methanol to 70% v/v, stirred for 1 hour and then re-centrifuged to an extract containing the active species.

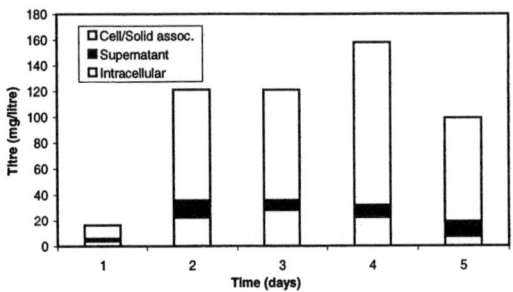

Figure 2   *Partitioning of Actives*

Analysis of this enriched fraction by LC-MS confirmed that a large number of active species had been produced (Figure 3, C8 HPLC analysis of cell extract). The actives fell into two classes, differentiated by chromophore and MS fragmentation characteristics. Within each class the various species differed in mass by 28 or 26 amu, suggesting that they incorporated varying lengths of fatty-acids. The minor class, series 'a', comprised an apparently structurally isomeric series which exhibited UV maxima at 200 and 280 nm.

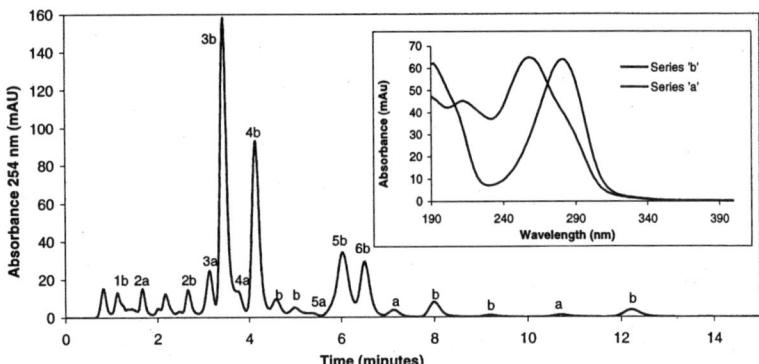

Figure 3   *C8 HPLC Analysis of Cell Extract*

Isolation of the most abundant actives from methanolic cell extract was carried out as depicted in Figure 4 (Isolation of inhibitors from cell extract). This procedure yielded the eight most abundant actives, the major one, **3b** (SB-253514) on gram scale.

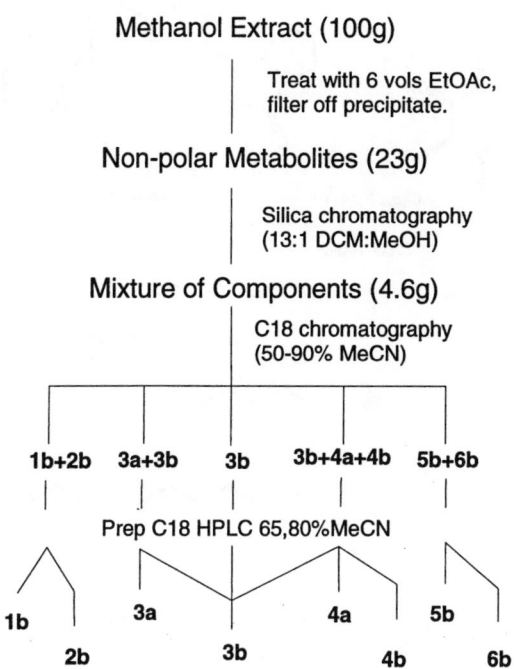

Methanol Extract (100g)

Treat with 6 vols EtOAc,
filter off precipitate.

Non-polar Metabolites (23g)

Silica chromatography
(13:1 DCM:MeOH)

Mixture of Components (4.6g)

C18 chromatography
(50-90% MeCN)

1b+2b **3a+3b** **3b** **3b+4a+4b** **5b+6b**

Prep C18 HPLC 65,80%MeCN

1b **3a** **4a** **5b**

2b **3b** **4b** **6b**

Figure 4 *Isolation of Inhibitors from Cell Extract*

5 STRUCTURAL ELUCIDATION OF ACTIVES

NMR studies on the major member of both series (**3a, 3b**) confirmed that these species were indeed closely related (Figure 5, Structures of isolated actives). The major series (series 'b') was based on a proline derived 5:5 bicyclic carbamate with an imide linked glycosylated fatty-acid derived chain. The minor series differs only in the geometry of closure of the carbamate to generate a 7:5 rather than a 5:5 bicycle. The IR stretch of the enol double bond was key in distinguishing the two structural types. Whilst the 7:5 system has precedent in the cyclocarbamides,[7] the 5:5 bicyclic carbamate system is novel. It was confirmed that within each series the species differ only in the length and/or degree of saturation of the sidechain, the rhamnose group being maintained for each member. The most striking difference between the two structural classes was the solution stability. Whilst the 5:5 series compounds were relatively stable between pH 4.0-7.5, the 7:5 compounds were unstable in aqueous solution even at neutral pH. For this reason it was decided not to progress these compounds.

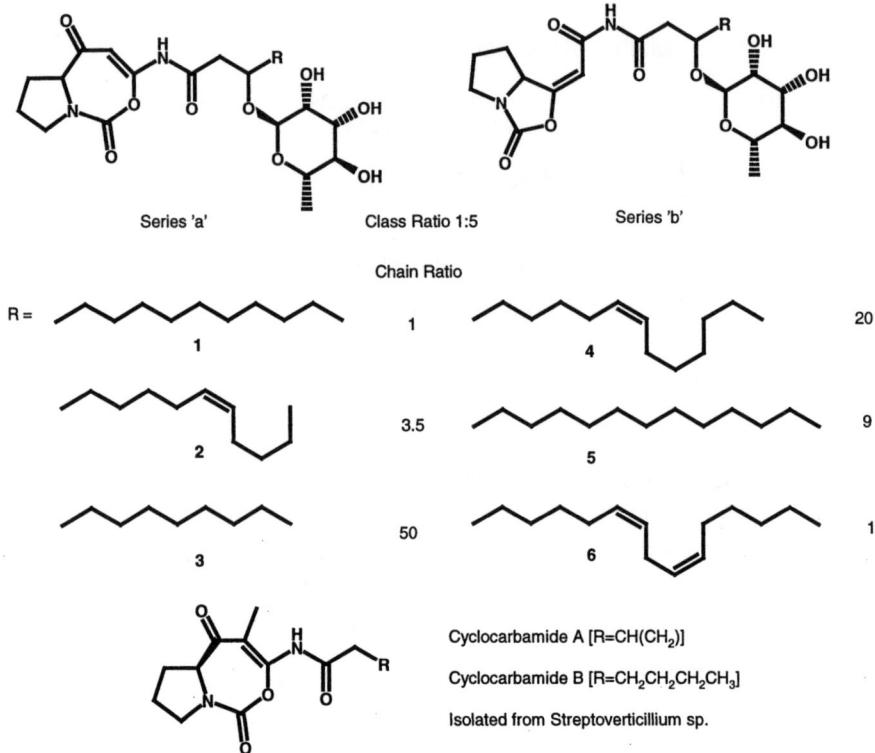

Figure 5   *Structures of Isolated Actives*

6   BIOLOGICAL ACTIVITY

6.1 *In vitro* Activity

Assaying the major component, SB-253514 (**3b**) against recombinant LpPLA$_2$ confirmed that it possessed potent activity. Interestingly, the inhibition was time-dependent, as potency was improved by pre-incubation of the inhibitor with the enzyme (Table 1, Activity of SB-253514 (3b)).

Pre-Incubation	IC$_{50}$ (nm)
0 min	530
10 min	51
30 min	18

Table 1   *Activity of SB-253514 (3b)*

Further studies indicated that the inhibition was competitive with respect to substrate, and that the rate of enzyme reactivation was very slow ($t_{1/2}$). These facts, together with the strained carbamate structure, suggested acylation of the active site serine by the carbamate. Indeed, LC-MS analysis of the inhibited enzyme indicated formation of a 1:1 covalent complex between **3b** and LpPLA$_2$. Assay of the other members of series 'b' (Table 2, Activity of isolated metabolites) showed a trend of increasing activity with increasing chain length (lipophilicity), the time dependence being exhibited throughout. One member of the 7:5 series (**3a**) was also assayed but exhibited very poor inhibition, possibly due to the instability of this compound under the assay conditions.

minutes pre-incubation	IC$_{50}$ R= C$_9$H$_{19}$ 1b	IC$_{50}$ R= C$_{11}$H$_{21}$ 2b	IC$_{50}$ R= C$_{11}$H$_{23}$ 3a/b	IC$_{50}$ R= C$_{13}$H$_{25}$ 4b	IC$_{50}$ R= C$_{13}$H$_{27}$ 5b	IC$_{50}$ R= C$_{15}$H$_{29}$ 6b
SB-	291071	291072	253514	253517	253518	291073
0	880 nM	1180 nM	530 nM	280 nM	96 nM	260 nM
10	30 nM	20 nM	51 nM	28 nM	6 nM	10 nM
SB-			315021			
0			3000 nM			
10			400 nM			

Table 2 *Activity of Isolated Metabolites*

It was critical that the presumed acylation activity of these compounds be specific to LpPLA$_2$ and not exhibited towards other lipases or hydrolytic enzymes. Accordingly **3b** was screened against porcine elastase, trypsin, chymotrypsin, thermolysin, fungal aspartic protease, bacterial type IX metalloprotease and the herpes protease enzymes (CMV, VZV, HSV-2). Gratifyingly, no inhibition was observed. Moderate inhibition of the functionally related enzyme, PAF acetylhydrolase II did however occur but only with an IC$_{50}$ of 485 nM. When tested against a panel of cytochrome P450 (CYP450) isozymes **3b** showed no inhibition.

6.2 *In vivo Activity*

Prior to animal dosing, **3b** was assayed against LpPLA$_2$ in plasma samples. The assay was conducted using human plasma and rabbit plasma, and in both cases potent, time dependent activity was observed (Figure 6, Activity of SB-253514 (3b) in plasma and *in vivo*). Encouraged by this result it was decided that this compound should be tested in a model animal for hyperlipidaemia, the Watanabe heritable hyperlipidaemic (WHHL) rabbit. This animal has no LDL receptor and thus accumulates very high levels of LDL providing a good model for hyperlipidaemia in humans. When dosed i.v., potent and long lived inhibition of LpPLA$_2$ was achieved (Figure 6, Activity of SB-253514 (3b) in plasma and *in vivo*). However when **3b** was dosed orally to WHHL rabbit no activity was achieved.

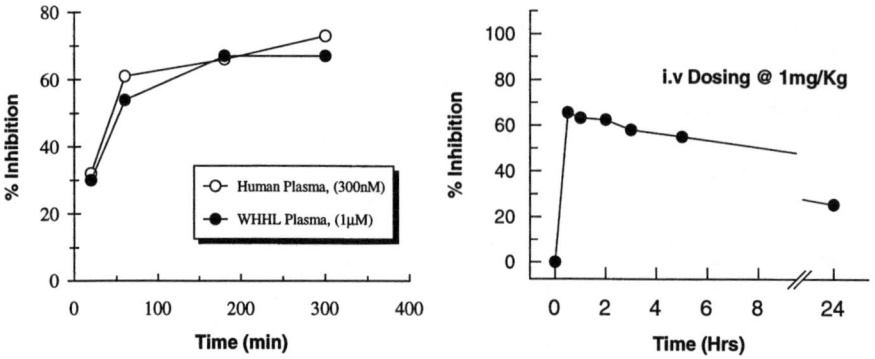

Figure 6   *Activity of SB-253514 (3b) in plasma and in vivo*

7   SYNTHETIC ANALOGUES

In order to attempt to address the issues of oral bioavailability and compound stability a number of synthetic analogues of the natural products were generated. It was found that long chain alkyl amides of the 5:5 bicycle system possessed comparable activity to their natural imide analogues. (Table 3, Activity of synthetic analogues). Whilst these compounds were more stable to pH extremes than the natural compounds, however, their solubility was lower, and they showed no oral bioavailability.

	min pre-incubation	IC_{50} $R = C_{12}H_{25}$	IC_{50} $R = C_{14}H_{29}$	IC_{50} $R = C_{16}H_{31}$ L D
		7	**8**	**9** | **10**
	0	2000 nM	>10 uM	9.6 uM | 7.0 uM
	10	170 nM	28 nm	25 nM | 12 nM

Table 3   *Activity of Synthetic Analogues*

8   BIOTRANSFORMATION OF SB-253514 (3b)

As a final attempt to gain an orally active compound it was decided that the sidechain sugar (believed to be rhamnose) should be removed in the hope that this would facilitate bioavailability and optimisation through derivatisation of the unmasked hydroxyl. It was additionally hoped that the resulting compound would be crystalline in order to allow determination of the stereochemistry by X-Ray crystallography.

A number of acid catalysed protocols were attempted to effect the hydrolysis but were unsuccessful due to concomitant hydrolysis of the carbamate so an enzymatic protocol was sought. Screening of a number of glycosidases identified that naringinase (a crude preparation from *Penicillium decumbens* which contains a rhamnosidase) possessed the desired activity, giving approximately 10% conversion to the deglycosylated compound SB-311009 (**11**). Optimisation of this reaction, which was carried out as a suspension in 10% methanolic buffer, produced a system with yields of 70-80%. In an attempt to increase levels of conversion by complexing the liberated rhamnose, bisulphite was added to the reaction system (as $Na_2S_2O_5$). Using this methodology conversions of >95% were obtainable after 16 hours of reaction (Figure 7, deglycosylation of SB-253514, (3b)).

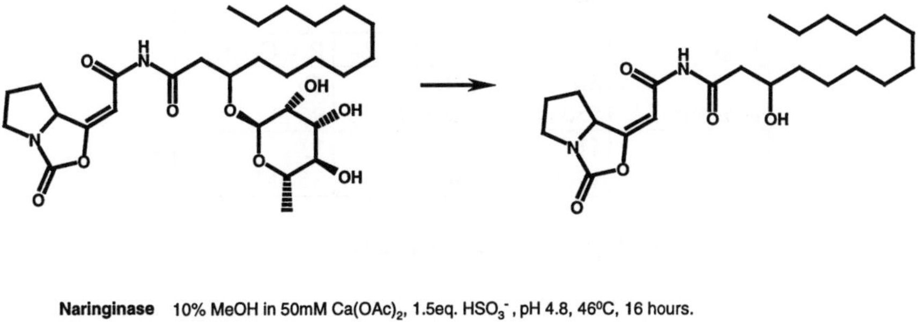

Naringinase   10% MeOH in 50mM Ca(OAc)$_2$, 1.5eq. HSO$_3^-$, pH 4.8, 46°C, 16 hours.

Figure 7   *Deglycosylation of SB-253514, (3b)*

Attempts to crystallise the deglycosylated compound were indeed successful (from MeOH) and yielded crystals suitable for X-Ray analysis. Crystals of two morphologies were obtained and one of these gave data of sufficient quality to determine the absolute stereochemistry. This surprisingly indicated that the stereochemistry at both the ring junction and side-chain hydroxyl was (*R*), implying an inversion of the proline chiral centre during biosynthesis (Figure 8, stereochemistry of SB-253514 (3b)).

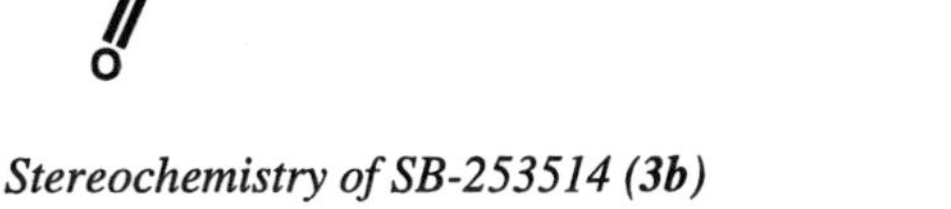

Figure 8   *Stereochemistry of SB-253514 (3b)*

9   BIOLOGICAL ACTIVITY OF DEGLYCOSYLATED COMPOUNDS

Three deglycosylated compounds were produced and assayed against recombinant LpPLA$_2$ to determine if there was a significant difference in activity compared to their glycosylated parents. Two species, with fully saturated sidechains (**3b, 5b**), showed a mild increase in activity, whilst the third (**4b**), possessing an unsaturated sidechain, showed a slight decrease in potency (Table 4, *In vitro* activity of deglycosylated compounds).

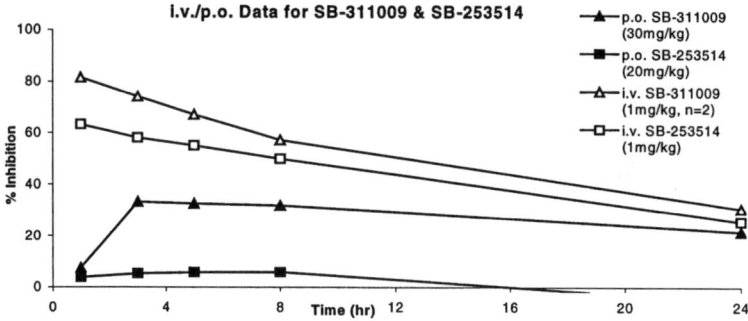

IC_{50}	R =$C_{11}H_{23}$	R = $C_{13}H_{25}$	R = $C_{13}H_{27}$
R'=	**3b**	**4b**	**5b**
Rhamnose	530 nM	280 nM	96 nM
R'= H	**11**	**12**	**13**
	170 nM	740 nM	35 nM

Table 4 *In vitro Activity of Deglycosylated Compounds*

In common with their parents, these compounds were selective, time dependent inhibitors of LpPLA$_2$. SB-311009 (**11**), derived from deglycosylation of the major fermentation species (**3b**), was tested *in vivo*. When dosed i.v. to WHHL rabbit, this compound demonstrated potent, long-lived activity, and was slightly more active than its parent, in line with the *in vitro* data (Figure 9, i.v./p.o Data for SB-311009 and SB-253514). Pleasingly, when **11** was dosed orally, in contrast to its parent, this compound showed significant, prolonged activity (Figure 9). This activity was, however, only obtained at relatively high dosing levels but it was anticipated that elaboration utilising the unmasked hydroxyl group could address this issue.

Figure 9 *i.v./p.o. Data for SB-311009 and SB-253514*

10   CONCLUSIONS

Eight potent inhibitors representing two novel structural classes were isolated from fermentation broths of DSM 11579 following rapid fermentation development. This development was key in both increasing the titre of the major component and the number of active species present and was thus crucial in allowing quick progression of these hits by providing large amounts of a number of active species in a short time frame. The ability to biotransform the isolated compounds was also key in maximising the activity of the hit compounds, and additionally led to later downstream isolation improvements. Thus the close integration of advanced microbiological methods, isolation processes and biotransformation was crucial in maximising the value of the initial screening hit.

Acknowledgements

The author gratefully acknowledges the other members of the team involved in this work;
Screening: Steve Elson, Emilio Diez, Alfonso Rivera, Juan Antonio Hueso, Jose Maria Sanchez-Pulles, Marisa Morata.
Ferm. Development: Craig Gershater, Sarah Ready, Steve Warr, Rosie Sheridan.
Isolation: Anna Stefanska, Mark Fulston, Keith Robins, Jagrup Aujla, Simon Spear.
Biological Data: Colin McPhee, Helen Boyd, David Tew, Kevin Millner, Maxine Taylor.
Discovery Chemistry: Simon Readshaw, Roy Copley, Ivan Pinto, Deirdre Hickey.

References

1.   D.G. Tew, C. Southern, S.Q.J. Rice, M.P. Lawrance, H. Li, H.F. Saul, K. Moores, I.S. Gloger and C.H. MacPhee, *Arteriocler. Thromb. Vasc. Biol.*, 1996, **16**, 591.
2.   M.T. Quinn, S. Parthasarathy and D. Steinberg, *Proc. Natl. Acad. Sci. U.S.A.*, 1988, **85**, 2805.
3.   M. Sakai, A. Miyazaki, H. Hakamata, Y. Sato, T, Matsumara, S. Kobori, M. Shichiri and S. Horiuchi, *Arteriocler. Thromb. Vasc. Biol.*, 1996, **16**, 600
4.   K. Kugiyama, S.A. Kerns, J.D. Morrisett, R. Roberts and P.D. Henry, *Nature (London)*, 1990, **344**, 160.
5.   C. Cowan and R.P. Steffen, *Arteriocler. Thromb. Vasc. Biol.*, 1995, **15**, 2290
6.   D.G. Tew, H.F. Boyd, S. Ashman, C. Theobold and C.A. Leach, *Biochemistry*, 1998 **37**, 10087.
7.   A. Isogai, S. Sakuda and K. Shindo, *Tetrahedron Lett.*, 1986, **27**, 1161.

3 Marine Natural Products

MARINE ORGANISMS AS A SOURCE OF NOVEL LEAD STRUCTURES FOR DRUG DEVELOPMENT

Amy E. Wright

Harbor Branch Oceanographic Institution, Inc, 5600 US 1 North, Fort Pierce, Florida USA 34946; e-mail wright@hboi.edu

1 INTRODUCTION

Screening of extracts obtained from terrestrial plants and animals has yielded a large number of novel natural products. Many of these have been developed into useful medicinal agents either directly as the natural product or through synthetic modification. It has been estimated that for the period 1983-1994, 78% of the antibacterial agents and 61% of the 31 anticancer agents approved for use, were derived from terrestrial natural products.[1] By comparison the marine environment remains relatively unexplored.

The oceans cover over 70% of the Earth's surface. Estimates of the number of marine invertebrate and algal species are at least 200,000, while those for marine microbial strains range into the millions. Many of the invertebrates are sessile as adults and have evolved in habitats which are often more extreme and self limiting compared to their terrestrial counterparts. This increased evolutionary pressure has resulted in organisms which produce chemically diverse compounds that have a wide variety of possible ecological roles. These include, but are not limited to: toxins, which can reduce predation, larval settlement and overgrowth by neighboring organisms; compounds which reduce palatability or nutrient uptake in predators; and compounds which direct larval settlement and reproduction.

Over the past quarter-century more than 10,000 compounds have been reported from marine-derived organisms. These compounds encompass a wide variety of chemical structures including acetogenins, polyketides, terpenes, alkaloids, peptides and many compounds of mixed biosynthesis. A number of excellent books and reviews document the diversity of both structures and bioactivities which have been observed for marine-derived compounds.[2-7]

Although many compounds have been discovered, only a few have progressed to drug candidates which have been tested in the clinic. Early studies on the Caribbean sponge *Tethya crypta* led to the isolation of the unusual nucleosides, spongothymidine and spongouridine.[8] The discovery of these novel nucleosides prompted the synthesis of a new class of arabinosyl nucleoside analogs, one of which, arabinosyl cytosine (ara-c), displayed in vivo antitumor activity. Ara-c is currently used in the treatment of acute myelocytic leukemia and non-Hodgkin's lymphoma.[9]

Four compounds are currently under clinical investigation for their use as anticancer agents (Figure 1). These are ecteinascidin 743 (**1**), from the tunicate *Ecteinascidia turbinate;*[10] aplidine (dehydrodidemnin B, **2**) from the ascidian *Aplidium albicans;*[11] dolastatin 10 (**3**) from the sea hare *Dolabella auratium;*[12] and bryostatin 1 (**4**) from the bryozoan *Bugula neritina.*[13] The marine environment has also been the source of a number of anti-inflammatory compounds such as pseudopterosin E, (**5**). The pseudopterosins constitute the major active components in a popular line of skin care products.[14]

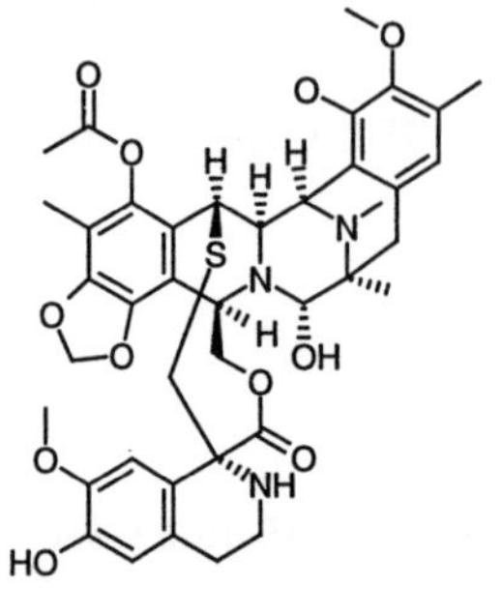

1

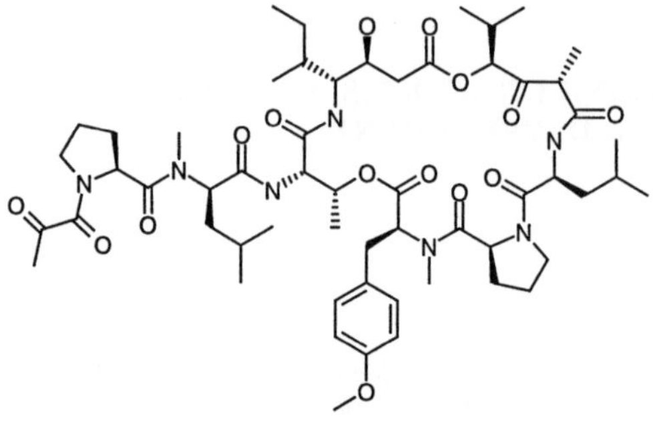

2

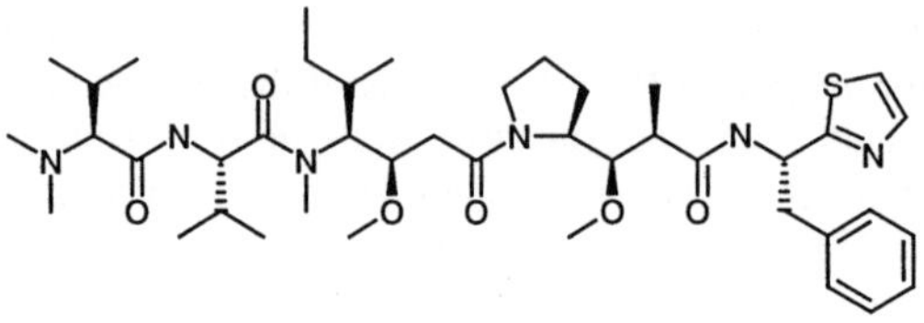

3

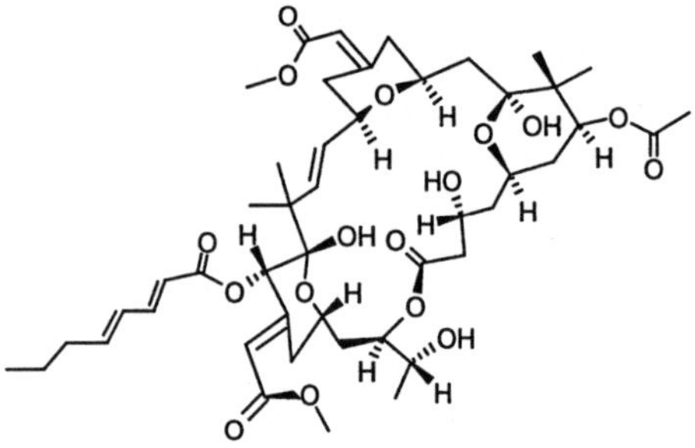

4

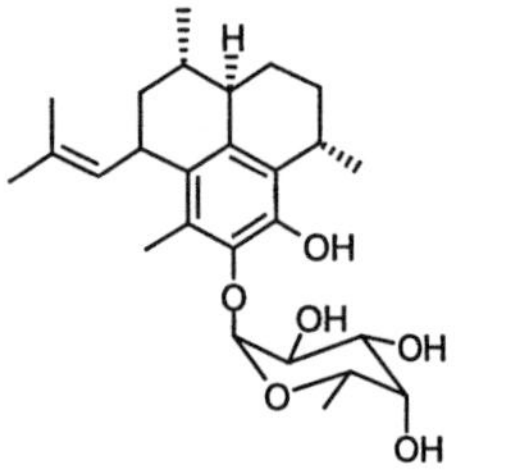

5

Figure 1 *Structures of marine-derived secondary metabolites which have entered clinical trials or which are used commercially. **1** (ecteinascidin 743), **2** (aplidine), **3** (dolastatin 10), **4** (bryostatin 1), **5** (pseaudopterosin E)*

2 DISCOVERY OF BIOACTIVE NATURAL PRODUCTS AT HBOI

In 1984, Harbor Branch Oceanographic Institution (HBOI) began a program aimed at the discovery of marine natural products with biomedical potential. HBOI operates three research vessels: the 204 ft. R/V Seward Johnson, the 168 ft. R/V Edwin Link, and the 114 ft. R/V Sea Diver. All three vessels are equipped as submersible support vessels and can carry one of HBOI's three submersibles: the Johnson-Sea-Link I or II, both of which are capable of diving to depths of 915 m; or the Clelia, which is capable of diving to depths of 305 m. These submersibles are uniquely equipped for the retrieval and documentation of both benthic and mid-water samples. Tools developed at HBOI include specialized manipulator arm attachments ("clam scoop", claw hand and suction adapter) and a rotating, conveyor belt work platform which maintains the specimens in separate acrylic buckets during the dive. Specimens are documented with taxonomic and environmental data, video and *in situ* 35 mm photographs. A PC-based integrated mission profiler/navigation system coupled with differential global positioning system allows for the documentation of the exact latitude and longitude for all samples collected by submersible. This capability allows us to return to the same site for re-collection of bioactive samples if necessary. We have found this system to have accuracy to better than ± 5 m. Specimens are also collected by scuba diving, wading and snorkeling. In these cases a handheld GPS is used to document latitude and longitude of the collection site and *in situ* photographs are taken whenever possible.

These tools have allowed HBOI to compile a collection of approximately 20,000 frozen marine invertebrate and algal specimens and 13,000 microbial isolates. Samples have been collected from around the world and at depths down to 915 m. All collections have been made with informed consent of the host countries. The wide diversity of marine habitats which have been sampled include: deep continental and island slopes; deep and shallow reefs; coral banks and walls; rhodolith banks; caves; grass beds; oyster and mussel beds; kelp forests; mangrove lagoons, inland salt lakes; shipwrecks; pilings and jetties. Table 1 shows a taxonomic distribution of organisms in the HBOI repository while Table 2 shows a distribution of samples by depth of collection.

Table 1 *Distribution of Organisms in the HBOI Repository by Phylum or Division*

Phylum	% of Total Samples
Porifera	59 %
Cnidaria	14 %
Urochordata (Ascidians)	5%
Echinodermata	5%
Mollusca	2 %
Bryozoa	1%
Crustacea	1%
Miscellaneous	1%
Cyanophyta	1%
Chlorophyta	4%
Phaeophyta	3%
Rhodophyta	5%

Two strategies are used in collecting new samples. The first is to maximize taxonomic diversity. In this process, an emphasis is placed on collecting specimens related to, but differing from, those known to contain bioactive natural products. The second emphasis is to evaluate ecological factors such as consumer pressure, growth form (e.g. thin encrusting vs. massive), level of resource competition, presence or absence of biofouling, *etc.*, and relate this to expression of secondary metabolism. For example, we have observed a higher percentage of cytotoxic agents in thin encrusting sponges. Cytotoxic compounds may provide an adaptive advantage in reducing overgrowth and larval settlement allowing for the successful colonization of relatively large surface areas. We have also observed a greater percentage of active compounds in invertebrate consortia (e.g. where two invertebrates grow in association with each other). For example, sponges of the genus *Spongosorites* are commonly found in association with vermetid gastropod molluscs of the genus *Siliquaria*. These sponge consortia are rich in bioactive compounds. Interestingly, we have also observed that organisms which are conspicuously colored that live in habitats with limited or no ambient light often possess interesting secondary metabolites (e.g. dercitins from the red-purple *Dercitus* sp. and topsentins from the yellow *Spongosorites* sp.). Such organisms are targeted for collection.

Table 2 *Distribution of Samples in the HBOI Repository by Depth of Collection*

Depth range	% of Total Samples
0 - 29 m	63 %
30 - 149 m	6 %
150 - 299 m	7 %
300 - 599 m	20 %
600 - 915 m	4%

Ethanol extracts are prepared of all specimens and sent through a battery of assays run both in house and in collaboration with partners in academia and industry.   Selected extracts and enriched fractions have been screened through approximately 20 assays over the past 15 years resulting in the isolation of over two hundred bioactive compounds. As with most programs the assay panel has evolved as new assays and screening strategies are developed. Examples of the types of assays run include whole cell assays (tumor cell lines, fungal pathogens, viral pathogens), enzyme assays (e.g. cdc25a, cpp32, chitin synthase) as well as receptor binding assays (C5a, Angiotensin II, LTB4). Representative hit rates are given in Table 3.[15, 16] Active compounds are purified using bioassay-guided fractionation and the structures defined through spectroscopic methods.

Table 3 *Representative Hit Rates for HBOI Marine Extracts*

Assay	% Considered Active
Inhibition of *in vitro* proliferation of P388 murine leukemia	10 %
Inhibition of *in vitro* proliferation of *Candida albicans*	10 %
C5a receptor antagonists	2.4 %
Inhibition of cdc25A enzyme	0.5 %
Endothelin-A receptor antagonists	0.1 %

3 HURDLES TO THE DEVELOPMENT OF MARINE-DERIVED NATURAL PRODUCTS

One of the greatest challenges to development of marine natural products as pharmaceutical, cosmetic or agrochemical products is the cost-effective production of the material.[17,18] In some instances organic synthesis may be a viable route for production while in others emerging technologies such as aquaculture, cell culture or genetic manipulation of the producing organism may be possible. Wild populations of *Ecteinascicida turbinata*, *Bugula neritina* and *Trididemnum candidum* have been harvested to provide the supply of compound needed for clinical trials of ecteinascidin 743, bryostatin 1 and didemnin B, respectively. To date, only dolastatin 10 was produced synthetically to provide the material required for clinical evaluation. Syntheses of ecteinascidin 743, bryostatin 1 and the didemnins have been published, but none of them are yet commercially viable. Pilot scale aquaculture of *E. turbinata*[19], *B. neritina*[19] and *Lissodendoryx* sp.[20] have been conducted, and hold promise for possible larger scale production of the compounds. The lack of supply can often place the development of compounds on hold as has happened for the promising compound eleutherobin.[17] At HBOI a multifaceted approach including aquaculture, cell culture, fermentation and synthesis is being used to address these supply issues. The following examples seek to illustrate this process.

3.1 Discodermolide

Lithistid sponges of the genus *Discodermia* have been shown to produce a variety of bioactive compounds. For example, Caribbean specimens have been the source of the cytotoxic compound polydiscamide A (6); [21] the antifungal compounds discobahamin (7),[22] and discodermide (8); [23] the indole derived compound discodermindole (9); [24] and the antitumor agent, discodermolide, (10) [25] (Figure 2).

Discodermolide was first reported as an immune suppressive agent isolated from a shallow water specimen of the sponge *Discodermia dissoluta* collected in the Bahama Islands. Continued investigation of the compound showed that it is a potent inducer of tubulin polymerization and stabilizes the microtubule matrix in a fashion similar to that of the clinically useful agent paclitaxel.[26] At low concentrations discodermolide induces apoptosis in A549 human adenocarcinoma cells through an undefined mechanism. Although the first specimen of *Discodermia* to yield discodermolide was collected by scuba at 37 m, the sponge is much more common at depths of 125-160 m and can be routinely collected throughout the Caribbean using the HBOI submersibles. Based upon its potent *in vitro* and *in vivo* antitumor activity, the compound was licensed to Novartis Pharmaceuticals in 1998. Clinical development will require a large supply of material and this remains the major stumbling block to development. Although discodermolide is a fairly small molecule (MW 593) it has 13 asymmetric centers and remains a challenge to synthetic organic chemists. A number of syntheses have been published[27-34] and very recently the Smith group[35] has reported a gram scale synthesis of discodermolide which holds promise for large-scale synthetic production of the material.

An alternative supply route may be achieved through fermentation of an associated microorganism. It has been suggested that many of the compounds produced by lithistid sponges such as *Discodermia* may have their true origin in microbial associates.[36] In the lithistid sponge, *Theonella swinhoei*, Bewley[36] demonstrated that the secondary metabolites, swinholide A and theopalauamide, could be localized into the microbial cell populations rather than the sponge cells suggesting a microbial origin of the compounds.

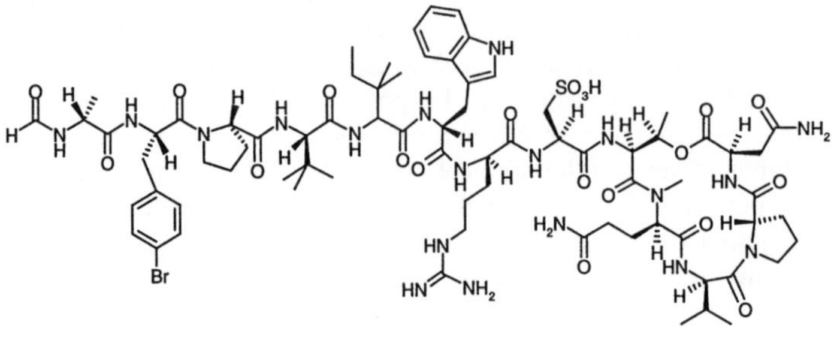

6

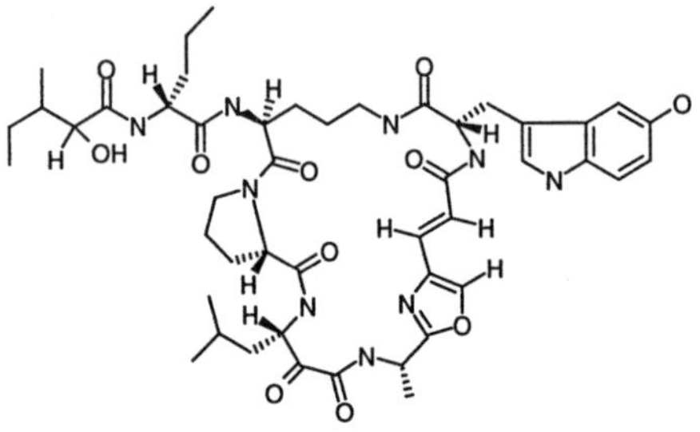

7

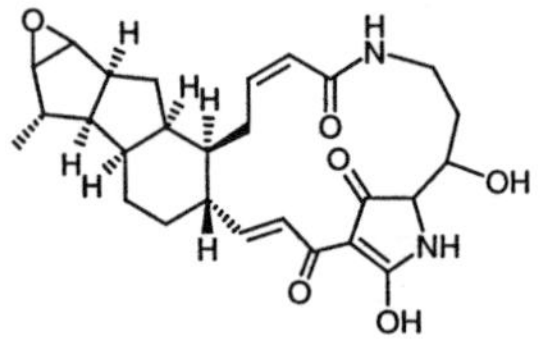

8

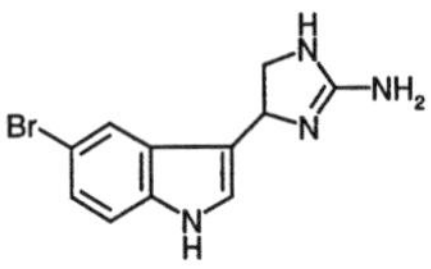

9

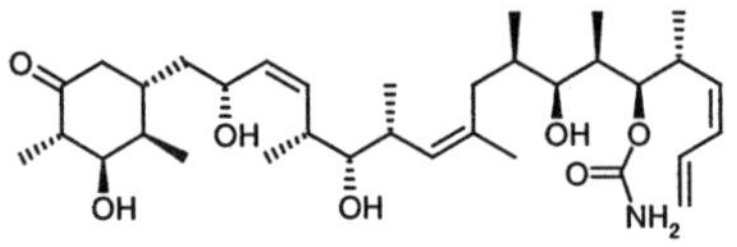

10

Figure 2   *Structures of compounds reported from Caribbean specimens of* Discodermia *spp.* **6** *(polydiscamide A),* **7** *(discobahamin A),* **8** *(discodermide),* **9** *(discodermindole),* **10** *(discodermolide)*

Investigation of the microbial population associated with *Discodermia* using 16s-rRNA analysis showed the presence of significant microbial diversity.[37] Although the case for microbial production is still speculative, we have begun a project aimed at the development of novel isolation and culture media in order to try to harness this potential supply of discodermolide. To date, over 800 microbial isolates have been isolated from *Discodermia* spp. and these are under investigation for production of discodermolide, discodermide and polydiscamide a.

Taking this a step further, a PO1 collaborative project with Professor David Sherman of the University of Minnesota aims to extract the genes responsible for biosynthesis of interesting compounds from marine-derived microorganisms and transfer them to a stable expression vector. The ultimate goal of this project will be both to produce a steady supply of important marine-microorganism derived natural products as well to produce novel compounds through combinatorial biology studies.

3.2   The Lasonolides

Lasonolide A (**11**, Figure 3) was first discovered in a specimen of the sponge, *Forcepia,* which was collected by scuba at a depth of 19 m in the British Virgin Islands.[38] Lasonolide A was found to inhibit the *in vitro* proliferation of A549 human lung carcinoma cells with an IC_{50} of 40 ng/ml and to inhibit cell adhesion in the EL4.IL-2 cell line with an IC_{50} of 19 ng/ml. The latter assay has been shown to be indicative of signal transduction activity.[39] Assay of the compound in the US National Cancer Institute (NCI) 60 cell line panel indicated that lasonolide A had a unique pattern of bioactivity and did not match any compounds of known mechanism of action when analyzed with the COMPARE algorithm.[40] This led the NCI to select the compound for further investigation. Unfortunately, the yield of compound was low (approximately 23 mg/kg) and the original specimen was small (400 g) and did not provide sufficient material to conduct the *in vivo* and mechanism of action studies.

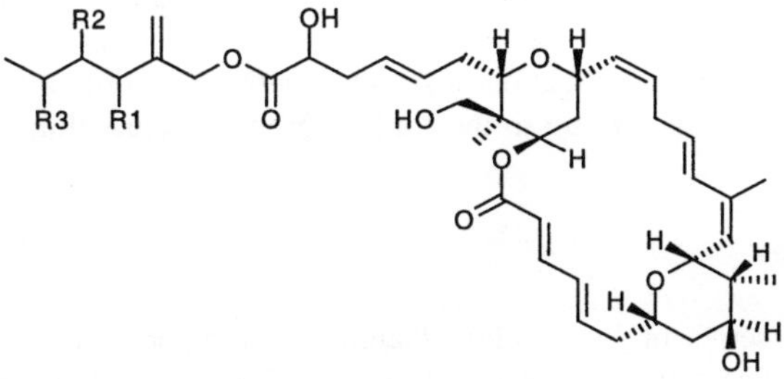

11 R_1=H, R_2=H, R_3=CH$_3$

12 R_1=CH$_3$, R_2=OH, R_3=H

Figure 3 *Structures of lasonolide A, 11, and lasonolide B, 12*

Analysis of the HBOI specimen database allowed for the identification of two additional specimens of the sponge, which had been collected as by-catch during a trawling mission aboard a commercial shrimp trawler operating in the US Gulf of Mexico. Both specimens were collected at approximately 70 m and although not identified as *Forcepia*, had similar morphological descriptions to the first specimen (spherical, red to orange in color, soft, thin ectosome, consolidating sediment). They also showed activity in the *in vitro* P388 murine leukemia assay. Taxonomic evaluation allowed for the classification of the sponges as *Forcepia* sp. and chemical investigation revealed the presence of the lasonolide class of compound. Although the combined weight of the samples was approximately 1.2 kg, the major compound present in these specimens was a different lasonolide, denoted as lasonolide B (**12**, Figure 3),[41] and there was still insufficient supply for *in vivo* antitumor evaluation. The NCI assisted with funding of a trawling expedition to the same area for recollection of the sponge. Unfortunately, after five days of trawling, only about eight small specimens of approximately 1.4 kg total weight were collected and once again a mixture of lasonolides A and B was found. Trawling does not allow one to observe the bottom topography and it was possible that the growth form of the sponge did not allow for facile collection by trawling. As the activity was still of interest and a need for additional material remained, HBOI returned to the site with the Johnson-Sea-Link II submersible in 1999. Using the submersible it was observed that the sponge was extremely common at the 70 m depth contour and over 15 kg of the sponge was successfully collected with minimal impact on other species. Upon observing the habitat in which the sponge grows, we propose that the reason that trawling failed to collect large amounts of the sponge is that the upper portions of the sponge are fragile while the base is packed with sediment and is more dense. As the sponge is picked up by the trawl, the larger fragile top is broken off and the dense base remains. Those specimens without a very dense base are lost entirely. It is possible that during the earlier trawling mission, the sea floor was actually "seeded" with small pieces of *Forcepia* which we were then able to harvest at the later date. Fragmentation is a common method of sponge propagation and it is likely that *Forcepia* will be an excellent candidate for aquaculture. The fact that the compounds have been isolated from both shallow and deeper water specimens, argues that transplantation and culture similar to that achieved for *Lissodendoryx*[20] (a source of the halichondrins) may be possible. Currently, *in vivo* testing and investigation of the mechanism of action of the lasonolides is in progress. If the compounds live up to their initial promise, aquaculture studies may be initiated.

3.3   The Topsentins

The topsentins are a series of bis(indole) alkaloids first reported from a specimen of *Topsentia genetrix.*[42] The parent compounds topsentin (**13**) and bromotopsentin (**14**) were isolated simultaneously from a specimen of *Spongosorites ruetzleri* collected in deep water in the Bahamas.[43] Related compounds in the series which have been isolated at HBOI include the nortopsentins A-C (**15-17**),[44] dragmacidin D (**18**),[45] 2,2-bis(6'-bromo-indol-3'-yl)ethyl amine (**19**),[46] dragmacidin (**20**)[47] and the hamacanthins (**21-22**),[48] (Figure 4).

Topsentin and bromotopsentin were shown to be potent anti-inflammatory agents.[49] At doses of 50 µg/ear they completely inhibit phorbol myristate acetate (PMA)-induced inflammation in the mouse ear with have ED_{50}s of 15 and 30 µg/ear for topsentin and bromotopsentin, respectively. Both compounds were also shown to inhibit PLA_2 with IC_{50}s of 0.5 and 6 µM, respectively. Topsentin and related bis(indole) compounds have also been shown to inhibit resinaferatoxin (RTX) induced inflammation with ED_{50}s ranging from 20-1.5 µg/ear (Table 4). This latter activity led to an interest in the development of the compounds for use in skin care products.

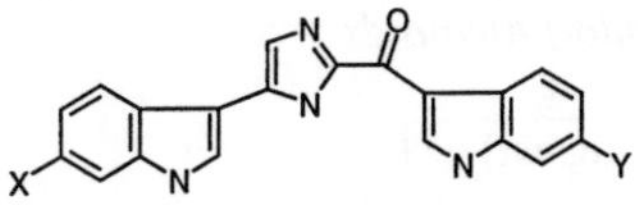

13 X=H Y=OH
14 X=Br Y=OH

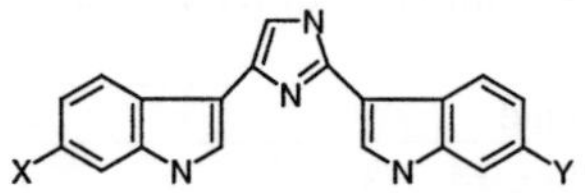

15 X=H Y=H
16 X=H Y=Br
17 X=Br Y=H

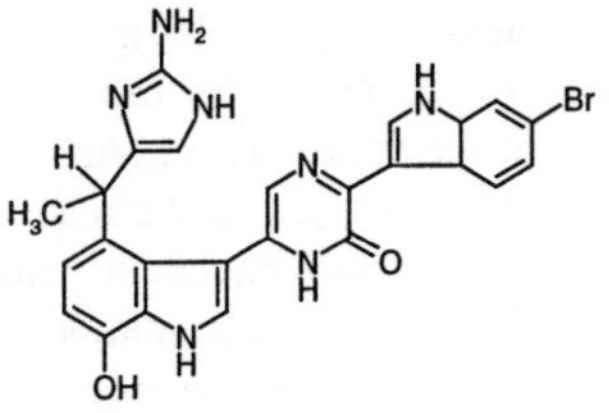

18

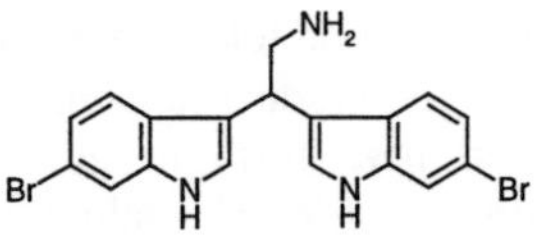

19

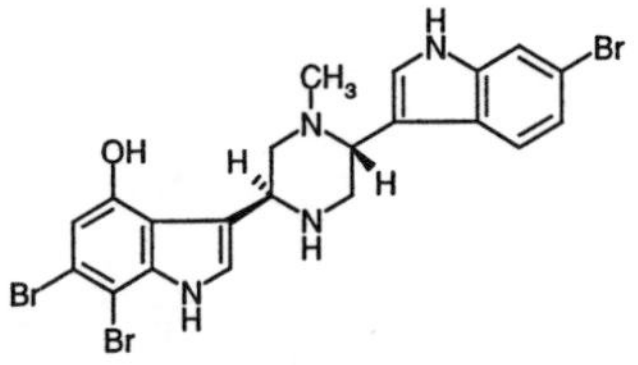

20

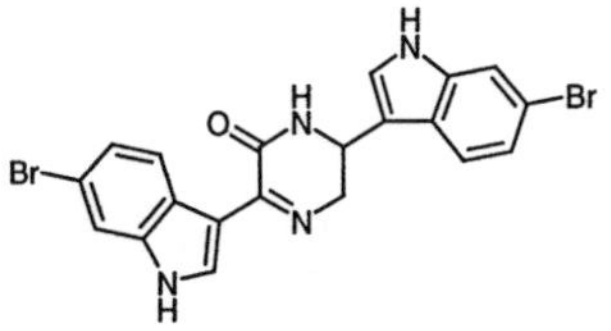

21

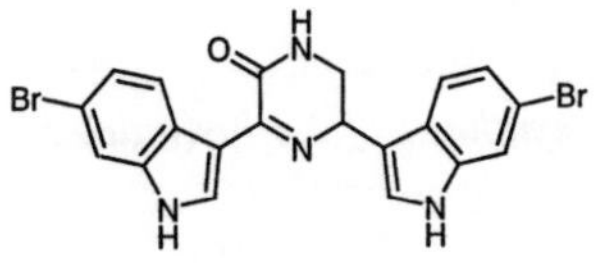

22

Figure 4 *Structures of topsentin class metabolites isolated at HBOI **13** (topsentin), **14** (bromotopsentin), **15** (nortopsentin A), **16** (nortopsentin B), **17** (nortopsentin C), **18** (dragmacidin D), **19** (dragmacidin), **20** (2,2-bis(6'-bromoindol-3'-yl)ethylamine), **21** (hamacanthin A), **22** (hamacanthin B)*

The planar nature and lack of asymmetric centers argued that a synthetic process could be developed to produce the compounds at a cost low enough for industrial production. A number of syntheses have been published[50-53] and in fact, 1 g of deoxytopsentin has been produced for additional biological testing using the Tsujii synthesis. Unfortunately the aldehyde intermediate is unstable and a number of purification steps were required making the synthesis of the compound less commercially feasible. The 2,2-bis(6'-bromo-indol-3'-yl) ethylamine analog, although not the most active in the series, seemed like a likely

Table 4 *Inhibition of RTX-Induced Inflammation by Bis(indol) Alkaloids*

Compound	ED_{50} (ug/ear)
Topsentin, **13**	20
Deoxytopsentin, **18**	17
Nortopsentin c, **19**	8
Hamacanthin b, **20**	1.5
2,2-bis(6'bromo-indol-3'yl)ethyl amine, **16**	18.1

candidate for synthesis. Coupling of indole with glyoxylic acid would yield the 2,2-bis(indol-3yl)-acetate, **23**,[51] which could in turn be converted to the amine by treatment with thionyl chloride followed by ammonia and then reduction of the amide with LiAlH4 to yield the desired product (Scheme 1). Testing of the intermediate **23**, which can be readily produced in pure form in gram quantities, indicated that it has an excellent anti-inflammatory profile. The inhibition of PMA-induced or RTX-induced inflammation of mouse ear edema inflammation gave ED_{50}s of 1.3 ug/ear and 5.1 ug/ear, respectively. Although the natural products are still of interest, this simple synthetic analog which can be produced in large quantities at low cost is the compound of choice for further development as an anti-inflammatory agent.

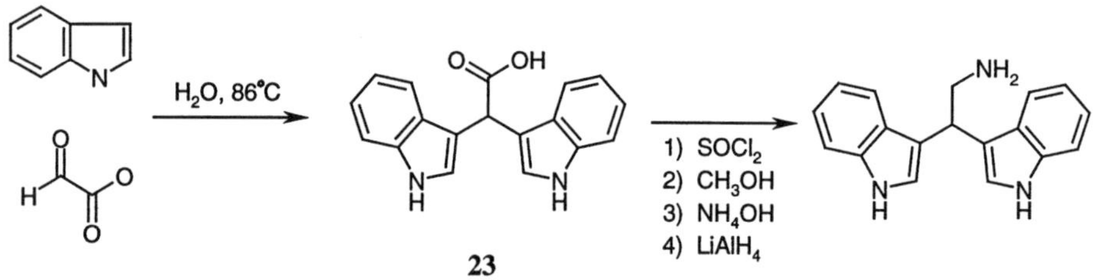

Scheme 1 *Proposed scheme for the synthesis of 2,2-bis(indol-3'-yl)ethylamine*

4 SUMMARY

In summary, the marine environment contains a wealth of plants, animals and microorganisms. Due to their unique adaptations to their ocean habitat, they contain a wide diversity of natural products. These compounds have shown activity in a variety of assays which have relevance to human diseases. As our understanding of the molecular basis of disease expands, these compounds and ones yet to be discovered will provide   lead compounds for human therapeutic treatment.   Innovations in synthesis, fermentation of symbionts as well as in manipulation of biosynthetic genes will allow us to produce sufficient material for clinical use of the compounds. Marine organisms provide a unique opportunity for access to chemical diversity.

Acknowledgements

I would like to thank Dr. Peter McCarthy and Dr. Shirley Pomponi for their assistance in editing this manuscript. This is HBOI Contribution Number 1345.

References

1.   G. M. Cragg, D. J. Newman and K. M. Snader, *J. Nat.Prod.*, 1997, **60**, 52.
2.   a) D. J. Faulkner, *Nat. Prod. Rep.*, 1984, **1**, 251; b) D. J. Faulkner, *Nat. Prod. Rep.*, 1984, **1**, 551; c) D. J. Faulkner, *Nat. Prod. Rep.*, 1986, **3**, 1; d) D. J. Faulkner, *Nat. Prod. Rep.*, 1987, **4**, 539; e) D. J. Faulkner, *Nat. Prod. Rep.*, 1988, **5**, 613; f) D. J. Faulkner, *Nat. Prod. Rep.*, 1990, **7**, 269; g) D. J. Faulkner, *Nat. Prod. Rep.*, 1991, **8**, 97; h) D. J. Faulkner, *Nat. Prod. Rep.*, 1992, **9**, 323; i) D. J. Faulkner, *Nat. Prod. Rep.*, 1993, **10**, 497; j) D. J. Faulkner, *Nat. Prod. Rep.*, 1994, **11**, 355; k) D. J. Faulkner, *Nat. Prod. Rep.*, 1995, **12**, 223; l) D. J. Faulkner, *Nat. Prod. Rep.*, 1996, **13**, 75.
3.   *Marine Biotechnology, Volume 1: Pharmaceutical and Bioactive Natural Products*, D. H. Attaway, and O. R. Zaborsky (eds), Plenum Press, New York, 1993.
4.   B. Carté, *Bioscience*, 1996, **46**, 271.
5.   F. Pietra, *Nat. Prod. Rep.* 1997, **14**, 453.
6.   W. Fenical, W. *Trends in Biotechnology* 1997, **15**, 339.
7.   D. J. Faulkner and all articles in *Chem. Rev.* 1993, **5**, 1671.
8.   a) W. Bergmann and D. C. Burke, *J. Org. Chem.*, 1955, **20**, 1501
9.   Physicians' Desk Reference, Medical Economics Company, 1999.
10.   a) A. E. Wright, D. Forleo, G. Gunawardana, S. Gunasekera, F. Koehn and O.McConnell, *J. Org. Chem.*, 1990, **55**, 4508.; b) K. L. Rinehart, T. Holt, N. Fregeau, J. Stroh, P. Keifer, F. Sun, D. Li, and G. Martin, *J.Org. Chem.*, 1990, **55**, 4512.
11.   K. L. Rinehart, L. S. Shields and M. Cohen-Parsons. In *"Marine Biotechnology, Volume 1: Pharmaceutical and Bioactive Natural Products"*, D. H. Attaway and O. R. Zaborsky (ed.) , Plenum Press, New York, 1993, 309.
12.   G.R. Pettit, S. B. Singh, F. Hogan, P. Lloyd-Williams, C. L. Herald, D. D. Burkett and P.J. Clewlow *J. Am. Chem. Soc.*, 1989, **111**, 5463.
13.   G.R. Pettit, C. L. Herald, D.L. Doubek and D. L. Herald, *J. Am. Chem. Soc.*,1982, **104**, 6846.
14.   a) S. A. Look, W. Fenical, G. K. Matsumoto, J. Clardy, *J. Org. Chem.*, 1986, **51**, 5140; b) S. A. Look, W. Fenical, R. S. Jacobs, J. Clardy, *Proc. Natl. Acad. Sci. USA*, 1986, **83**, 6238; c) *Scripps Institute of Oceanography Explorations*, 1997, **4** , 11.
15.   J. K. Reed, S. H. Sennett, P. J. McCarthy, T .A. Pitts, A. E. Wright and S. A. Pomponi, *Proceedings of the American Academy of Underwater Scientists, 18th Annual Symposium*, Vancouver, BC, 1998, 50.
16.   M. A. Sills, Netsci.org http://www.netsci.org/Science/Screening/feature10.html.
17.   A.J.S. Rayl, *The Scientist* 1999 **13**[20]: Oct 11, 1999

18. G.M. Cragg, S.A. Schepartz, M. Suffness, and M.R. Grever, *J. Nat. Prod.*, 1993, **56**, 1657.
19. D. Mendola, Abstract O-40, *American Society of Pharmacognosy 37th Annual Meeting*, July 19-24, 1998, Orlando, Florida
20. A. R. Duckworth, C. N. Battershill, D. R. Schiel and P. R. Bergquist, Proceedings of the 5th International Sponge Symposium -Origin and Outlook, *Memoirs of the Queensland Museum* 1999, **44**, 155.
21. N.K. Gulavita, S. P. Gunasekera, S. A. Pomponi, E.V. Robinson, *J. Org. Chem.*, 1992, **57**, 1767.
22. S. P. Gunasekera, S. A. Pomponi, P.J. McCarthy, *J. Nat. Prod.*, 1994, **57**, 79.
23. S. P. Gunasekera, M. Gunasekera, M.; P.J. McCarthy, *J. Org. Chem.*, 1991, **56**, 4830
24. H.H. Sun, S.J. Sakemi, *J. Org. Chem.*, 1991, **56**, 4307.
25. S. P. Gunasekera, M. Gunasekera, R. Longley and G. Schulte, *J. Org. Chem.*, 1990, **55**, 4912.
26. E. Ter Haar, R.J. Kowalski, E. Hamel, C. Lin, R. Longley, S. Gunasekera, H. Rosenkranz and B. Day, *Biochem.*, 1996, **35**, 243.
27. J.B. Nerenberg, D. T. Hung, P.K. Somers and S. L. Schreiber, *J. Am. Chem. Soc.*, 1993, **115** 12621.
28. A. B. Smith, Y.P. Qiu, D. R. Jones and K. Kobayashi, *J. Am. Chem. Soc.*, 1995, **117**, 12011.
29. D. T. Hung, J. Chen and S. L. Schreiber *J. Am. Chem. Soc.*, 1996, **118**, 11054.
30. D. T. Hung, J.B. Nerenberg and S. L. Schreiber *Chemistry and Biology* 1996, **3**, 287
31. S. S. Harried, G. Yang, M. A. Strawn and D. C. Myles. *J. Org. Chem.*, 1997, **62**, 6098.
32. J. A. Marshall and B.A. Johns, *J. Org. Chem.*, 1998, **63**, 7885.
33. D. P. Halstead *Diss Abstr. Int.* B 1999, **60 (3)**, 1087.
34. I. Paterson, G.J. Florence, K. Gerlach and J.P. Scott, *Angew. Chem., Intl. Ed. Engl.*, 2000, **39**, 377.
35. A. B. Smith, M.D. Kaufman, T.J. Beauchamp, M.J. LeMarche and H. Arimoto, *Org. Lett.* 1999, **1**, 1823.
36. a) D. J. Faulkner, M.D. Unson and C. A. Bewley, *Pure & Appl. Chem.*, 1994, **66**, 1983; b) C.A. Bewley, N.D. Holland, and D. J. Faulkner, D.J. *Experientia*, 1996, **52**, 716.
37. J. V. Lopez, P. J. McCarthy, K. E. Janda, R. Willoughby and S.A. Pomponi. Proceedings of the 5th International Sponge Symposium -"Origin and Outlook". *Memoirs of the Queensland Museum*. 44: 329-341. Brisbane ISSN 0079-8835.
38. P.A. Horton, F. E. Koehn, R. E. Longley and O. J. McConnell, *J. Am. Chem. Soc.*, 1994, **116**, 6015.
39. R. E. Longley and D. Harmody, *J. Antibiot.*, 1991, **44**, 93.
40. M. R. Boyd, K.D. Paull, *Drug Dev. Research* 1995, **34**, 91.
41. P.A. Horton, F. Koehn, R. E. Longley, O.J. McConnell and S. A. Pomponi, US Patent 5,478,861, Cytotoxic Macrolides and Methods of Use. 1997.
42. K. Bartik, J.-C. Braekman, D. Daloze, C. Stoller, J. Huysecom, G.. Vandevyver and R. Ottinger, *Can. J. Chem.*, **1987**, *65*, 2118.

43. S. Tsujii, K.L. Rinehart, S.P. Gunaskera, Y. Kashman, S. Cross, M.S. Lui, S.A. Pomponi and M. C. Diaz, *J. Org. Chem.*, 1988, **53**, 5446.

44. S.H. Sakemi and H. H. Sun, *J. Org. Chem.*, 1991, **56**, 4304.

45. A. E. Wright, S. S. Cross, S. A. Pomponi and P. McCarthy, *J. Org. Chem.*, 1992, **57**, 4772.

46. a) E. Fahy, B. C. M. Potts, D. J. Faulkner, K. Smith, *J. Nat. Prod.*, 1991, **54**, 564 b) G. Bifulco, I. Bruno, R. Riccio, J. Lavayre, G. Bourdy, *J. Nat. Prod.*, 1995, **58**, 1254.

47. S. Kohmoto, Y. Kashman, O.J. McConnell, K. L. Rinehart, A. Wright and F. E. Koehn, *J. Org. Chem.*, 1988, **53**, 3116.

48. S. P. Gunasekera, P.J. McCarthy and M. Kelly-Borges, *J. Nat. Prod.*, 1994, **57**, 1437.

49. B. L. Wylie, N. B. Ernst, K. J. S. Grace and R. S. Jacobs, in *Phospholipase A$_2$. Basic and Clinical Aspects in Inflammatory Diseases.* (W. Uhl, TJ. Nevalainen, M.W. Buchler eds.) *Prog Surg.* Basel, Kurger 1997 **24**, 146.

50. Braekman, J.C., D. Daloze and C. Stoller, *Bull. Soc. Chim. Belg.*, 1987, **96**, 809.

51. Achab, S. *Tetrahedron Lett.*, 1996, **37**, 5506.

52. I. Kawasaki, M. Yamashita and S. Ohta, *J. Chem. Soc., Chem. Comm.*, 1994, 208.

53. W. E. Noland and M.R. Venkiteswaran, *Cyclizative Condensations*, 1961, IV. 4263.

THE ANTICANCER DOLASTATINS AS CYANOBACTERIAL METABOLITES

George G. Harrigan,[1]* Hendrik Luesch,[2] Richard E. Moore[2] and Valerie J. Paul[3]

1. Searle Discovery Research, 700 Chesterfield Parkway North, Chesterfield, MO 63198, USA
2. Department of Chemistry, University of Hawaii at Manoa, Honolulu, HI 96822, USA
3. University of Guam Marine Laboratory, UOG Station, Mangilao, Guam 96923, USA
*To whom correspondence should be addressed. Phone: (314) 737-6663. FAX: (314) 737-7300
E-mail:george.g.harrigan@monsanto.com

Several analogues of the dolastatins, potent cytotoxic compounds originally derived from the Indian Ocean seahare, *Dolabella auricularia*, have now been found in the marine cyanobacteria, *Lyngbya majuscula* and *Symploca hydnoides*. These discoveries support the proposal that the dolastatins isolated from *D. auricularia* are of dietary origin. The implications of such an observation to natural products programmes is discussed. Aspects of the structure elucidation of these complex molecules are also presented.

1 INTRODUCTION

The dolastatins are a series of remarkable cytotoxic compounds isolated from the Indian Ocean seahare, *Dolabella auricularia*.[1] Most are peptide derivatives although other *Dolabella* metabolites are represented in a wide range of biosynthetic chemotypes. The impetus for investigations of *D. auricularia* as a source of new anticancer compounds derived from the initial discovery in 1972 of potent cytotoxic extracts of this seahare.[2] Many dolastatins with pronounced anticancer activity have now been isolated. The most important of these are dolastatins 10 (**1**)[3] and 15 (**2**),[4] the structures of which are shown in Figure 1, analogues of which are in phase I trials as anticancer agents.[5]

However the exceedingly low yields of the dolastatins[6] and of other metabolites obtained from *D. auricularia* (typically 10^{-6} -10^{-7}%) imply that this mollusk is not the true producer of these compounds. *D. auricularia* is a known generalist herbivore and many metabolites originally isolated from seahares are of dietary origin.[7] We present here results obtained from investigations of anticancer compounds isolated from the marine cyanobacteria (blue-green algae), *Lyngbya majuscula* and *Symploca hydnoides*, and also highlight the many biosynthetic similarities between compounds isolated from *D. auricularia* and those of unambiguous cyanobacterial or other algal origin.

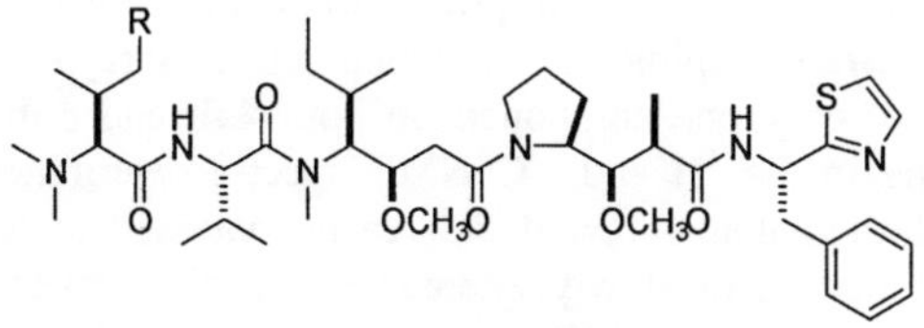

1 R = H

10 R = CH3

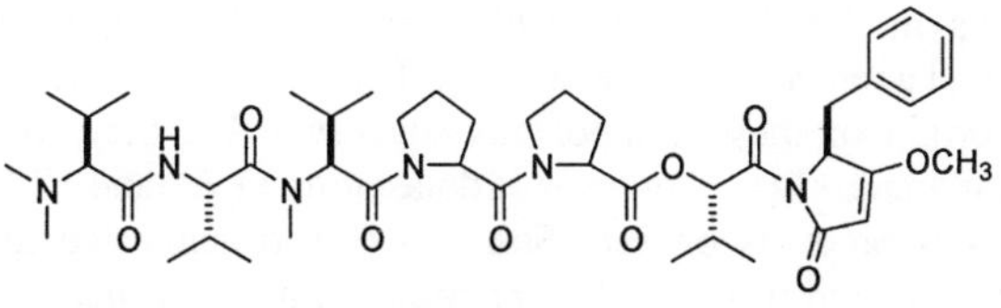

2

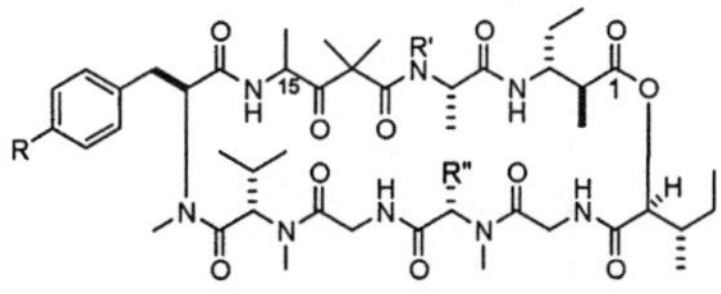

	R	R'	R"
3	OCH₃	H	(CH₃)₂CHCH₂- S at C-15
4ab	OCH₃	CH₃	(CH₃)₂CHCH₂- epimeric at C-15 (a = 15-S, b = 15-R)
5ab	H	CH₃	(CH₃)₂CHCH₂- epimeric at C-15 (a = 15-S, b = 15-R)
6	OCH₃	H	CH₃CH₂(CH₃)CH-

Figure 1 *Structures of compounds **1** (dolastatin 10), **2** (dolastatin 15), **3** (dolastatin 11), **4a** (lyngbyastatin 1), **4b** (epilyngbyastatin 1), **5a** (dolastatin 12), **5b**, **6** (majusculamide C) and **10** (symplostatin 1).*

The purpose of this review is i) to demonstrate that most, if not all, metabolites currently isolated from *D. auricularia* are of cyanobacterial (blue-green algal) origin and ii) to discuss the implications of such an observation to industrial natural products programs.

2 RECENT DISCOVERIES OF DOLASTATINS FROM MARINE CYANOBACTERIA

Our interest in the dolastatins was stimulated by our recent isolation of a C-15 epimeric mixture of an N-methylated analog of dolastatin 11 (**3**), which we termed lyngbyastatin 1 (**4a**) and epilyngbyastatin 1 (**4b**), and dolastatin 12 (**5a**) (also as a C-15 epimeric mixture) from a *L. majuscula/Schizothrix calcicola* assemblage collected near Guam.[8] Dolastatins 11 (**3**) and 12 (**5a**) were originally isolated as potent cytotoxic compounds from *D. auricularia*.[9]

The structure elucidation of **4ab** and **5ab** proved to be an interesting challenge due to their isolation as inseparable epimers.[8] Although electrospray mass spectroscopy suggested the presence of only one component in both **4ab** and **5ab**, extensive signal doubling and broadening in the ^1H and ^{13}C NMR spectra determined in a variety of deuterated solvents indicated that **4ab** and **5ab** each contained at least two isomeric species or existed as a mixture of slowly interconverting conformers in solution. This greatly hampered any attempt to use NMR spectroscopy as a significant tool in structure elucidation. It was evident from these spectral data however, that both **4ab** and **5ab** contained N-methylated depsipeptides and that **4ab** was a methoxyl analogue of **5ab**. This was supported by HRFABMS, which indicated a 30 mass unit difference between these two; the components in **4ab** had a molecular formula of $C_{51}H_{82}N_8O_{12}$, whereas the components in **5ab** had a molecular formula of $C_{50}H_{80}N_8O_{11}$. The molecular formulae and the cyanobacterial origin of these components suggested that they were N-methylated analogues of majusculamide C {**6**}[10] or even of dolastatins 11 {**3**} and 12 {**5a**}.

The base hydrolysis products of **4ab**, **5ab** and **6** were subjected to FABMS. The fragmentation patterns shown in Figure 2 supported the proposal that **4ab** and **5ab** were almost identical to majusculamide C {**6**} and dolastatin 11 {**3**} and 12 {**5a**}.

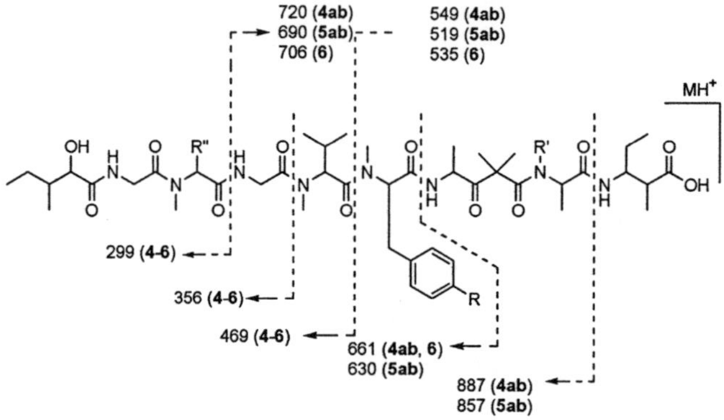

	R	R'	R"
4ab	OCH3	Ch3	(CH3)2CHCH2-epimeric at C-15 (a = 15-S, b = 15-R)
5ab	H	CH3	(CH3)2CHCH2- epimeric at C-15 (a = 15-S, b = 15-R)
6	OCH3	H	CH3CH2(CH3)CH-

Figure 2 *FABMS Spectral Fragmentation pattern of Base Hydrolysates of **4ab**, **5ab** and majusculamide C(6). Numbers in parentheses indicate to which hydrolysate a particular mass fragment refers.*

Supporting evidence for this proposal was obtained from amino acid analyses which demonstrated that glycine was the only typical or non N-methylated amino acid present in **4ab** and **5ab**. Marfey analysis of the acid hydrolysates of **4ab**, **5ab** and majusculamide C {**6**} confirmed that **4ab** and **5ab** differed from majusculamide C (**6**) in possessing an N-methyl-L-alanine residue rather than an L-alanine residue. Moreover, **4ab** and **5ab** contained an N-methyl-L-leucine residue rather than an N-methyl-L-isoleucine residue, indicating a closer structural relationship to dolastatin 11 (**3**) and dolastatin 12 (**5a**) than to majusculamide C (**6**).

These were the only differences however, and even after comparison with the ^{1}H- and ^{13}C-NMR spectra of a known sample of dolastatin 12 (**5a**), this did not immediately indicate the causes of the extensive broadening and doubling of signals in the NMR spectra of **4ab** and **5ab**. Chromatographic peaks in the Marfey profile corresponding to racemic 2-amino-4-methylpentanone derived from the 4-amino-2,2-dimethyl-3-oxopentanoic acid residue offered the prospect that **4ab** and **5ab** each existed as a pair of C-15 epimers. However, proof of this possibility is made problematic by the acid/base sensitivity of the 4-amino-2,2-dimethyl-3-oxopentanoic acid residue. Racemic mixtures 2-amino-4-methylpentanone are obtained in the acid hydrolysis of majusculamide C (**6**) where the chirality at C-15 has been assigned as *S* on the basis of synthetic endeavours.[11] This stereochemistry also exists in dolastatin 11 (**3**) and dolastatin 12 (**5a**).[11]

Epimerisation of majusculamide C (**6**) could be effected with 5% TFA in CH$_3$CN on overnight standing at room temperature. This resulted in doubling and broadening of signals in its ^{1}H NMR spectrum. No changes, however, were evident in the ^{1}H NMR spectra of **4ab** and **5ab** on similar acid treatment indicating that epimerisation at C-15 had already occurred.

Isolation of complex molecules in low yields presents challenges in structure elucidation. In all cases, as illustrated above, our structure elucidation of cyanobacterial-derived dolastatins would have been greatly hampered without their isolation in good yields or prior structural proofs of the seahare-derived metabolites. This clearly highlights the importance of correctly identifying the true biological origin of metabolites of interest.

One significance of the results presented above is that it represented the first time that a dolastatin had been isolated from cyanobacteria and implied that other dolastatins, including those of clinical significance, would prove to be of dietary origin. Indeed, recent investigations of a lyngbyastatin 1 (**4ab**) producing *L. majuscula* collection from Guam afforded lyngbyastatin 2 (**7**),[12] a closely related analogue of dolastatin G (**8**) (Figure 3).[13] Structure elucidation of this new metabolite was based on NMR comparisons including extensive 2-D NMR experiments with dolastatin G (**8**).[14] Dolastatin G required total synthesis for its structure proof. Interestingly, another dolastatin of mixed peptide-polyketide biogenesis is dolastatin 14 (**9**),[15] which has yet to be isolated from a cyanobacterial source.

Our most intriguing finding to date was the isolation, from *Symploca hydnoides*, of symplostatin 1 (**10**),[16] a derivative only one methylene unit different than dolastatin 10 (**1**).[3] The structure of **10** is shown in Figure 1. Analogues of dolastatin 10 (**1**) are currently in phase I trials in Europe and Japan as anticancer agents.[1] Biosynthetic similarities between dolastatin 10 (**1**) with barbamide (**11**),[17] a metabolite from *L. majuscula*, had previously suggested that dolastatin 10 was most probably a cyanobacterial metabolite and we were extremely encouraged to find it in our own collections.

Structure elucidation and NMR spectral analyses of symplostatin 1 (**10**) was complicated by extensive broadening and considerable overlap of several signals in the ^{1}H NMR spectrum and also by the presence of a minor conformer[18] which doubled the number of signals. Despite these analytical difficulties, direct comparison of the NMR spectra indicated that symplostatin 1 (**10**) differed from **1** at only one site in the molecule. The HRFABMS of **10** established the molecular formula as C$_{43}$H$_{70}$N$_6$O$_8$S, one methylene unit greater than that for **1**. ^{13}C NMR chemical shift differences clearly showed that one of the isopropyl groups in **1** had been replaced by a *sec*-butyl group in **10** while HMBC NMR spectroscopy indicated that **10** differed from **1** in the presence of a terminal *N,N*-dimethylisoleucine residue instead of a *N,N*-dimethylvaline residue.

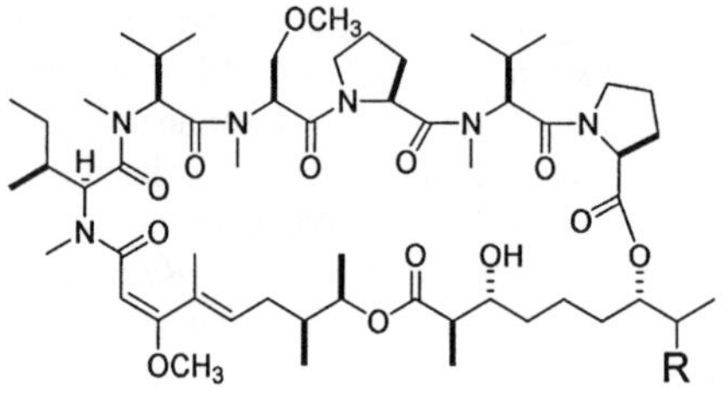

7 R = H
8 R = CH₃

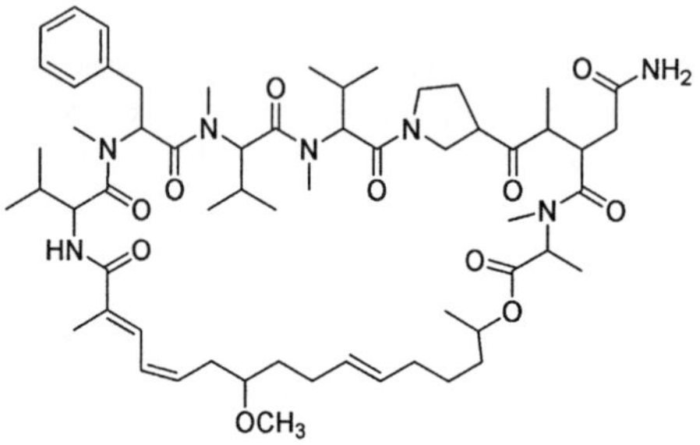

9

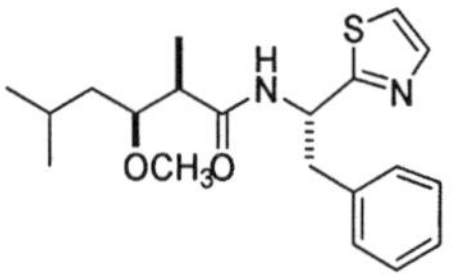

11

Figure 3 *Structures of compounds 7 (lyngbyastatin 2), 8 (dolastatin G), 9 (dolastatin 14) and 11 (barbamide).*

Dolastatin 10 (**1**) appears to be one of the most potent antineoplastic compounds known to date.[1] The isolation of a closely related analogue from a cultivable source is significant, as this potentially allows study of its biosynthesis.

Also isolated from *Symploca hydnoides* was symplostatin 2 (**12**),[19] a cyclic depsipeptide with close structural analogy to dolastatin 13 (**13**) (Figure 4).[20] It had in fact been suggested earlier that dolastatin 13 (**13**) most probably had a cyanobacterial dietary origin.[21] This proposal was based on its structural similarity with the nostopeptins,[22] micropeptins,[23] oscillapeptin[24] and A90720A.[25] These latter metabolites have all been isolated from freshwater cyanobacterial whereas symplostatin 2 (**12**) represents the first such analogue from a marine source.

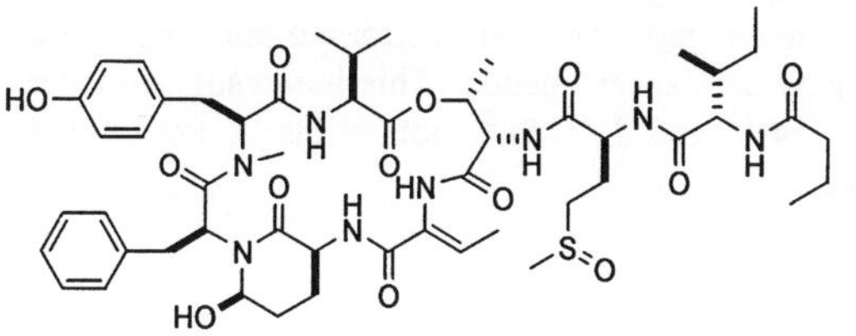

12

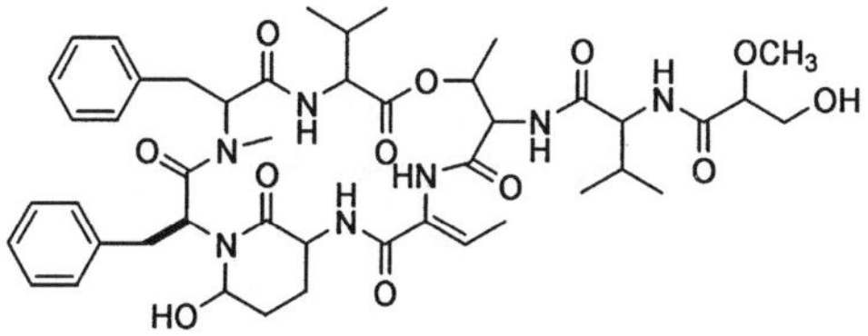

13

Figure 4 *Structures of compounds 12 (symplostatin 2) and 13 (dolastatin 13).*

Again, structure elucidation of **12** presented some challenges due to broadened signals in NMR spectra. However, the amino acid content, particularly the presence of the rarely found 3-amino-6-hydroxy-2-piperidone (AHP) and 2-amino-2-butenoic acid (ABU) residues, clearly suggested a close structural relationship to dolastatin 13 (**13**). A cyclic core as shown in structure **12** could therefore be proposed based on 1-D and 2-D NMR spectral analyses. Because of signal broadening in the NMR spectra only a few of the amide proton to adjacent amino acid carbonyl carbon connectivities, which are generally needed for peptide sequencing by HMBC, were discernible. Nevertheless the following sequences could be established from HMBC: *N*-Me-tyrosine-phenylalanine-AHP-ABU and threonine-methioninesulfoxide-isoleucine-butanoic acid. ROESY correlations were required to confirm the placement of the valine residue. The absolute stereochemistries at the α-position of the amino acid units were determined by Marfey analysis. Chiral GC-MS was required to unambiguously establish the absolute stereochemistries at the β-positions of the isoleucine and threonine residues. The absolute stereochemistry has yet to be established for dolastatin 13 (**13**) and this is obviously attributable to the exceedingly low yields in which it was isolated.

3 OTHER *DOLABELLA* METABOLITES OF PROBABLE CYANOBACTERIAL OR OTHER ALGAL ORIGIN

The extraordinary technical achievement and success of the Pettit group in isolating exceedingly small amounts of highly bioactive metabolites from vast excesses of *D. auricularia* extracts has encouraged other investigators to pursue *D. auricularia* and other seahares as sources of new anticancer agents. This has resulted in the isolation of, for example, dolastatins C (**14**)[26] and D (**15**),[27] doliculiols A and B (**16, 17**),[28] and the dolabelides (**18-21**) (Figures 5 and 6).[29]

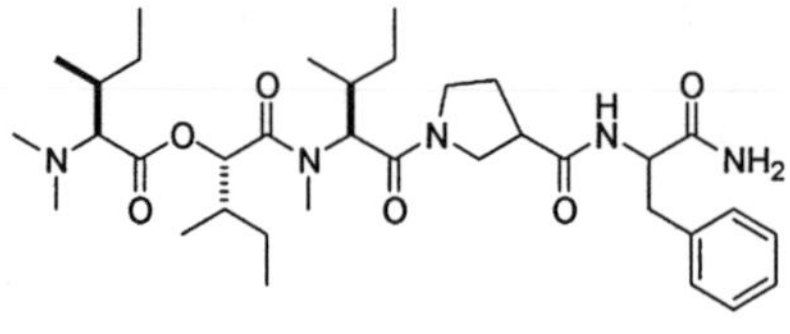

14

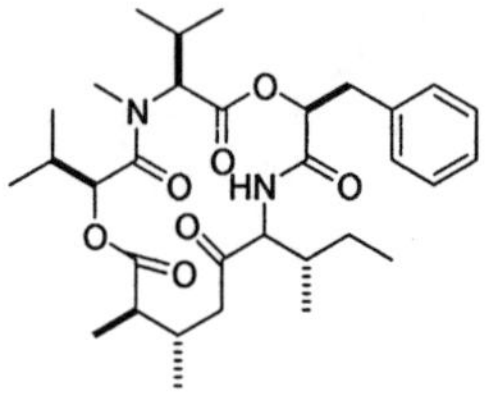

15

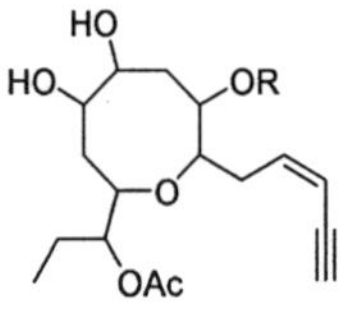

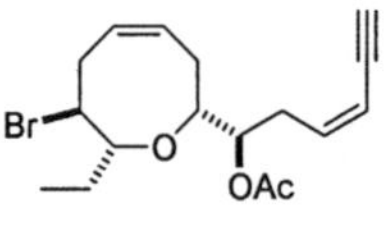

16 R = H
17 R = Ac

22

Figure 5   *Structures of compounds 14 and 15 (dolastatins C and D), 16 and 17 (doliculiols A and B), and 22 (laurencin).*

Dolastatins C (**14**) and D (**15**) clearly show structural resemblance to dolastatins 10 (**1**) and 15 (**2**). The doliculiols (**16, 17**) show striking similarities to laurencin (**22**)[30] and other closely related analogues from the red alga, *Laurencia*. Significantly, many *Laurencia* metabolites have been found in the seahare, *Aplysia*.[7] The dolabelides (**18-21**) are structurally related to the scytophicins C-E[31] (**23-25**) and also to aplyronine A (**26**), a potent antitumor compound with demonstrated *in vivo* efficacy against certain cancer cell lines.[32] Aplyronine A was originally isolated from the seahare *Aplysia kurodai* but is almost certainly an algal metabolite. The scytophycins were originally found in several species of the terrestrial cyanobacterium, *Scytonema*.[31]

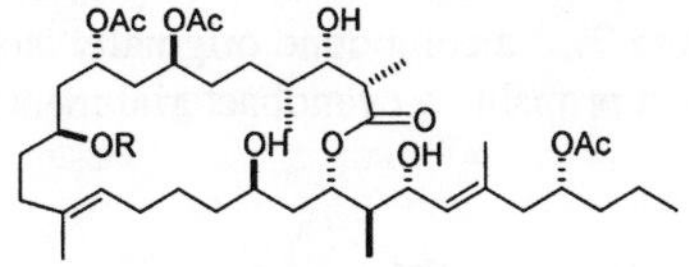

18 R = Ac
19 R = H

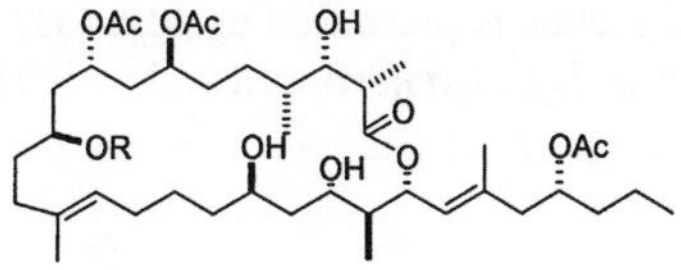

20 R = Ac
21 R = H

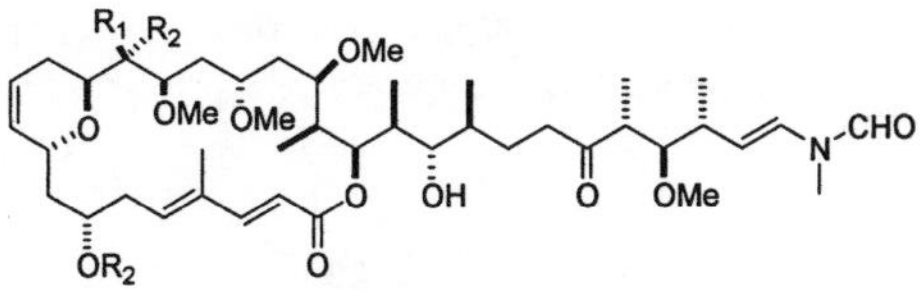

23 R₁ = CH₃ R₂ = H
24 R₁ = CH₃ R₂ = OH
25 R₁ = CH₂OH R₂ = H

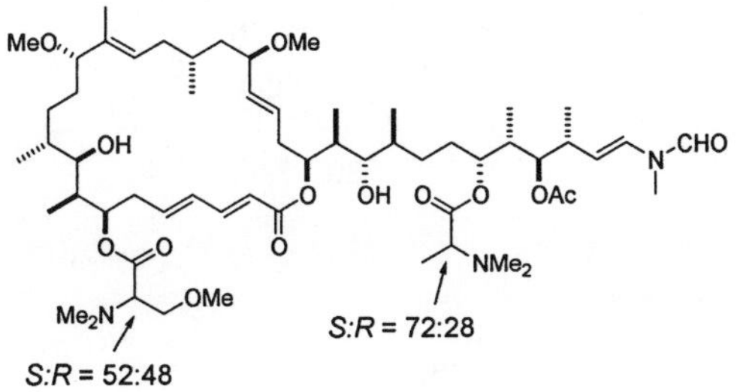

26

Figure 6   *Structures of compounds **18-21** (the dolabelides), **23-25** (scytophicins C-E), and **26** (aplyrone A).*

4   CYANOBACTERIAL BIOSYNTHETIC SIGNATURES AND *DOLABELLA* METABOLITES

We consider that all of the dolastatins will prove to be of cyanobacterial or other dietary origin. There are some important biosynthetic signatures, for example, which suggest that dolastatin 15 (**2**) (Figure 1) is most probably derived from *L. majuscula*. These signatures include the presence of the N,N-dimethylvaline residue, a valine-N-methylvaline-proline sequence also found in the microcolins (**27, 28**)[33] (Figure 7) and the presence of a pyrrolidone ring presumably derived from condensation of phenylalanine and acetate. A pyrrolidone ring derived from glycine and acetate is present in malyngamide A (**29**)[34] while the microcolins (**27, 28**)[33] and majusculamide D (**30**)[35] (Figure 7) contain a pyrrolidone ring derived from alanine and acetate. A pyrrolidone ring derived from valine and acetate is present in dysidin (**31**) (Figure 7),[36] a compound originally isolated from the sponge *Dysidea herbacea* and which is most probably a cyanobacterial metabolite.

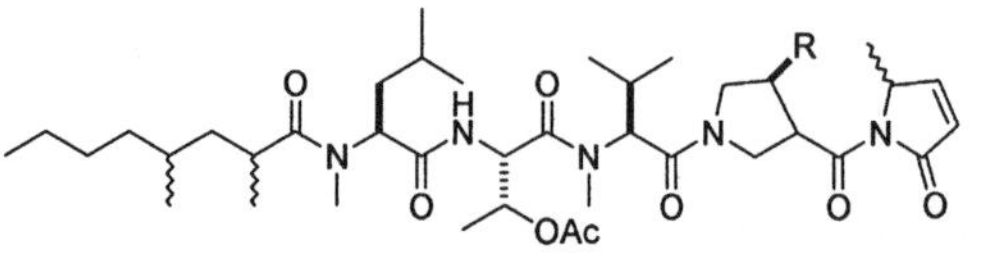

27 R = OH
28 R = H

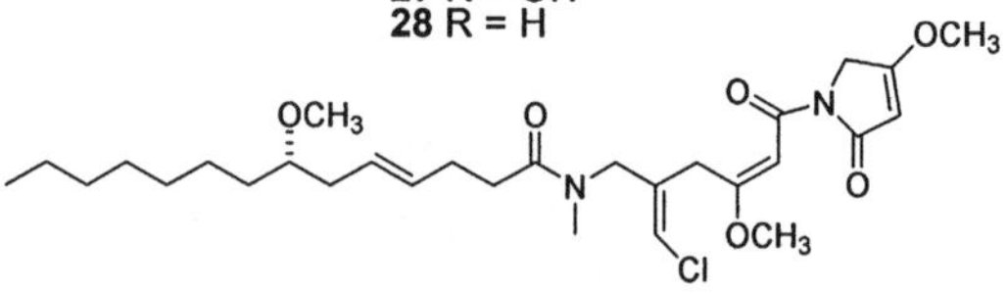

29

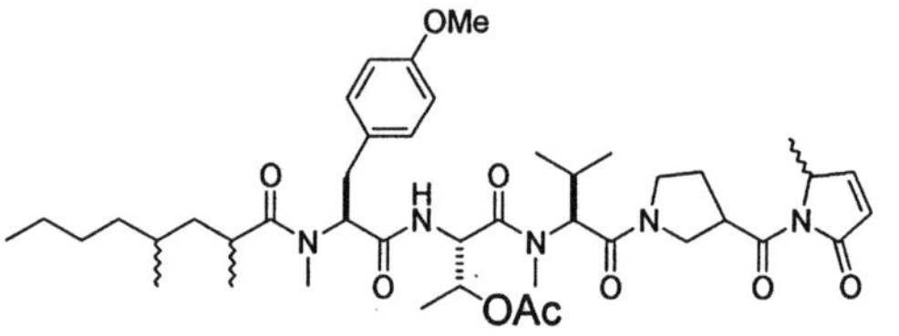

30

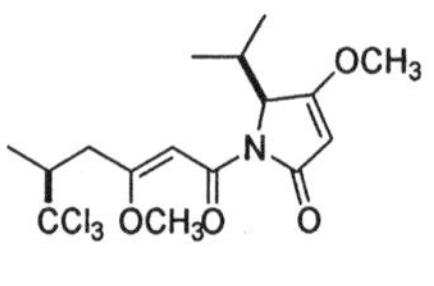

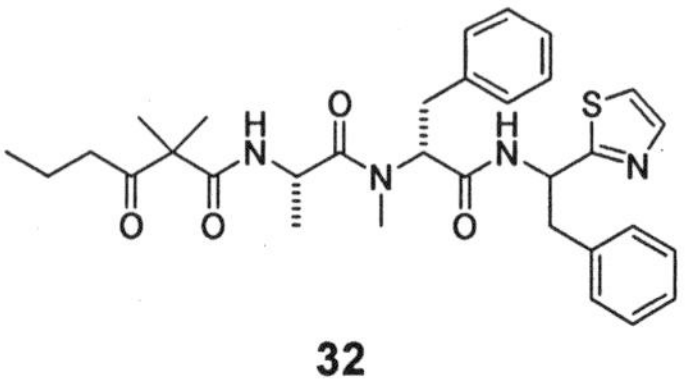

31   **32**

Figure 7   *Structures of compounds 27 and 28 (the microcolins), 29 (malyngamide A), 30 (majusculamide D), 31 (dysidin) and 32 (dolastatin 18).*

Dolastatin 18 (**32**)[37] (Figure 7) also contains significant biosynthetic signatures that imply that it is of cyanobacterial origin. These include a 2,2-dimethyl-1,3-dicarbonyl moiety found in lyngbyastatin 1 (**4a**) and majusculamide C (**6**) and a dolaphenine residue found in symplostatin 1 (**10**) and barbamide (**11**).

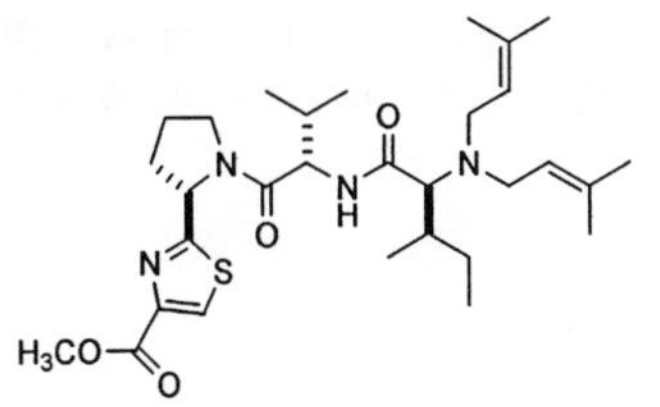

33

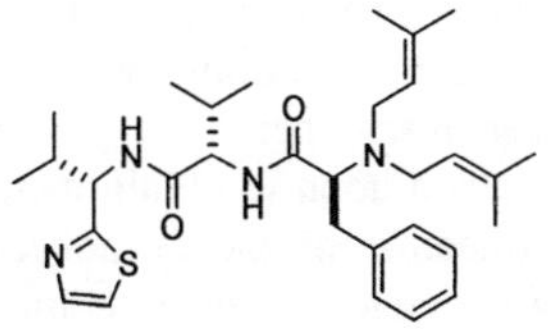

34

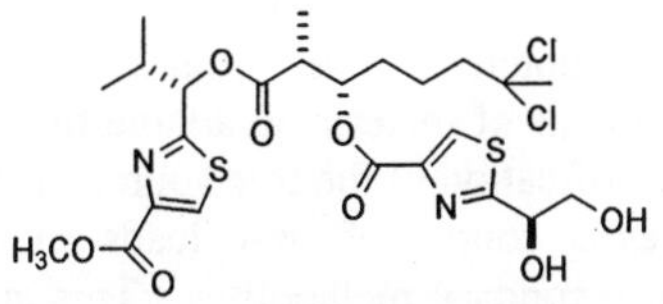

35

Figure 8 *Structures of compounds **33** (aeruginosamide), **34** (dolabellin) and **35** (virenamide A).*

Another interesting example where biosynthetic signatures can be used to imply cyanobacterial origin of compounds isolated from seahares (and other marine macroorganisms) is that based on the recent isolation of aereuginosamide (**33**)[38] (Figure 8) from a *Microcystis aeruginosa* bloom. The diisoprenylamine and carboxylated thiazole groups of this linear peptide had not previously been encountered in cyanobacterial metabolites. However dolabellin (**34**),[39] initially discovered from *D. auricularia* does contain a carboxylated thiazole group whereas the diisoprenylamine group has been found in virenamide A (**35**) (Figure 8).[40] The virenamides were originally isolated from *Diplosoma virens*, an ascidian known to harbor blue-green algae in its cloacal cavity.[40]

Clearly, as the structural diversity of cyanobacterial metabolites is further elucidated, many compounds attributed to marine macroorganisms will prove to be derived from these prolific prokaryotes.

5 CONCLUDING REMARKS

We have attempted in this review to demonstrate that most, if not all, *Dolabella* metabolites are of dietary origin. The dietary origin of many compounds isolated from seahares has been suggested since the 1970s. Indeed this fact has been recognized, although not acted upon, by most of the current investigators in *Dolabella* research. The ecological implications of such large scale field collections have also been considered, albeit deemed negligible, by many of these researchers.[41] It is our hope that a well documented ecological phenomenon can be acted upon to obtain metabolites not through excessive collections of seahares but by collection of the cyanobacteria or algae upon which they feed. This will result in higher yields with concomitant greater ease of structure elucidation and shorter time-lines from identification of active components to their evaluation in efficacy trials. It also allows the exploitation of culturable resources, allowing smaller field collections and less, potentially deleterious, ecological impact.

The most important aspect of drug discovery in industry today is speed. Natural product extracts containing low yields of complex unidentified metabolites are inimical to this process. Aspects of the structure elucidation of the complex dolastatins we encountered in our research on cyanobacterial metabolites were presented to illustrate this point. All these molecules presented required extensive 2-D NMR analyses, chemical derivatisations or were based on structures which had been proven by prior synthetic endeavors. Many of the problems encountered in the initial structure elucidation of the dolastatins were related to the low amounts of material available for structural analyses, a clear consequence of the initial misidentification of the true source of these metabolites.

Considerations of biodiversity as a source of new leads must address the true biological origin of desired bioactive secondary metabolites. This is most certainly true when considering the marine environment as a source of new leads for the pharmaceutical and agricultural industries.

The poor discovery rate for new leads from combinatorial chemistry approaches indicate that natural products chemistry will continue to warrant support. However attention to source material, particularly the true biological origin of desired metabolites, direct assessments of chemical diversity and effective dereplication strategies, is essential.

Acknowledgements

Funding for the original research on the isolation of dolastatin analogs from cyanobacteria carried out in the laboratories of REM and VJP was provided by a NCNPDDG Grant CA53001 from the National Cancer Institute and by NIH grants GM 38624 and GM 44796,

References And Notes

1. G. R. Pettit, Y. Kamano, C. L. Herald, Y. Fujii, H. Kizu, M.R. Boyd, F. E. Boettner, D. L. Doubek, J. M. Schmidt, J.-C. Chapuis and C. Michel, *Tetrahedron*, 1993, **49**, 9151. G. R. Pettit, *J. Nat. Prod.*, 1996, **59**, 812. G. R. Pettit. *'Progress in the Chemistry of Organic Natural Products'*, W. Herz, G. W. Kirby, R. E. Moore, W. Steglich and C. Tamm, (Eds.), Springer -Verlag Wien, New York, 1997; Volume 7, p. 1

2. G. R. Pettit, R. H. Ode, C. L. Herald, R. B. Von Dreele and C. Michel, *J. Am. Chem. Soc.*, 1976, **98**, 4677

3. G. R. Pettit, Y. Kamano, C. L. Herald, A. A. Tuinman, F. E. Boettner, H. Kizu, J. M. Schmidt, L. Baczynskyj, K. B. Tomer and R. J. Bontems, *J. Am. Chem. Soc.*, 1987, **109**, 6883

4. G.R. Pettit, Y. Kamano, C. Dufresne, R.C. Cerny, C.L. Herald and J.M. Schmidt, *J. Org. Chem.*, 1989, **54**, 6005

5. G. R. Pettit, C .L. Herald and Y. Kamano, US Pat. 4, 816, 444, Mar 28, 1989, 26 pages. G.R. Pettit, J.K. Srirangam and D. Kantoci, US Pat. 5,663,149, Sept 2, 1997, 50 pages. K. Sakakibara, M. Gondo and K. Miyazaki, US Pat. 5,654,399, Aug 5, 1997, 48 pages

6. For example, it required a collection of 1600 Kg of seahares to obtain only 20 mg of dolastatin 10 (**1**)[3]

7. T. H. Carefoot, *Oceanogr. Mar. Biol. Ann. Rev.*, 1987, **25**, 167. D. J. Faulkner, *Tetrahedron*, 1977, **33**, 1421. D. J. Faulkner, *Nat. Prod. Rep.*, 1984, **1**, 251. D. J. Faulkner, *Nat. Prod. Rep.*, 1984, **1**, 551. D. J. Faulkner, *Nat. Prod. Rep.*, 1986, **3**, 1. D. J. Faulkner, *Nat. Prod. Rep.*, 1987, **4**, 539. D. J. Faulkner, *Nat. Prod. Rep.*, 1988, **5**, 613. D. J. Faulkner, *Nat. Prod. Rep.*, 1990, **7**, 269. D. J. Faulkner, *Nat. Prod. Rep.*, 1991, **8**, 97. D. J. Faulkner, *Nat. Prod. Rep.*, 1992, **9**, 323. D. J. Faulkner. *'Ecological Roles of Marine Secondary Metabolites'*, V. J. Paul, (Ed.) Cornell University Press, Ithaca, New York, 1992, p.119. D. J. Faulkner, *Nat. Prod. Rep.*, 1993, **10**, 497. D. J. Faulkner, *Nat. Prod. Rep.*, 1994, **11**, 355. D. J. Faulkner, *Nat. Prod. Rep.*, 1995, **12**, 223. D. J. Faulkner, *Nat. Prod. Rep.*, 1996, **13**, 75

8. G. G. Harrigan, W. Y. Yoshida, R. E. Moore, D. G. Nagle, P. U. Park, J. Biggs, V. J. Paul, S. L. Mooberry, T. H. Corbett and F. A. Valeriote, *J. Nat. Prod.*, 1998, **61**, 1221

9. G. R. Pettit, Y. Kamano, H. Kizu, C. Dufresne, C. L. Herald, R. J. Bontems, J. M. Schmidt, F. E. Boettner and R. A. Nieman, *Heterocycles*, 1989, **28**, 553

10. D. C. Carter, R. E. Moore, J. S. Mynderse, W. P. Niemczura and J. S. Todd, *J. Org. Chem.*, 1984, **49**, 236

11. R. B. Bates, K. G. Brusoe, J. J. Burns, S. Caldera, W. Cui, S. Gangwar, M. R. Gramme, K. J. McClure, G. P. Rouen, H. Schadow, C. C. Stessman, S. R. Taylor, V. H. Vu, G. V. Yarick, J. Zhang, G. R. Pettit and R. J. Bontems, *J. Am. Chem. Soc.*, 1997, **119**, 2111

12. H. Luesch, W. Y. Yoshida, R. E. Moore and V. J. Paul, *J. Nat. Prod.*, submitted for publication

13. T. Mutou, T. Kondo, M. Ojika and K. Yamada, *J. Org. Chem.*, 1996, **61**, 6340

14. T. Mutou, T. Kondo, T. Shibata, M. Ojika, H. Kigoshi and K. Yamada, *Tetrahedron Lett.*, 1996, **40**, 7299

15. G. R. Pettit, Y. Kamano, C. L. Herald, C. Dufresne, R. E. Bates and J. M. Schmidt, *J. Org. Chem.*, 1990, **55**, 2989

16. G. G. Harrigan, H. Luesch, W. Y. Yoshida, R. E. Moore, D. G. Nagle, V. J. Paul, S. L. Mooberry, T. H. Corbett and F. A. Valeriote, *J. Nat. Prod.*, 1998, **61**, 1075

17. J. Orjala, J. and W. H. Gerwick, *J. Nat. Prod.*, 1996, **59**, 427

18. Two conformers were observed in a ratio of about 3:1 in CD_2Cl_2. This is the same as reported for dolastatin 10 (**1**)[3]

19. G. G. Harrigan, H. Luesch, W. Y. Yoshida, R. E. Moore, D. G. Nagle and V. J. Paul, *J. Nat. Prod.*, 1999, **62**, 655

20. G. R. Pettit, Y. Kamano, C. L. Herald, C. Dufresne, R. L. Cerny, D. L Herald, J. M. Schmidt and H. Kizu, *J. Am. Chem. Soc.*, 1989, **111**, 5015

21. R. E. Moore, *J. Industr. Microbiol.*, 1996, **16**, 134

22. T. Okino, S. Qi, H. Matsuda, M. Murakami and K. Yamaguchi, *J. Nat. Prod.*, 1997, **60**, 158

23. K. Ishida, H. Matsuda, M. Murakami, and K. Yamaguchi, *J. Nat. Prod.*, 1997, **60**, 184

24. H. J. Shin, H. Murakami, H. Matsuda, K. Ishida and K. Yamaguchi, *Tetrahedron Lett.*, 1992, **36**, 5235

25. A. Y. Lee, T. A. Smitka, R. Bonjouklian and J. Clardy, *Chem. Biol.*, 1994, **1**, 113

26. H. Sone, T. Nemoto, M. Ojika and K. Yamada, *Tetrahedron Lett.*, 1993, **34**, 8445

27. H. Sone, T. Nemoto, H. Ishiwata, M. Ojika and K. Yamada, *Tetrahedron Lett.*, 1993, **34**, 8449

28. M. Ojika, T. Nemoto and K. Yamada, *Tetrahedron Lett.*, 1995, **36**, 3461. K. Suenaga, T. Nagoya, T. Shibata, H. Kigoshi and K. Yamada, *J. Nat. Prod.*, 1997, **60**, 155

29. M. Ojika, T. Nagoya and K. Yamada, *Tetrahedron Lett.*, 1995, **36**, 7491

30. T. Irie, M. Suzuki and T. Masamune, *Tetrahedron*, 1968, **24**, 4193. A. F. Cameron, K. K. Cheung, G. Ferguson and J. M. Robertson, *J. Chem. Soc. (B)*, 1969, 559

31. S. Carmeli, R. E. Moore and G. M. L. Patterson, *J. Nat. Prod.*, 1990, **53**, 1533

32. K. Yamada, M. Ojika, T. Ishigaki, Y. Yoshida, E. Ekimoto and M. Arakawa, *J. Am. Chem. Soc.*, 1993, **115**, 11020

33. F. E. Koehn, R. E Longley and J. K. Reed, *J. Nat. Prod.*, 1992, **55**, 613

34. J. H. Cardellina, F.-J. Marner and R. E. Moore, *J. Am. Chem. Soc.*, 1979, **101**, 240

35. R. E. Moore and M. Entzeroth, *Phytochemistry*, 1988, **27**, 3101

36. W. Hofheinz and W. E. Oberhansli, *Helv. Chim. Acta*, 1977, **60**, 660

37. G. R. Pettit, J. Xu, M. D. Williams, F. Hogan., J. M. Schmidt and R.L. Cerny, *Bioorg. Med. Chem. Lett.*, 1997, **7**, 827

38. L. A. Lawton, L. A. Morris and M. Jaspers, *J. Org. Chem.*, 1999, **64**, 5329

39. H. Sone, T. Kondo, M. Kiryu, H. Ishiwata, M. Ojika and K. Yamada, *J. Org. Chem.*, 1995, **60**, 4774

40. A. R. Carroll, Y. Feng, B. F. Bowden and J. C. Coll, *J. Org. Chem.*, 1996, **61**, 4059

41. In their paper on the isolation of dolabelides from 175 Kg of *D. auricularia*, Suenaga *et al* (ref. 28) remark "Although a large number of seahares had to be collected for this work, *D. auricularia* is a very common animal, and plenty of them were observed the following year at the collection location. Our collection thus posed no significant ecological threat to the species."

Note Added in Proof

Further "*Dolabella*" compounds have now been isolated from collections of the marine cyanobacterium, *Lyngbya majuscula*. These include dolastatin 3 (S. S. Mitchell, D. J. Faulkner, K. Rubins and F. D. Bushman, *J. Nat. Prod.*, 2000, **63**, 279) and lyngybyabellin A, a close analogue of dolabellin A (H. Luesch, W. Y. Yoshida, R. E. Moore, V. J. Paul and S. L. Mooberry, *J. Nat. Prod.*, 2000, in press).

A Cu^{2+} SELECTIVE MARINE METABOLITE.

Linda A. Morris[1] and Marcel Jaspars

Marine Natural Products Laboratory, Department of Chemistry, University of Aberdeen, Old Aberdeen, AB24 3UE, Scotland

1 INTRODUCTION AND BACKGROUND

1.1 General

This chapter deals with the unique metal binding properties and potential ecological significance of modified cyclic octapeptides isolated from the Indo-Pacific ascidian *Lissoclinum patella* (Order *Enterogena,* Family *Didemnidae). L. patella* belongs to the phylum Chordata, as do vertebrates. Ascidians are also known as sea squirts due to their ability to squirt water out of their inner hollow cavities *via* siphons, or tunicates as their body is covered in a tunic, made of tunicin, which forms their outer layer. Adult ascidians are either solitary or colonial sessile filter feeders. Their morphology is diverse with solitary species ranging from 1-15 cm in length while colonial species like *L. patella* resemble sponges. Unlike other marine invertebrates, most tunicates have a well-developed immune system.[2] Many species are host to symbionts and in the case of *L. patella* the symbiont is the prokaryote *Prochloron*. This is a photosynthetic symbiont and phagocytosis of such a symbiont is thought to be the possible origin of chloroplasts in green algae and higher plants.[3] In *L. patella* the *Prochloron* is found residing within the tunic of the ascidian.

Prochloron is found in and on a variety of ascidian species, and also other marine invertebrates, sometimes along with other symbionts, but in *L. patella* it is the only species of symbiont that is found. It is not known whether other symbionts are deterred by some chemical means, but the symbiosis is beneficial to both symbiont and host. Larval stages of *L. patella* bud from the adult already containing some *Prochloron* cells and research has shown that the settling of these larvae on new substrates is controlled by light conditions being suitable for efficient photosynthesis by *Prochloron*.[4] *L. patella* receives between 30 and 50% of its carbon requirements and a large proportion of its nitrogen from this symbiont. Much of the carbon is transferred from *Prochloron* to *L. patella* as glycolate, which is an inhibitor of photosynthesis. The mechanism of this transfer is as yet unknown, as is that of the nitrogen, but if ^{13}C labelled CO_2 is provided the final fate of the carbon transferred is in the lipids, nucleic acids and proteins of the host organism. In return for this transfer the symbiont has an exclusive habitat where the translucency of its host allows photosynthesis. The host actively encourages optimum photosynthesis by taking up glycolate and filtering out green to orange wavelengths with pigments in the tunic so that the light reaching *Prochloron* is of the optimum wavelengths required for photosynthesis.

1.2 Compounds Isolated from *L. patella*

L. patella is known for the yielding of several families of closely related cyclic peptides, many of which have significant biological activity.[5] These peptides have been grouped into four main structural types - ulithiacyclamides, patellamides, lissoclinamides and tawicyclamides - according to the number of amino acids and inclusion of thiazole, thiazoline and oxazoline rings within their structure. The ulithiacyclamides all possess a disulphide bridge. To date eight patellamides (A-G and ascidiacyclamide) (**1-8**),[6,7,8,9] eight lissoclinamides (1-8) (**9-16**),[10,11,12] four ulithiacyclamides (A-D) (*e.g.* **17, 18**),[13] and two tawicyclamides (A and B)(**19, 20**)[14] have been isolated. The structures of these are shown in Figure 1 to illustrate the range of compounds isolated from *L. patella*. There is significant evidence to suggest that these active compounds may be produced by the symbionts rather than the ascidians.[2] In the case of *L. patella*, when *Prochloron* cells were removed from the host, the amounts of peptides extracted from them, on a weight for weight basis, were equal to or greater than the amounts in the host tissue alone.

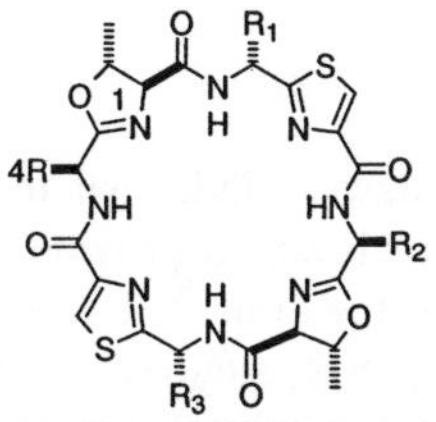

(1) Patellamide A R1=R3=D-Val R2=R4=L-Ile Ring 1-noMe
(2) Patellamide B R1=D-Phe R2=L-Leu R3=D-Ala R4=L-Ile
(3) Patellamide C R1=D-Phe R2=L-Val R3=D-Ala R4=L-Ile
(4) Patellamide D R1=D-Phe R2=R4=L-Ile R3=D-Ala
(5) Patellamide E R1=D-Phe R2=R3=L-Val R4=D-Ile
(6) Patellamide F R1=D-Phe R2=R3=Val R4=Ile
(7) Patellamide G R1=D-Phe R2=L-Ile R3=D-Ala R4=L-Leu
(8) Ascidiacyclamide R1=R3=D-Val R2=R4=L-Ile

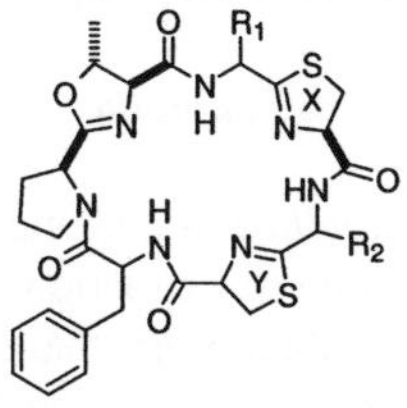

(9) Lissoclinamide 1 R1=L-Val R2=D-Ile X=Y=thiazole
(10) Lissoclinamide 2 R1=D-Ile R2=D-Ala X=thiazolineY=thiazole
(11) Lissoclinamide 3 R1=D-Ile R2=L-Ala X=thiazolineY=thiazole
(12) Lissoclinamide 4 R1=L-Val R2=D-Phe X=thiazolineY=thiazole
(13) Lissoclinamide 5 R1=L-Val R2=D-Phe X=Y=thiazole
(14) Lissoclinamide 6 R1=D-Val R2=D-Phe X=thiazolineY=thiazole
(15) Lissoclinamide 7 R1=D-Val R2=D-Phe X=Y=thiazoline
(16) Lissoclinamide 8 R1=Val R2=Phe X=thiazolineY=thiazole

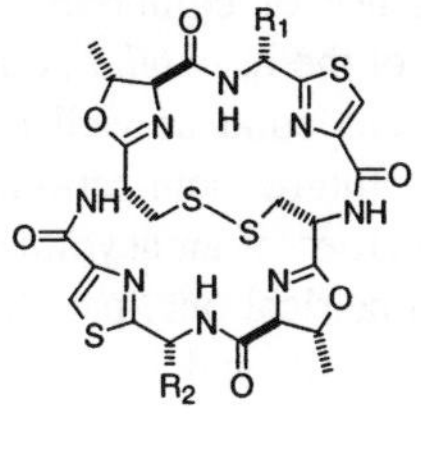

(17) Ulithiacyclamide A R1=R2=D-Leu
(18) Ulithiacyclamide B R1=D-Ile R2=D-Phe

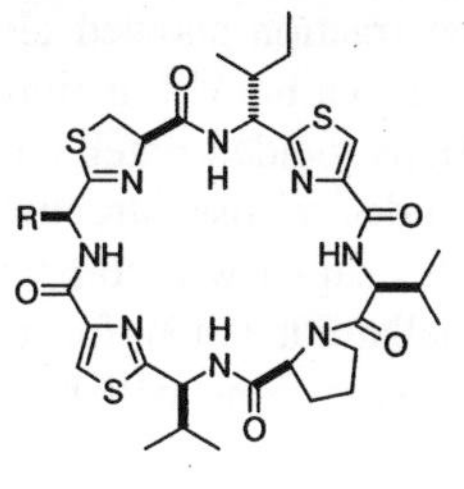

(19) Tawicyclamide A R = L-Phe
(20) Tawicyclamide B R= L-Leu

Figure 1. *Structures of* Lissoclinum patella *cyclic peptides*

As previously stated, these cyclic peptides have shown significant bioactivity in different screening regimes and a strong structure-activity relationship has been noted by several workers, with the disulphide bridge in the ulithiacyclamides (*e.g.* **17, 18**) making these the most cytotoxic of the compounds isolated from *L. Patella*. It is thought that the bridge fixes the conformation of the molecule and that this is the important factor in their high cytotoxicity. The inclusion of an oxazoline moiety in a compound was shown by Shioiri and co-workers[15] to give much higher levels of activity than other residues. However, a comparison of natural and synthetic lissoclinamides by Wipf and co-workers showed that the replacement of thiazoline rings with oxazolines decreased activity to a greater extent than replacement of oxazoline rings with thiazolines.[16] This study further showed that it was not only the individual components of the macrocycle that conferred high activity but rather the overall conformation of the molecules. The structure activity relationship is also demonstrated when comparing lissoclinamides 4 (**12**) and 5 (**13**). These compounds differ only in the oxidation state of a single thiazole unit but this difference makes lissoclinamide 5 (**13**) two orders of magnitude less cytotoxic than lissoclinamide 4 (**12**) against bladder carcinoma (T24) cells.[17] In addition to their cytotoxic properties, patellamides B and C have been shown to reduce multi drug resistance (MDR) *in vitro* of drug resistant lymphoblasts.[12] The peptides from *L. patella* are also of interest for their metal-binding properties. In structure they resemble 21-azacrown-7 and 24-aza-crown-8 macrocycles and several members of the patellamide group have been found to bind copper and zinc atoms within their central cavities.[18,19]

1.3 Metals Present in *L. patella*

Our sample of *L. patella* was collected in the Molucca Sea, Sulawesi, Indonesia in 1996, and underwent a solvent extraction followed by a solvent partition to provide water, butanol, methanol, dichloromethane and hexane partition fractions.[20] In order to investigate whether metal chelation by peptides from this species occurred *in vivo* or only under laboratory conditions, the solvent partition fractions were screened by inductively coupled plasma mass spectrometry (ICP-MS) for their metal content. This analysis revealed that the dichloromethane extract was rich in zinc and copper (see Table 1). The concentration of copper and zinc in this extract is roughly 10^4 x greater than that found in the surrounding seawater. The fact that the metals were found in non-polar fractions suggested that they must be complexed. The ^1H and ^{13}C spectra of the dichloromethane fraction showed clear evidence of the presence of compounds with thiazole, amide and carbonyl functional groups - characteristic of the *L. patella* peptides. Further separations yielded patellamide C (**3**) as the major compound as well as two new lissoclinamides whose structure will be detailed in a future publication. The remainder of this chapter will compare the metal binding properties of an asymmetrical patellamide, patellamide C (**3**), from our collection and a symmetrical one, patellamide A (**1**). Several suggestions will be made as to the possible ecological role of these peptides.

Table 1. *Metal content of solvent extracts of L. patella*

Partition fraction	Mass (mg)	Metals found	conc. (ppm)
CH$_2$Cl$_2$	1420	Zn	2.5
		Fe	1.2
		Cu	2.5
MeOH	5840	none	-
Hexane	2430	Fe	1.4
n-BuOH	710	none	-

2 PREVIOUS STUDIES ON THE PATELLAMIDES

2.1 Conformations of the Patellamides

Molecular modelling and NOE restrained molecular dynamics studies by Ishida *et al*[9] have defined the different conformations adopted by the patellamides (Figure 2). These studies and others[7,11,21] have indicated that symmetrically substituted patellamides (patellamide A (**1**) and ascidiacyclamide (**8**)) adopt the type I 'square' structure both in the solid state and in solution. Our own NOE restrained molecular dynamics studies, on patellamide A (**1**) indicate that in non-polar solvents such as CDCl3 it takes up a type III conformation[22] (see Section 5.2) whereas Doi *et al.*[21] show that in more polar solvents such as DMSO-d6 symmetrical patellamides take up a type I conformation. For the asymmetrically substituted patellamides (patellamides B-G, (**2-7**)) the predominant conformation in the solid state is the 'figure of eight' type IV, held closed by hydrogen bonds with those from NH to C=O forming a type II β turn mimic (Figure 2). In solution these structures relax to give the type III 'figure of eight' conformations which are held closed mainly by the conformational constraints imposed by the thiazole and oxazoline ring systems.

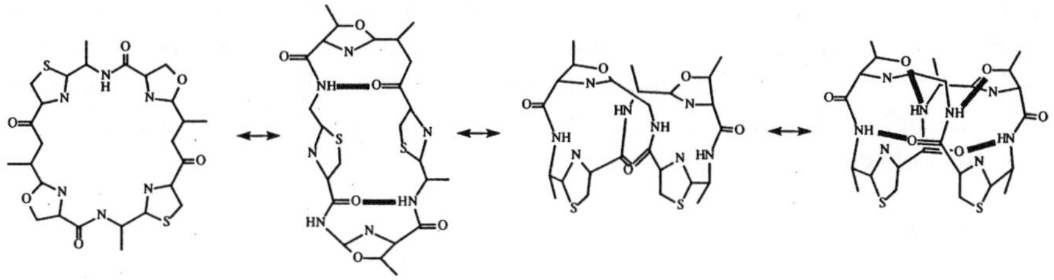

I   **II**   **III**   **IV**

Figure 2. *Conformations of the patellamide backbone*

2.2 Metal Binding of the Patellamides

The ability of the cyclic peptides from *L. patella* to bind metals was first investigated by Hawkins who monitored the formation of a 1 : 1 complex with Cu(II) by patellamide D (**4**) using circular dichroism (CD).[23] Hawkins and co-workers then studied this family of peptides further, examining copper bound ascidiacyclamide (**8**), potassium binding by its hydrolysis product, and copper bound patellamide D (**4**), using MS, EPR and CD.[19,24,25] After deprotonation with base they found that both these peptides bound two equivalents of copper with a bridging carbonate ion incorporated from atmospheric carbon dioxide. Using X-ray crystallography they found that each copper atom co-ordinated to the thiazole, intervening amide and oxazoline nitrogens (the 'TAO' motif), as well as H2O and a bridging carbonate ion.[19] Comba *et al* then compared the copper binding of patellamide A (**1**) and a synthetic analogue which contained threonine instead of the oxazoline rings.[26] Using EPR and molecular mechanics calculations they found both molecules bound to two atoms of copper and that the complex existed in solution in a 'saddle' conformation. However despite attempts using ^{13}C labelled carbonate they found no evidence for the formation of a complex with a bridging carbonate ion.

Subsequently Freeman *et al* utilised CD to look at the conformation and binding constants of patellamides A (**1**), B (**2**) and E (**5**) with calcium, magnesium, zinc and copper.[18] This work found no evidence of binding to either calcium or magnesium for these compounds but that all three formed 2 : 1 complexes with zinc with no conformational change occurring. Binding to copper was found to be more complex with patellamides A (**1**) and E (**5**) forming 1 : 1 complexes and patellamide B (**2**) a 2 : 1 complex. They also reported that patellamides B (**2**) and E (**5**) underwent a conformational change from the figure of eight to the square form on complexing to copper. Binding constants were calculated for the formation of all the peptide-metal complexes. Most recently, the interaction of zinc with ascidiacyclamide (**8**) and a synthetic analogue was studied using CD and NMR by Grondahl *et al.*[27] This work showed the formation of a 1 : 1 complex of zinc with the peptides both with and without the presence of base to deprotonate the peptides. In the absence of base a large excess of zinc was required to completely form the complex.

3 CIRCULAR DICHROISM STUDIES ON PATELLAMIDE A AND C

Circular dichroism (CD) is the ideal technique for the study of small changes in the chiral environment of UV chromophores.[28] Previous studies have taken advantage of the conformational change concomitant with the binding of the patellamides to divalent metal ions.[18,23] One obvious advantage of using CD is that only very small amounts of sample (<1 mg) are required making it an ideal technique for the small amounts of compounds isolated from *L. patella*.

3.1 CD of Free Peptides

In order to compare the solution conformation of patellamide A and C, spectra were obtained at the same concentration, 0.2 mg/mL in MeOH. The CD spectrum of patellamide C (**3**) has two prominent positive maxima centred around 250 and 205 nm, whilst that of patellamide A (**1**) only displays a maximum at 205 nm (Figure 3). Freeman *et al.* suggest that π- π interactions of the stacked thiazole rings in the figure of eight conformation are the main contribution to the maximum at 250 nm.[18] It is more likely to be due to the 'figure of eight' conformation of the patellamides in which a type II β turn mimic is present (Figure 2). In larger peptides and proteins these turns produce a CD maximum around 230 nm but this can be shifted to 250 nm if there is conjugation present in the molecule.[29] The standard CD absorbance values for a type II β turn are 220-230 nm [θ] = 4 x 10^{-3} and 195-205 nm [θ] = 20 x 10^{-3}. In the 'square' patellamide structures the maximum is due to the conjugated peptide bond similar to the random coil values observed in larger peptides and proteins (217 nm [θ] = 3 x 10^{-3} and 195 nm [θ] = 30 x 10^{-3}). In other words, the maximum at 250 nm is associated with the 'figure of eight' conformation and the maximum at 205 nm is associated mainly with the 'square' form of the patellamides. Studies by Freeman *et al* showed that at very low temperature (-74°C) patellamide A (**1**) changes from the 'square' to the 'figure of eight' form, whereas little change in the conformation of the asymmetric patellamides B (**2**) and E (**5**) was observed.[18] This is in keeping with our molecular mechanics calculations which indicate that the energy of the 'figure of eight' form is considerably lower than that of the 'square' conformation.

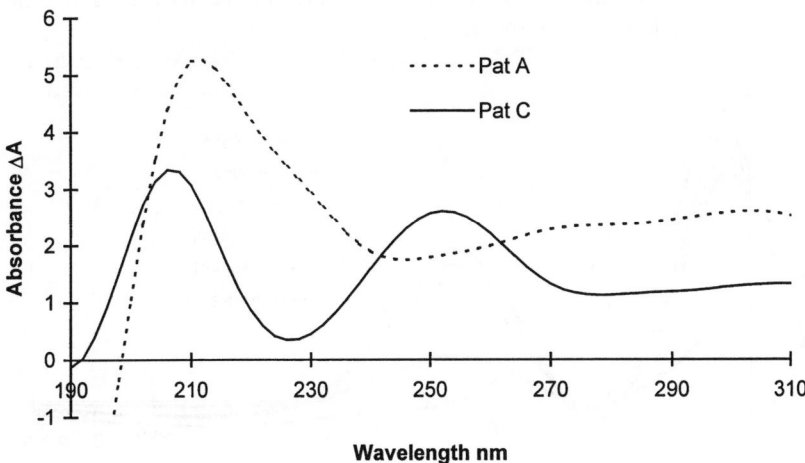

Figure 3. *CD spectra of patellamides A and C at 0.2 mg/mL in methanol*

3.2 CD of Metal Bound Patellamides

As there was a significant amount (50 mg) of patellamide C (**3**) available, all the preliminary studies were carried out on this. The metal ions chosen were zinc and copper, as we had found them present in the crude extracts of *L. patella* by ICP-MS analysis. Nickel and cobalt were also investigated as they have the same oxidation state (+2) and are of a similar size. Nickel is also known to bind to copper sites in biological systems and we were interested to see if the copper site of these peptides could also bind nickel.[30] We found that changing the anion had no effect on the binding. Copper and zinc were first added separately to two portions of patellamide C (**3**) to give solutions with concentrations of 0.02 M with respect to the metal ion and 0.5 mg/mL of the peptide (6×10^{-4} M). The CD spectrum obtained on addition of copper showed a significant change from the unbound spectrum with a decrease in ΔA of 4.5 at the λ_{max} at 250 nm and a shift to 211 nm of the other maximum at 205 nm combined with an increase in ΔA of 1.5 (Figure 4). This suggests that on binding to copper, patellamide C changes conformation to one approximating the 'square' form as the spectrum closely resembles that measured for patellamide A. This change in conformation is probably necessary in order to provide the optimum binding site for copper as it prefers a square planar environment which is not available with the peptide in the figure of eight form (see Section 5).

There was an observable difference between the copper/peptide spectrum and that of the peptide when zinc was added as the latter solution gave no shift of the 205 nm peak, only an increase in ΔA of 1.2, nor was there any collapse of the λ_{max} at 250 nm, only a decrease in ΔA of 1.1, suggesting there was little or no conformational change when patellamide C binds to zinc (Figure 4). Zinc has a more adaptable co-ordination sphere and can bind in a distorted tetrahedral environment which may be partly available in the 'figure of eight' form. A spectrum was also recorded of the peptide with 5 equivalents of zinc added. On comparison of the two patellamide C/zinc spectra a concentration effect was observed, *i.e.* a measurable change in absorbance with increasing amount of zinc added, thus showing the system to be suitable for the measuring of binding constants by titration with metal solutions.

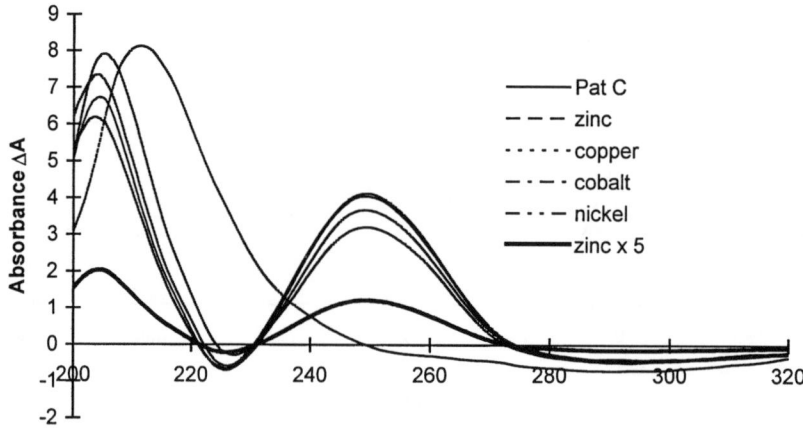

Figure 4. *CD spectra of patellamide C at 0.2 mg/mL in methanol plus 2 eq. of each metal and also with 5 eq. of zinc*

Spectra of solutions containing patellamide C with two equivalents of nickel and with cobalt were also measured, with addition of cobalt causing no overall change to the shape of the spectrum apart from an increase ΔA of 0.4 at the λ_{max} at 205 nm and a decrease in ΔA of 0.3 at the λ_{max} 250 nm. Nickel caused negligible changes, there was a decrease in ΔA of 0.8 at 205 nm and a decrease in ΔA of 0.3 at 250 nm. This indicated there was very little or no binding to nickel and maybe a small amount of non-specific binding to cobalt. Mercury can sometimes bind to copper binding sites and therefore CD spectra of solutions containing patellamide C plus two equivalents and also with an excess of mercury were also measured, but there was no visible binding effect. Therefore any interaction between the mercury and the peptide was a non-specific association.

Addition of zinc caused no overall change to the spectrum of patellamide A (Figure 5) apart from a steady slow increase in ΔA at the λ_{max} at 209 nm with increasing zinc, similar to that seen with patellamide C and zinc. After the addition of one equivalent the increase in ΔA becomes even slower up to two equivalents when the increase stops. This would indicate the formation of a complex containing two zinc atoms. The first addition of copper to patellamide A caused a shift of the λ_{max} at 209 to 211 nm, similar to that seen with addition of copper to patellamide C, and also an increase in the ΔA up to one equivalent of Cu (see Figure 5). Further addition of copper then caused a significant broadening of the λ_{max} up to the addition of two equivalents but no change thereafter. This suggests the formation of the 1 : 1 complex causes no conformational change but the 2 : 1 complex does give rise to a slight change in conformation. This is corroborated by the X-ray crystal structure of ascidiacyclamide which shows that one equivalent of copper can be bound easily in the square form by the 'TAO' motif.[31] However in order to bind the second equivalent it must adapt this to the saddle form.[19] An interesting point to note is that in the work by Freeman *et al* they did not detect the formation of the 2 : 1 complex possibly due to their monitoring the titration of patellamide A with copper at 209 and 250 nm only whereas during our study a full spectrum was recorded after each addition of metal solution.[18] Addition of two equivalents of nickel to a solution of patellamide A caused no apparent change in the spectrum, therefore we suspect that no binding is taking place.

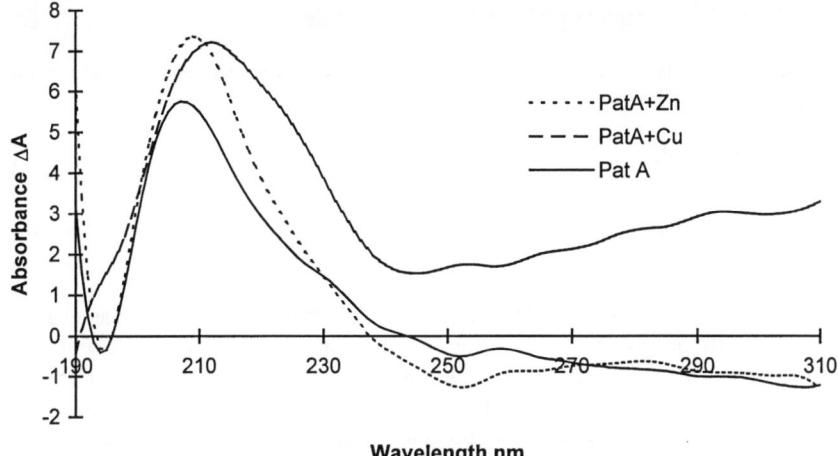

Wavelength nm

Figure 5. *CD spectra of patellamide A at 0.2 mg/mL in methanol plus 2 eq. of zinc and copper*

3.3 Measurement of Binding Constants

In order to calculate the binding constants for Zn and Cu they were initially added at three concentrations, 1/2, 1 and 5 equivalents, to patellamide C (**3**) in order to observe the range in which most change in the spectra occurred. In the case of copper there was no change after the addition of one equivalent suggesting only the formation of a 1 : 1 complex. The spectra for the addition of zinc showed marked differences between the spectra for the addition of one equivalent and of five equivalents suggesting the formation of at least a 2 : 1 complex. The next step involved starting with metal-free patellamide C (**3**), at a concentration of 0.02 mg/mL (0.025 mM), and adding the metal in small aliquots (0.04 eq) in order to ascertain the binding constant. This was carried out using copper over the concentration range 0 - 3 eq. The spectrum was recorded at each addition and change in ΔA at 211 nm and 250 nm plotted against metal concentration in order to calculate binding constants. Unbound peptide was also titrated with zinc solution in the same manner up to five equivalents and absorbance at 205 and 250 nm recorded after each addition. The copper and zinc titrations were repeated with patellamide A (**1**) under the same conditions and binding constants calculated.

Calculation of the binding constants for patellamide A (**1**) was quite straightforward as only three sequential states had to be considered. These were unbound, singly bound and doubly bound and it was assumed that this took place in the manner shown below and that there was no difference in the two binding sites available. Although deprotonation of the peptide occurs prior to inclusion of the metal at each step, as shown in equations 1-3 (2nd deprotonation not included), this could not be incorporated into the calculation as there are no deprotonation constants for amides available.[27]

[Peptide] $\rightleftharpoons$ [Peptide - H]$^-$ + H$^+$ Eq. 1

[Peptide - H]$^-$ + M^{2+} $\rightleftharpoons$ [Peptide - H + M]$^+$ Eq. 2

[Peptide - H + M]$^+$ + M^{2+} $\rightleftharpoons$ [Peptide - 2H + 2M]$^{2+}$ + H$^+$ Eq. 3

Equations 1-3 were simplified to give Eq. 4 as deprotonation constants are unknown.

[Peptide] + M^{2+} $\rightleftharpoons$ [Peptide + M]$^+$ + M^{2+} $\rightleftharpoons$ [Peptide + 2M]$^{2+}$ Eq. 4

This then gives the following equations for calculation of approximate binding constants.

$$K_1 = \frac{[\text{Peptide} + \text{M}]^+}{[\text{Peptide}]}$$ Eq. 5

$$K_2 = \frac{[\text{Peptide} + 2\text{M}]^{2+}}{[\text{Peptide} + \text{M}]}$$ Eq. 6

As the free metal has no CD absorbance, and there is a measurable change in absorbance with increasing concentration of metal complex formed, we may utilise the Beer-Lambert law to calculate the concentration of the species present from the measured change in absorbance and thus calculate these constants. It was felt that iterative calculations as carried out by Freeman *et al* were not necessary due to the vast difference between the values of K_1 and K_2.[18]

However, when calculating the binding constant for patellamide C with copper the change in conformation, from the figure of eight (∞) to the square ($\square$), that takes place must also be taken into consideration, as shown below, as this causes changes to the CD signal as well as the formation of the copper complex.

[Pat C ∞] $\rightleftharpoons$ [Pat C $\square$] $\rightarrow$ [Pat C + Cu]   Eq. 7

From this we can calculate a constant for the change in conformation and the constant for binding to copper again using the Beer-Lambert law in a similar manner to previous calculations. This was found to be in the order of $K_{\infty \text{-} \square} = 2 \times 10^5$. Values obtained during our studies are tabulated along with those obtained by Freeman *et al*.[18] for comparison in Table 2.

Table 2. *Binding constants of patellamide A and C with copper and zinc*

Peptide	*K* values for Zn	*K* values for Zn (Freeman)	*K* values for Cu	*K* values for Cu (Freeman)
Patellamide A	$K_1 = 3.0 \times 10^4$ $K_2 = 1000$	$K_1 = 3.0 \times 10^4$ $K_2 = 16$	$K_1 = 2.0 \times 10^4$ $K_2 = 780$	$K_1 = 2.0 \times 10^4$ -
Patellamide C	$K_1 = 1.8 \times 10^4$ $K_2 = 806$	- -	$K_1 = 6.8 \times 10^4$ -	- -

3.4 Competition Experiments

The fact that the binding constants were the same order of magnitude seemed to suggest that patellamide C would show little selectivity between Cu^{2+} and Zn^{2+}. A competitive binding study was therefore carried out between Cu^{2+} and Zn^{2+} with patellamide C (Figure 6). The initial experiment involved the addition of 2 equivalents of Cu^{2+} to patellamide C (**3**), after which 2 equivalents of Zn^{2+} were added. The CD spectrum of $3/Cu^{2+}$ was essentially identical to that of $3/Cu^{2+}/Zn^{2+}$ except for a small change in ΔA of the λ_{max} at 211 nm. A further 2 equivalents of Zn^{2+} were added, and no further change was observed in the CD spectrum. This indicates that the binding is selective for Cu^{2+} in the presence of an excess of Zn^{2+}. Even taking into account the binding constant differential, an observable degree of binding to Zn^{2+} would be expected, and this should manifest itself in an increase of ΔA at λ_{max} 250 nm. In addition, the $3/Cu^{2+}/Zn^{2+}$ and $3/Cu^{2+}$ spectra have their other λ_{max} at 211 nm and not at 205 nm as observed for $3/Zn^{2+}$. To determine whether the order of addition was important the competition experiment was carried out by adding 2 equivalents of Zn^{2+} to patellamide C after which 2 equivalents of Cu^{2+} was added. The resulting dramatic change is shown in Figure 6, and the $3/Zn^{2+}/Cu^{2+}$ CD spectrum is identical to that of $3/Cu^{2+}$. This again indicates the strong selective binding for Cu^{2+}. The binding constants determined by titration suggest that there should be binding of both Zn^{2+} and Cu^{2+} to patellamide C. Initial thoughts were that the selectivity was kinetic in nature, but re-recording the CD spectrum of the $3/Cu^{2+}/Zn^{2+}$ sample after 3 months showed that no change had taken place indicating that no binding of Zn^{2+} to **3** can take place in the presence of Cu^{2+}. The selectivity of patellamide C for Cu^{2+} could be explained in terms of the different binding environments favoured by Zn^{2+} and Cu^{2+}.[32] Zn^{2+} favours a tetrahedral binding environment, not available in either the 'square' or 'figure of eight' form of **3**. Cu^{2+} favours square planar environments, and this is afforded readily by the 'TAO' motif in the 'square' form, but not in the 'figure of eight' form. This explanation does not give any indication as to why patellamide C does not bind to Ni^{2+} whereas many Cu^{2+} selective binding agents do.[30] Addition of nickel to patellamide C (**3**) causes a small change in the spectrum but on addition of copper the copper bound spectrum is obtained. Addition of nickel to the copper bound peptide causes no change. Any binding with nickel is therefore non-specific and does not compete with the binding of copper.

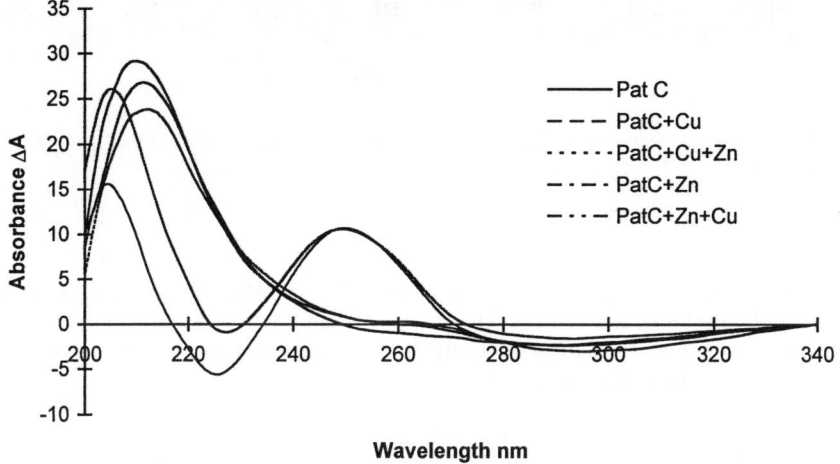

Figure 6. *Patellamide C (at 0.2 mg/mL in methanol) zinc and copper competition experiment*

A further competition experiment was carried out using copper bound patellamide C and EDTA. Two equivalents of EDTA were added to the copper bound peptide and the spectrum measured and compared with that of copper/patellamide C but there was no measurable difference. A further two equivalents of EDTA were added but no change to the spectrum was observed. This shows that the EDTA could not displace the copper from patellamide C despite the formation constant for copper-EDTA being 6×10^{18}, many orders of magnitude greater than that measured for patellamide C.

Addition of copper to a solution of zinc/patellamide A caused no major change in the spectrum and similarly addition of zinc to copper bound patellamide A caused little change. This would suggest no preferential binding of either metal takes place. A third experiment was carried out by adding a mixture containing two equivalents of copper and two equivalents of zinc to a solution of this peptide. The resulting spectrum appeared to be a mix of the zinc bound and copper bound spectra with a large increase in the λ_{max} at 209 nm and also distinct broadening, suggesting the presence of a mixture of species (Figure 7).

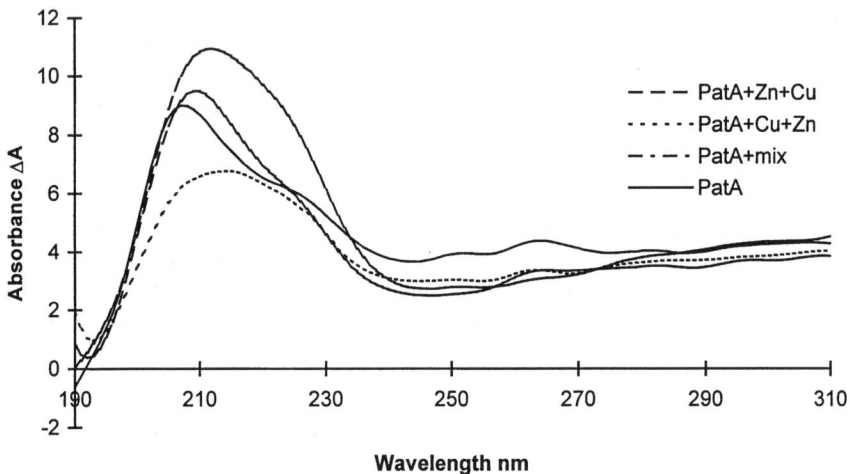

Figure 7. *Patellamide A (at 0.2 mg/mL in methanol) zinc and copper competition experiment*

3.5 Discussion

The results obtained using CD gave insights into the specificity and strength of binding of metals by patellamide A and C. The Irving-Williams series of complex stability for divalent ions bound to conformationally flexible ligands puts divalent ions in the following order independent of ligand.[33]

$$(Mg^{2+}, Ca^{2+}) < Mn^{2+} < Fe^{2+} < Co^{2+} < Ni^{2+} < Cu^{2+} > Zn^{2+}$$

This would lead us to expect selectivity for copper over zinc as observed as well as preferential binding to copper over nickel. It must be noted that the symmetrical patellamide A is non-selective for Cu^{2+}, even though the Irving-Williams series would lead us to expect this. It appears that the change of symmetry between patellamides A and C is responsible for this change in selectivity. The order of the Irving-Williams series can be changed by altering the binding geometry of the ligand, *i.e.* using conformationally rigid ligands. Therefore an explanation for this selectivity must be the different binding environments favoured by the two metals. Cu^{2+} is known to prefer a square planar environment and this is provided by the 'TAO' motif when the patellamides adopt the 'square' form. Zn^{2+} favours a tetrahedral binding environment, which is not available in either conformation of patellamide C, and although binding of zinc does take place it is probably less ideal than that with copper so this could be an explanation for the observation that zinc can be displaced from patellamide C on the addition of copper. Nickel, however, prefers an octahedral binding environment, which is perhaps unavailable in these peptides. The preferred geometries of relevant divalent metal ions is shown in Table 3.

Table 3. *Preferred geometries in simple co-ordination compounds*

Metal ion	Preferred geometries
Cu^{2+}	Tetragonal > 5 co-ordination > tetrahedral
Ni^{2+}	Octahedral > others
Co^{2+}	Octahedral > tetrahedral > others
Zn^{2+}	Tetrahedral > octahedral > 5 co-ordination

Grondahl *et al.*[27] found that binding of zinc to ascidiacyclamide (**8**) was a very slow process. If this is the case with all of the peptides and zinc it would also help explain the observed selectivity for copper over zinc, but when solutions of patellamide C (**3**) and a mixture of copper and zinc were measured after three months no change had occurred and only the copper complex was in evidence.

A deciding factor in the specificity of binding of copper to patellamide C (**3**) compared to patellamide A (**1**) may be the difference in their respective side chains. The presence of the phenylalanine may distort the planarity of the 'square' conformation of patellamide C, suggesting that this distortion changes the geometry of the macrocycle sufficiently to provide the ideal binding site for copper in the 'TAO' motif. At this point bonding in metal complexes must also be considered, in particular, Jahn-Teller effects. Complexes of Zn^{2+} are tetrahedral as they involve sp^3-hybrid bonds, all the $3d$-orbitals being fully occupied. Four sp^3-hybrid orbitals can be formed leading to a tetrahedral complex. Complexes of Cu^{2+} involve dsp^2-hybrid orbitals, resulting in a co-planar square structure, which is better described as a distorted octahedron. This hybridisation of orbitals leads to an asymmetric arrangement of electrons, resulting in two short bonds along the z-axis and four longer bonds along the x- and y-axes. This is the Jahn-Teller distortion, and this may also explain how copper is more easily accommodated within the macrocycle than zinc and nickel.

4 MASS SPECTROMETRIC STUDIES OF THE METAL BINDING OF PATELLAMIDE A AND C

4.1 Introduction

As all the cyclic peptides isolated from *Lissoclinum patella* had been found to bind both copper and zinc in the circular dichroism study it was decided to further investigate these complexes by mass spectrometry in order to ascertain their exact nature and which, if any, counter-ions were incorporated into the complexes. Mass spectrometry was ideally suited for this purpose as accurate mass measurements and analysis of isotope ratio patterns of peaks in a spectrum allow definite assignment of the components of a complex. The copper complexes of ascidiacyclamide (**8**) and patellamide D (**4**) were extensively studied using mass spectrometry by van den Brenk *et al.*[24,19] Both the peptides were shown to complex two copper atoms by deprotonation of amide nitrogens with the metals bridged by a carbonate ion. The first stage of the mass spectrometry experiments involved studying the uncomplexed patellamide A and C to acquire accurate mass data and to observe the fragmentation patterns of the uncomplexed molecules. The copper and zinc complexes of each were then studied in detail, followed by competition and titration experiments.

4.2 Mass Spectrometry of the Metal Complexes

Initial work was again performed on patellamide C, as this was the most abundant compound. One equivalent of Cu^{2+} was added to patellamide C in methanol and the mass spectrum was observed. The relative intensities and identity of species formed are shown in Table 4. The fully protonated peptide is denoted $PatCH_4$, singly deprotonated as $PatCH_3$, *etc.* for clarity. In the spectrum of copper bound patellamide C, the main signals are due to singly charged species. These species are $[PatCH_4 + H]^+$ at m/z 763 (minor) and its sodium adducts (m/z 785), $[PatCH_3 + Cu]^+$ (m/z 824) its HCl adduct (m/z 860), $[PatCH_3 + CuClNa]^+$ at m/z 882 and two copper complexes at m/z 921 and 975/7. The cluster at m/z 975/7 was initially thought to be $[PatCH_3 + 2Cu + C_2O_4]^+$ but later identified as $[PatCH_2 + 2Cu + C_3H_5O_3]^+$ by HRMS and a neutral loss of 90.0314 ($\Delta 1.1$ ppm from calcd. for $C_3H_6O_3$). There was no significant occurrence of doubly charged species. The appearance of two copper species was unexpected as formation of this had not been detected by CD. This suggests that only the binding of the first copper causes any measurable change in the conformation of the molecule and once this change has occurred a further copper may be complexed with no further conformational change required, which is corroborated by structural studies (see Section 5). When copper acetate was used in place of copper chloride the base peak observed was m/z 975/7, compared to 10% intensity when copper chloride was used, with only minor peaks for

Table 4. *Main species detected in the MS of patellamide C plus 2eq. Cu^{2+} solution*

m/z	Intensity %	Species	Formula	HRMS Δ ppm from calcd.
763	10	$[PatCH_5]^+$	$C_{37}H_{47}N_8O_6S_2$	763.3059 $\Delta 1.1$ppm
785	20	$[PatCH_4 + Na]^+$	$C_{37}H_{46}N_8O_6S_2Na$	-
824	100	$[PatCH_3 + Cu]^+$	$C_{37}H_{45}N_8O_6S_2Cu$	824.2198 $\Delta 1.0$ ppm
860	24	$[PatCH_4 + CuCl]^+$	$C_{37}H_{46}N_8O_6S_2CuCl$	861.1961 $\Delta 1.0$ ppm
882	10	$[PatCH_3 + CuClNa]^+$	$C_{37}H_{45}N_8O_6S_2CuClNa$	-
921	5	$[PatCH_2 + Cu_2Cl]^+$	$C_{37}H_{44}N_8O_6S_2Cu_2Cl$	-
975/7	10	$[PatCH_2 + Cu_2 + C_3H_5O_3]^+$	$C_{40}H_{49}N_8O_9S_2Cu_2$	975.1767 $\Delta 12.5$ ppm

other copper species suggesting the formation of the m/z 975/7 species is greatly enhanced in the presence of acetate. When a three-month old solution of patellamide C and copper (added as copper chloride) was analysed the only species present was the two copper complex at m/z 975/7, suggesting that this is the preferred copper complex.

Zinc complexed patellamide C produces a much more complicated spectrum with many doubly charged species such as $[PatCH_4 + Zn]^{2+}$ at m/z 413. It would seem from this that zinc may be complexed by patellamide C without the abstraction of a proton which always accompanied the complexation of copper although $[PatCH_3 + Zn]^+$ is also seen at m/z 825. This indicates that the binding to zinc is probably of a less specific nature than that to copper as was seen in the CD competition study. In the spectrum there were also peaks for many singly and doubly charged species many of which could be assigned using the general formula $PatCH_{4-x} + xZn^{2+} + (x + 1)Cl$. Also present were water adducts of these species. Some of these species are assigned in Table 5 using the same notation as for patellamide C and copper.

Table 5. *Main species detected in the MS of patellamide C plus 2eq. Zn^{2+} solution*

m/z	Intensity %	Species	Formula	HRMS Δ ppm from calcd.
413.3	75	$[PatCH_4 + Zn]^{2+}$	$C_{37}H_{46}N_8O_6S_2Zn$	413.1091 Δ9.9 ppm
481.7	20	$[PatCH_4 + Zn_2Cl_2]^{2+}$	$C_{37}H_{47}N_8O_6S_2ZnCl$	-
549.0	7	$[PatCH_4 + Zn_3Cl_4]^{2+}$	$C_{37}H_{46}N_8O_6S_2Zn_3Cl_4$	-
763.6	20	$[PatCH_5]^+$	$C_{37}H_{47}N_8O_6S_2$	763.3059 Δ1.2 ppm
781.6	15	$[PatCH_5 + H_2O]^+$	$C_{37}H_{49}N_8O_7S_2$	-
785.3	20	$[PatCH_4 + Na]^+$	$C_{37}H_{46}N_8O_6S_2Na$	-
795.2	20	$[PatCH_5 + MeOH]^+$	$C_{38}H_{51}N_8O_7S_2$	-
825.5	5	$[PatCH_3 + Zn]^+$	$C_{37}H_{45}N_8O_6S_2Zn$	-
863.3	100	$[PatCH_4 + ZnCl]^+$	$C_{37}H_{46}N_8O_6S_2ZnCl$	863.1932 Δ1.2 ppm (for ^{66}Zn)
994.9	5	$[PatCH_4 + Zn_2Cl_3]^+$	$C_{37}H_{46}N_8O_6S_2Zn_2Cl_3$	-
1130.5	5	$[PatCH_4 + Zn_2Cl_3 + 2H_2O]^+$	$C_{37}H_{50}N_8O_8S_2Zn_2Cl_3$	-

The mass spectrum of patellamide A with two equivalents of Cu^{2+} was dominated by the signal at m/z 804 for $[PatAH_3 + Cu]^+$ with a smaller cluster at m/z 840/2 for its HCl adduct. One other interesting signal present was a two copper complex at m/z 901 for $[PatAH_2 + 2Cu + Cl]^+$. The species detected are shown in Table 6.

After addition of two equivalents of Zn^{2+} to patellamide A, a significant proportion of the peptide remained uncomplexed, either reflecting slow binding, as proposed by Grondahl et al.[27], in the case of ascidiacyclamide (**8**) which, of all the patellamides most resembles patellamide A (**1**) or non-specific binding that was unstable in the MS environment. An excess of zinc was then added (4 equivalents) to force formation of more zinc bound species. As was seen with patellamide C and zinc, zinc-bound patellamide A also produced a large number of doubly charged species and even species with charges up to five were seen in the region below m/z 200 of the spectrum. Signals with four charges were of very low intensity but those with double and triple charges

Table 6. *Main species detected in the MS of patellamide A plus 2eq. Cu^{2+} solution*

m/z	Intensity %	Species	Formula	HRMS Δ ppm from calcd.
743.6	5	$[PatAH_5]^+$	$C_{35}H_{51}N_8O_6S_2$	743.3372 $\Delta1.3$ ppm
779.5	5	$[PatAH_5 + HCl]^+$	$C_{35}H_{52}N_8O_6S_2Cl$	779.3154 $\Delta7.0$ ppm
804.5	100	$[PatAH_3 + Cu]^+$	$C_{35}H_{49}N_8O_6S_2Cu$	804.2512 $\Delta1.2$ ppm
822.6	5	$[PatAH_3 + Cu + H_2O]^+$	$C_{35}H_{51}N_8O_7S_2Cu$	-
842.5	45	$[PatAH_4 + CuCl]^+$	$C_{35}H_{50}N_8O_6S_2CuCl$	842.2277 $\Delta3.0$ ppm (for ^{65}Cu)
901.2	5	$[PatAH_2 + Cu_2Cl]^+$	$C_{35}H_{48}N_8O_6S_2Cu_2Cl$	901.1440 $\Delta3.4$ ppm

were quite intense. These multiply charged species were not evident in the copper complexes. Many adducts were also present with as many as three molecules of HCl included whilst water adducts were less abundant. This reflects the less specific binding of zinc by patellamide A. The formation of the larger aggregates such as $[PatAH_2 + Zn_3Cl_3 + H_2O]^+$ is more likely to be an artefact of the mass spectrometer conditions than a reflection of actual complexes formed in solution.

4.3 Competition Experiments

Competition experiments were carried out in a similar manner to those performed using CD. Zinc was added to copper bound patellamide and copper to zinc bound patellamide and the resulting solutions analysed. A mixed solution of copper and zinc was also added to unbound peptide in some cases to ensure that observed selectivity of binding was real and not dependant on which metal was introduced first to the peptide.

As was seen in the CD competition study (Section 3.4) copper displaced zinc from patellamide C and when a mixture of the two metals was added only copper species were formed. When copper was added to a solution containing predominantly the two zinc complex $[PatCH_4 + Zn_2Cl_3]^+$ at m/z 995 formed by adding a large excess of zinc, this was displaced and the two copper complex $[PatCH_3 + Cu_2Cl_2]^+$ at m/z 957 became the dominant species present. A further competition experiment was carried out with copper bound patellamide C in the presence of acid. The solution of copper and patellamide C was diluted with 2% formic acid and then introduced into the mass spectrometer. No copper species of any kind were detected because the amide NH's were fully protonated at low pH so no binding to Cu^{2+} was possible.

The competition experiments with patellamide A produced mixed species in each case with some zinc only, some copper only and some copper/zinc doubly charged complexes. The peaks in the spectra for copper only species were of slightly greater intensity suggesting a small preference for copper but not on the scale of that displayed by patellamide C. These results agreed with those obtained by CD where a hybrid copper/zinc spectrum was obtained in the presence of both metals. Further evidence came from the HRMS of a solution of patellamide A to which a mixture of copper and zinc had been added which showed a peak at 902.1476 $\Delta 8$ ppm from calculated for $[PatAH_2 + Cu + Zn + Cl]^+$.

As an additional experiment, a 1 : 1 mixture of patellamides A and C was made up in methanol and to this was added 1 eq. of copper. The resulting spectrum showed as the 100% peak $[PatAH_3 + Cu]^+$ at *m/z* 805. The patellamide C was mainly in the form of $[PatCH_4 + Na]^+$ at *m/z* 785. This would suggest that patellamide A can out-compete patellamide C for copper, but further experiments would be necessary to confirm this.

4.4 Titration Experiments

These experiments were undertaken in a similar manner to those described by Brady and Sanders for measuring relative binding affinities of metals with steroid derivatives by ESI-MS.[34] Titrations with zinc and copper solutions were carried out on patellamides C and A by sequential addition of 0.25 equivalents of the metal solutions to the peptides (conc. 0.02 mg/mL in MeOH). These titrations were monitored by measuring the formation of the metal species as well as the loss of the uncomplexed peptides, using both the full spectrum and the selected ion mode of the instrument in parallel, as the ionisation efficiency of these species were very different. Once fully complexed zinc species were obtained for each peptide, copper solution was then titrated into these at a rate of 0.25 equivalents to again observe the competition effects.

By measuring the change in relative intensities of the bound and unbound species and plotting these against the amount of metal added, an estimate of their binding affinity could be made to compare with that measured by CD. Species chosen for monitoring were those present throughout the titration as many transient species were formed when zinc was added. The binding constants for zinc and copper calculated for patellamides A (1) and C (3) from the MS titrations were comparable with those obtained from the CD titrations. All other values were within the same order of magnitude apart from that for the formation of the 2Zn complex of both peptides. Both sets of results are shown in Table 7 for comparison.

Table7. *Binding constants of patellamides A and C*

Peptide	Species used for calculation	K values by MS	K values by CD
Patellamide C and copper	$[PatCH_3 + Cu]^+$	$K_1 = 1.2 \times 10^4$	$K_1 = 6.8 \times 10^4$
	$[PatCH_2 + Cu_2 + (C_3H_5O_3)]^+$	$K_2 = 6.0 \times 10^3$	not observed
Patellamide A and copper	$[PatAH_3 + Cu]^+$	$K_1 = 3.3 \times 10^4$	$K_1 = 3 \times 10^4$
	$[PatAH_2 + Cu_2Cl]^+$	$K_2 = 1.0 \times 10^4$	$K_2 = 1000$
Patellamide C and zinc	$[PatCH_3 + Zn]^+$	$K_1 = 2.4 \times 10^3$	$K_1 = 1.8 \times 10^4$
	$[PatCH_3 + Zn_2Cl_2]^+$	$K_2 = 2.6 \times 10^3$	$K_2 = 806$
Patellamide A and zinc	$[PatAH_3 + Zn]^+$	$K_1 = 2.8 \times 10^3$	$K_1 = 2 \times 10^4$
	$[PatAH_3 + Zn_2Cl_2]^+$	$K_2 = 3.9 \times 10^3$	$K_2 = 780$

From the plots of the relative intensities of the metal complexes formed it could be seen that the formation of many zinc species was independent of the concentration of metal salt added. This was the case for all the doubly charged zinc complexes of patellamide C which, after an initial fast increase in their intensities, stabilised at around 1 equivalent of zinc added and remained more or less constant thereafter. This was also the case with zinc complexes of patellamide A with only the intensities of the $ZnCl_2$ and $Zn_2Cl_2(H_2O)_2$ showing a linear dependence on concentration.

The formation of the Cu_2Cl complex of patellamide A became constant after the addition of 1 equivalent of copper suggesting that although two coppers could be bound by patellamide A this complex was much less stable than the single copper complex or the 2Cu complex formed by patellamide C. The formation of the $[PatAH_3 + Cu]^+$ complex of patellamide A was very clear with the intensity of that species reaching a maximum after the addition of one equivalent of the metal and remaining constant thereafter.

The appearance of the $[PatCH_2 + Cu_2 + (C_3H_5O_3)]^+$ complex of patellamide C appeared to be dependent on the concentration of metal solution added, with a steady increase in intensity with increasing metal concentration, but previous measurements of this complex had shown that its formation was very slow.

4.5 Discussion

The study of these peptides and their copper and zinc complexes by mass spectrometry revealed much information about their composition and binding characteristics. The most interesting results were those from the study of patellamide C and its copper complexes. This revealed the formation of complexes, such as $[PatCH_2 + 2Cu + (C_3H_5O_3)]^+$ which were not detected by CD. This species might be similar to the copper complex of ascidiacyclamide, which was shown by X-ray crystallography to have a bridging carbonate between the two copper atoms.[19]

Studying the zinc complexes by MS revealed that the binding of zinc is not a straightforward process with many intermediate species being formed, both singly and multiply charged, and with various combinations of counter ions and deprotonation of the peptides. This would suggest that the binding of zinc is predominantly non-specific and also that the binding constants calculated from the CD titration for patellamides A and C with copper and zinc, are only approximate as they do not take into account all the species formed. The binding affinities measured from the titration experiments monitored by MS were however comparable with those from the CD experiments and we may take these measurements as being representative of the specific binding values for these peptides.

5 NOE RESTRAINED MOLECULAR DYNAMICS STUDIES OF PATELLAMIDE A AND C

5.1 Introduction

As described in Section 2.1 and Figure 2, patellamides have been studied extensively by NOE restrained molecular dynamics. The compounds modelled in our study were patellamides A (**1**) and C (**3**), the zinc complex of patellamide C and singly bound copper/patellamide C. The NOE restraints were obtained from a T-ROESY NMR spectrum of the patellamide (in $CDCl_3$) or its complex (in CD_3OD) and assigning the intensity of each correlation observed as either strong (1.8 - 2.5 Å), medium (1.8 - 3.5 Å) or weak (1.8 - 5.0 Å) with an additional 0.5 Å allowed for correlations to methyls. The NOE restrained molecular dynamics was then carried out using Brunger's X-PLOR.[35]

5.2 Modelling of Free Peptides

In total 64 nOe restraints were obtained from the T-ROESY spectrum in CDCl₃ of patellamide C (**3**), and the most important of these are indicated in Figure 8. Usually 8-10 nOe restraints per amino acid residue are required to model a peptide, but the cyclic nature of the compounds under study reduces the number of degrees of freedom and therefore the number of nOes required for the resulting conformation to be reliable.

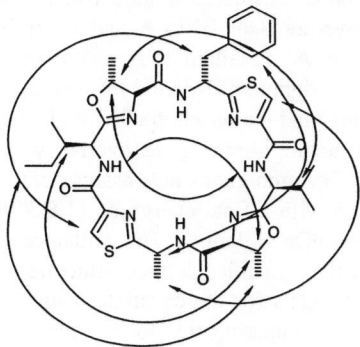

Figure 8. *Key long range nOe correlations of patellamide C in CDCl₃*

The consensus structure of patellamide C after refinement is shown in Figure 9. This solution state conformation has a 'figure of eight' form, which matches the crystal structure closely apart from the sidechains which would have more freedom of movement in solution. The structure has a final energy of 37.3 KJ/mol and the backbone RMSD is 0.10 ± 0.05 Å with only 3 minor NOE violations. We also calculated the structure using Ishida's data (only 28 nOe's),[9] which excludes sidechain restraints, to check our methodology and the figures for this are a final energy of 22.3 KJ/mol, a backbone RMSD of 0.11 ± 0.03 Å with only 3 minor NOE violations. The comparison of our calculated structure with that derived by Ishida was also favourable with a backbone RMSD of 0.20 ± 0.09 Å.

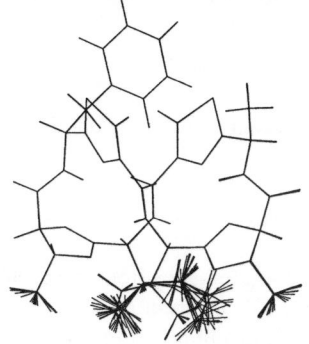

Figure 9. *Refined structure of patellamide C - ensemble of 14 lowest energy structures*

The modeling of patellamide A was complicated by the symmetry of this compound, especially the amide proton overlap, and this led to ambiguity in the assignment of the nOes. Correlations to the amide protons were unable to be distinguished and therefore the nOe restrained molecular modeling in X-PLOR was performed using 'ambiguous' nOe's. The program gives a different weight to the nOe's depending on the initial proximity of the protons in a randomised structure. By hand, using a flat structure, assignments would normally be set to the closest neighbour whereas this is not always correct when the third dimension is considered (Figure 10). The ambiguous assignments of the nOe's in the figure above are B or B' to A and B' or B to A'. The 'obvious' assignments are B to A and B' to A'. Patellamide A was modelled using both methods of assignment of the nOes. The assignments by hand resulted in a 'square' conformation with a final energy of 21.0 kJ/mol and the backbone RMSD is 0.22 ± 0.14 Å with only 3 minor nOe violations. This structure shows good overlap with the published crystal structure of ascidiacyclamide.[7] The ambiguous nOe assignments in X-PLOR resulted in a 'figure of eight' conformation with a final energy of 11.8 KJ/mol and the backbone RMSD is 0.22 ± 0.14 Å with no nOe violations. This structure shows excellent overlap with our nOe restrained molecular dynamics derived structure of patellamide C. Back-calculating the expected strong nOe's for the calculated 'square' and 'figure of eight' structures of patellamide A and comparing this to the observed nOe's indicated that patellamide A adopts a 'figure of eight' conformation in CDCl₃ solution (Figure 11) as opposed to the 'square' form adopted by ascidiacyclamide in DMSO-d6 solution.[21]

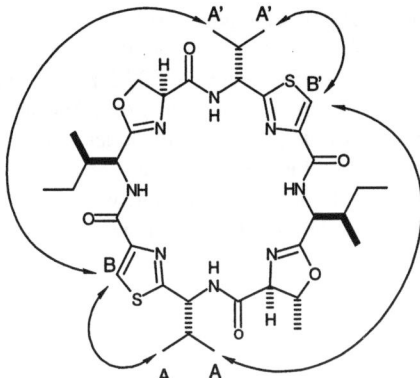

Figure 10. *Ambiguous nOe's for the symmetrical patellamide A*

5.3 Modelling of the Metal Complexes

A sample of patellamide C plus 2 eq. of anhydrous zinc chloride in CD₃OD was prepared under nitrogen. The T-ROESY spectrum of this complex produced a total of 26 nOes which were then used in the modelling of the zinc bound form using the same method as for patellamide C. As the spectrum was in CD₃OD there were no NH correlations. The resulting structure closely resembled that of the free peptide only slightly more opened (Figure 12).

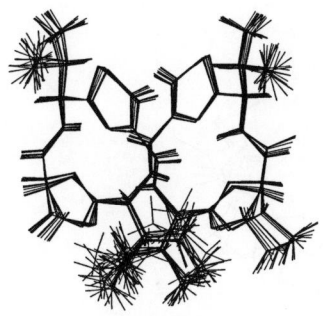

Figure 11. *Ensemble of the 14 lowest energy structures of patellamide A in the 'figure of eight' form*

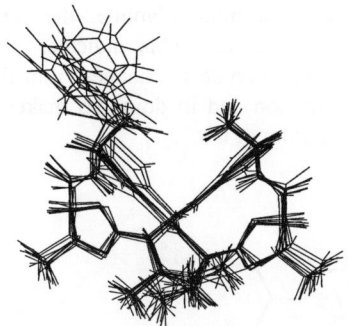

Figure 12. *Ensemble of the 10 lowest energy structures of zinc bound patellamide C*

This is in agreement with the data obtained by CD which showed that patellamide C underwent only slight conformational change when forming zinc complexes. The final energy of the structure was 11.8 kJ/mol and the backbone RMSD is 0.24 ± 0.09 Å with no nOe violations. An overlay of the zinc complex structure and the uncomplexed structure is shown in Figure 13 to demonstrate their similarity.

A solution of patellamide C and one eq. of copper chloride in CD$_3$OD was prepared under nitrogen and used to obtain nOe restraints. Due to paramagnetic line broadening in the spectrum, a consequence of the presence of Cu(II), a 2D T-ROESY spectrum of this compound proved to be of little value. Instead, a series of selective 1D-NOESY experiments was carried out by irradiating each proton in turn and measuring the intensities of the resulting nOe correlations. When these nOes were analysed it was found that those for the valine side of the molecule were well resolved but those for the other side of the molecule were unclear. From this it was supposed that the copper was bound mostly to the isoleucine side and its paramagnetic effects made it impossible to resolve the nOes to that portion of the molecule. The molecule was then modelled with 32 nOe restraints, and the resulting ensemble of structures is shown in Figure 14. The

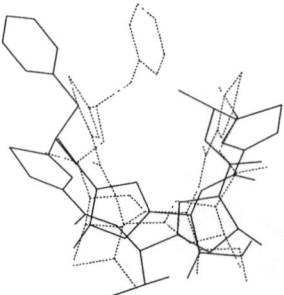

Figure 13. *Overlay of zinc bound (line) and uncomplexed (dashed line) patellamide C (heavy atoms only)*

final energy of this structure was 38.0 kJ/mol and the backbone RMSD is 0.13 ± 0.07 Å with two minor nOe violations. When the structure of patellamide C/Cu complex was compared to the X-ray crystal structure of ascidiacyclamide,[7] the overlay on the Val side was found to be extremely good (Figure 15). This indicates that upon binding 1 equivalent of Cu^{2+}, the structure of patellamide C changes from the 'figure of eight' conformation to the 'square' conformation and in doing so makes the second 'TAO' motif available for binding to another atom of Cu^{2+}.

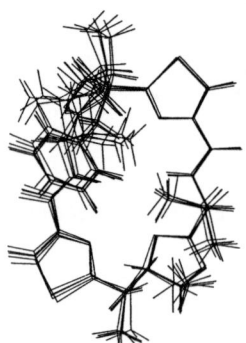

Figure 14. *Ensemble of the 6 lowest energy structures of 1 : 1 copper/patellamide C complex showing poor resolution on the Ile (left) side*

5.4 Discussion

The inter-atomic distances and relative orientations of the copper binding sites available within patellamides A and C were measured on the structures generated during this work. These were then compared to the sites present in copper-bound ascidiacyclamide (Figure 16). In the structure of patellamide C a nitrogen-lined cavity was evident, but the nitrogens of the thiazole, oxazoline and intervening amide, were not co-planar although the distances between them were similar to those measured in copper-bound ascidiacyclamide. However, in the single copper bound patellamide C these nitrogens on

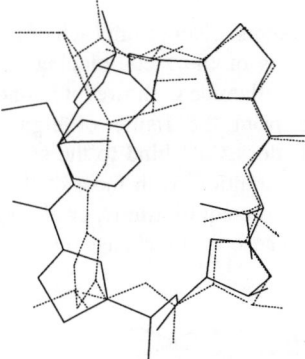

Figure 15 *Overlay of copper bound patellamide C (line) and X-ray crystal structure of ascidiacyclamide (dashed line) (heavy atoms only)*

the unbound side of the molecule were re-orientated to be co-planar thus providing a suitable binding site for the second atom of copper after the first has bound. This corresponds with the CD data, which showed a conformational change occurred only in binding the first equivalent of copper.

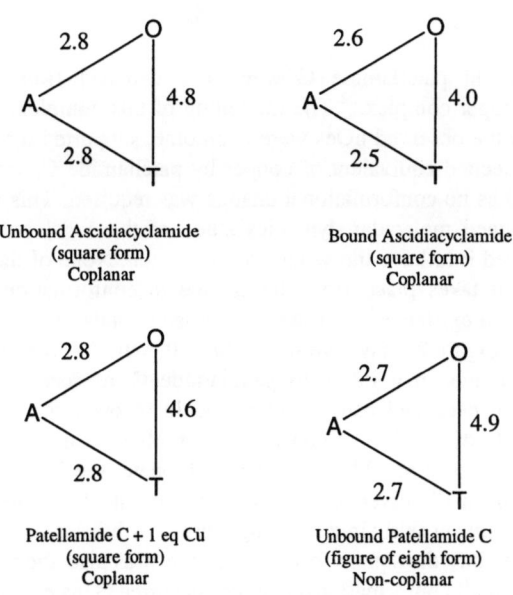

T = Thiazole nitrogen, A = Amide nitrogen, O = Oxazoline nitrogen

Figure 16 *Distances in Å between and relative orientation of nitrogen atoms available for copper binding in patellamides A and C*

6 DISCUSSION OF BINDING PROCESSES

In this section a brief summary of the main points arising from each of the studies above will be made and related to different aspects of the metal-binding properties of the compounds studied. The studies by CD on patellamide C showed it bound Cu^{2+}, and this binding of copper resulted in a change from the figure of eight to the square conformation. This raises two questions: 1. does Cu^{2+} bind to the small proportion of square form patellamide C that is present in solution, with more of the figure of eight form then changing to the square form to restore equilibrium?, or alternatively, 2. does the binding of copper to patellamide C cause it to change conformation? These alternatives are illustrated in Figure 17.

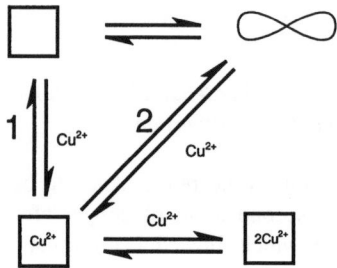

Figure 17. *Alternative modes of conformational change by patellamide C on binding copper (∞ denote 'figure of eight' form and $\square$ the 'square' form)*

Binding sites for copper in patellamide C were also deduced from the MS fragmentation of the 1 : 1 copper complex.[22] The modelling of this complex, once the problems in interpretation of the observed nOes were overcome, supported the location of these sites. Binding of a second equivalent of copper by patellamide C, seen in the MS, was not detected by CD as no conformational change was required. This was also supported by the nOe restrained molecular dynamics study of the 1 : 1 patellamide C/Cu^{2+} complex which showed it to have the square conformation. None of the studies however, tell us which event takes place first, the change in conformation, or, the binding of the first copper. The existence of a cage of hydrogen bonds seen in the type IV 'figure of eight' structure (Figure 2) may give some clues. It could be envisioned that as copper deprotonates an amide on bonding to patellamide C, as seen in the MS experiments, this disrupts the cage and causes the molecule to open out. A further question arises from the study of the 2 : 1 copper complex of patellamide C, which formed slowly except in the presence of added acetate or carbonate (see Section 4.2). Is carbonate, either added or from dissolved CO_2, necessary for the formation and/or stability of this complex? We are currently investigating these possibilities.

Patellamide A, adopting mainly the square form in solution, was also shown to bind copper, but in doing so only small conformational changes occurred. This can be related to the XRD structure of ascidiacyclamide which distorts only slightly, from the 'square' form to what has been termed the 'saddle' form, to accommodate two copper atoms.[19] No bridged 2Cu complexes of patellamide A were detected, indeed only small amounts of 2 : 1 copper complexes were detected by MS. The 2Cu bridged complex of ascidiacyclamide was formed after addition of large amounts of base (4 eq.), [19] and perhaps patellamide A also requires base to form 2 : 1 copper complexes.

Patellamides A and C were shown to form complexes with zinc in both the CD and MS studies. The formation of multiply charged, transient and aggregate species with zinc would suggest that zinc complexes are largely non-specific associations with various counter-ions included to satisfy the charge balance requirements. Looking at the 'square' conformation of the patellamide backbone and looking for possible tetrahedral sites for zinc, we can see that three nitrogen atoms might be suitably orientated to provide this. This may explain the lack of specificity between copper and zinc observed for patellamide A, which adopts the 'square' conformation in methanol, in both the CD and MS studies. In the 'figure of eight' form of patellamide C only two nitrogens are suitably positioned for the formation of tetrahedral complexes. This would lead to the less specific binding of zinc by patellamide C that was observed.

As discussed in Section 3.5, the Irving-Williams series for complex formation to divalent metals explains the observed selectivity shown for copper over zinc of patellamide C in both the CD and MS competition experiments.[33] At pH 7 copper can more easily deprotonate amides than zinc, so we could predict that copper would out-compete zinc for peptide ligands. However, from this series we would expect Ni^{2+} to also complex with patellamide C but no evidence of this was seen. This is unusual as nickel and copper are sometimes able to occupy the same binding sites.[30] Ni^{2+} complexes prefer an octahedral environment with symmetric bonds.[36] Within the macrocyclic cavity of patellamides the sixth site is blocked and only square pyramidal co-ordination is possible, and this might be why Ni^{2+} is unable to bind.

7 POSSIBLE ECOLOGICAL ROLES FOR THE PATELLAMIDES

In all extractions of *L. patella*, the major compounds isolated are always cyclic peptides, with one of the patellamides usually the most abundant of these.[6,8,11] There has been only one reported case of *L. patella* which did not have *Prochloron* as its symbiont.[10] When this sample was extracted no cyclic peptides were isolated. This supports the case for their being produced by *Prochloron* rather than the host. The production of these compounds, by either the host or symbiont, is costly to the internal metabolism of the producing organism and they must therefore play some ecological role. The selectivity and structural studies show these peptides are most suited to bind copper. Copper is therefore proposed to be the ecologically relevant metal for the patellamides. It is possible that the patellamide/copper complexes might be co-factors in enzymatic processes. We will discuss a few further possibilities here.

Are they involved in the activation and mobilisation of CO₂? Van den Brenk *et al.* suggested this from their studies of patellamide D and ascidiacyclamide, which formed carbonate bridged 2Cu complexes. They suggested that these complexes fix CO_2 for use in the formation of $CaCO_3$ (used in the internal skeleton of the tunicate).[24]. One argument against this is that marine photosynthetic organisms fix HCO_3^- which is plentiful in seawater (0.002 M). The proposed process, $CO_2 \rightarrow HCO_3^-$ is therefore the reverse of that expected.

Are they a form of carbonic anhydrase? Carbonic anhydrases (CAs) play an important role in photosynthetic carbon fixation, converting $HCO_3^-{}_{(aq)}$ in seawater to CO_2. It is possible that the copper complexes of these peptides perform this function. Evidence from the studies by van den Brenk *et al.*[24] show that patellamide D (**4**) forms $[PatDH + Cu_2 + CO_2]^+$ and $[PatDH_3 + Cu_2 + CO_3]^+$ complexes in the mass spectrometer lending credence to this proposal.

Do they have catalytic activity? Type 3 copper proteins have two copper atoms in close proximity (3.6Å), and have roles in oxygen transport and uptake, and as oxidising agents.[32] In ascidiacyclamide/Cu_2 the two copper atoms are 4.5 Å apart. The formation of the $C_3H_5O_3^-$ unit observed during our studies with patellamide C/Cu_2 might be the result of catalytic activity of this system.

Are they involved in carbon metabolism? There are many candidates in the carbon metabolism of *Prochloron* that could be converted to the $C_3H_5O_3^-$ moiety of the 2Cu patellamide C complex. Studies have shown that *L. patella* actively removes glycolate to encourage photosynthesis by *prochloron*.[4] Are the copper complexes of these cyclic peptides involved as enzymic cofactors in this process?

Are they involved in copper detoxification? Our sample of *L. patella* came from an area with high local concentrations of copper in the rocks and seawater.[37] Copper is toxic at high concentrations, so it is possible that the ecological role of the patellamide/copper complexes is for the removal of copper. A point against this would be the loss of expensive metabolites this would incur as these peptides only bind one or two atoms of metal, as opposed to metallothioneins which form aggregates of metal. We observed a precipitate of a blue green solid, possibly copper carbonate, when we combined patellamide C, copper, carbonate and acetate at pH 7. Van den Brenk *et al* found a similar precipitate formed during their preparation of $[PatDH_2 + Cu_2 + CO_3]$.[19] If the end product of copper complexation is indeed insoluble copper carbonate then the peptides would be recyclable and could be used in another round of detoxification.

8   ACKNOWLEDGEMENTS

Financial support for the collection of samples came from the Carnegie Trust and the Nuffield Foundation. MJ participated in a University of California, Santa Cruz marine natural products chemistry expedition partially supported by NIH grants CA47135 and CA52955. We would like to thank Prof Chris Ireland of the University of Utah for generously donating the sample of patellamide A used in our studies. LAM is the recipient of an EPSRC quota award. Dr Sharon Kelly and Prof Nick Price at the BBSRC CD centre in Stirling are acknowledged for their assistance in the acquisition of CD spectra. We would like to thank Dr Jantien Kettennes and Kees Versluis at the University of Utrecht for extensive help with the acquisition and interpretation of the mass spectra. Prof Albert Heck is thanked for allowing us access to the MS instrumentation at Utrecht. Dr Gary Thompson in Prof Steve Homans' group introduced MJ to the mysteries of X-PLOR.

References

1   The text of this was modified from the PhD thesis of L. A. Morris, 'Studies on the Cu(II) and Zn(II) binding properties of cyclic peptides from the ascidian *Lissoclinum patella*', Aberdeen University, Aberdeen, 1999.

2   H. L. Sings and K. L. Rinehart, *J. Ind. Microbio.*, 1996, **17**, 385.

3   V. A. Harris, 'Sessile animals of the seashore', Chapman and Hall, 1990.

4   R. A. Lewin and L. Chang, 'Prochloron', Chapman and Hall, 1989.

5   B. S. Davidson, *Chem. Rev.*, 1993, **93**, 1771.

6   M. A. Rashid, K. R. Gustafson, J. H. C. II, and M. R. Boyd, *J. Nat. Prod.*, 1995, **58**, 594.

7   Y. Hamamoto, M. Endo, M. Nakagawa, T. Nakanishi, and K. Mizukawa, *J. Chem. Soc., Chem. Commun.*, 1983, 323.

8   L. A. McDonald and C. M. Ireland, *J. Nat. Prod.*, 1992, **55**, 376-379.

9   T. Ishida, Y. In, F. Shinozaki, and M. Doi, *J. Org. Chem.*, 1995, **60**, 3944.

10 B. M. Degnan, C. J. Hawkins, M. F. Lavin, E. J. McCaffrey, D. L. Parry, A. L. v. d. Brenk, and D. J. Watters, *J. Med. Chem.*, 1989, **32**, 1349.

11 F. J. Schmitz, M. B. Ksebati, J. S. Chang, J. L. Wang, M. B. Hossain, D. v. d. Helm, M. H. Engel, A. Serban, and J. A. Silfer, *J. Org. Chem.*, 1989, **54**, 3463.

12 X. Fu, T. Do, F. J. Schmitz, V. Andrusevich, and M. H. Engel, *J. Nat. Prod.*, 1998, **61**, 1547.

13 D. E. Williams, R. E. Moore, and V. J. Paul, *J. Nat. Prod.*, 1989, **52**, 732.

14 L. A. McDonald, M. P. Foster, D. R. Phillips, C. M. Ireland, A. Y. Lee, and J. Clardy, *J. Org. Chem.*, 1992, **57**, 4616.

15 T. Shioiri, Y. Hamada, S. Kato, M. Shibata, Y. Kondo, H. Nakagawa, and K. Kohda, *Biochem. Pharmacol.*, 1987, **36**, 4181.

16 P. Wipf, P. C. Fritch, S. J. Geib, and A. M. Sefler, *J. Am. Chem. Soc.*, 1998, **120**, 4105.

17 C. J. Hawkins, M. F. Lavin, K. A. Marshall, A. L. v. d. Brenk, and D. J. Watters, *J. Med. Chem.*, 1990, **33**, 1634.

18 D. J. Freeman, G. Pattenden, A. F. Drake, and G. Siligardi, *J. Chem. Soc., Perkin Trans II*, 1998, **2**, 129.

19 A. L. v. d. Brenk, K. A. Byriel, D. P. Fairlie, L. R. Gahan, G. R. Hanson, C. J. Hawkins, A. Jones, C. H. L. Kennard, B. Moubaraki, and K. S. Murray, *Inorg. Chem.*, 1994, **33**, 3549.

20 M. Jaspars, T. Rali, M. Laney, R. C. Schatzman, M. C. Diaz, F. J. Schmitz, E. O. Pordesimo, and P. Crews, *Tetrahedron*, 1994, **50**, 7367.

21 M. Doi, F. Shinozaki, Y. In, T. Ishida, D. Yamamoto, M. Kamigauchi, M. Sugiura, Y. Hamada, K. Khoda, and T. Shioiri, *Biopolymers*, 1999, **49**, 459.

22 L. A. Morris, G. S. Thompson, and M. Jaspars, *Unpublished results*, 1999.

23 C. J. Hawkins, *Pure Appl. Chem.*, 1988, **60**, 1267.

24 A. L. v. d. Brenk, D. P. Fairlie, G. R. Hanson, L. R. Gahan, C. J. Hawkins, and A. Jones, *Inorg. Chem.*, 1994, **33**, 2280.

25 A. L. v. d. Brenk, D. P. Fairlie, L. R. Gahan, G. R. Hanson, and T. W. Hambley, *Inorg. Chem.*, 1996, **35**, 1095.

26 P. Comba, R. Cusack, D. P. Fairlie, L. R. Gahan, G. R. Hanson, U. Kazmaier, and A. Ramlow, *Inorg. Chem.*, 1998, **37**, 6721.

27   L. Grondahl, N. Sokolenko, G. Abbenante, D. P. Fairlie, G. R. Hanson, and L. R. Gahan, *J. Chem. Soc., Dalton Trans.*, 1999, 1227.

28   A. Rodger and B. Norden, 'Circular dichroism and linear dichroism', ed. R. G. Compton, S. G. Davies, and J. Evans, Oxford University Press, 1997.

29   G. D. Fasman, in 'Circular dichroism and the conformational analysis of biomolecules', New York, 1996.

30   C. Harford and B. Sarkar, *Acc. Chem. Res.*, 1997, **30**, 123.

31   T. Ishida, M. Tanaka, M. Nabae, and M. Inoue, *J. Org. Chem.*, 1988, **53**, 107.

32   W. Kaim and B. Schwederski, 'Bioinorganic Chemistry: Inorganic Elements in the Chemistry of Life', John Wiley and Sons, 1995.

33   R. J. P. Williams and J. J. R. F. de Silva, 'The biological chemistry of the elements', Clarendon Press, 1991.

34   P. A. Brady and J. K. M. Sanders, *New J. Chem.*, 1998, 411.

35   A. T. Brunger, in 'X-PLOR, a system for X-ray crystallography and NMR', New Haven, 1992.

36   G. I. Brown, 'A new guide to modern valence theory', Longman Group Ltd., 1974.

37   J. C. Carlile and A. H. G. Mitchell, *J. Geochem. Explor.*, 1994, **50**, 91.

4   Plant Natural Products

WATER SOLUBLE BIOACTIVE ALKALOIDS

Alison A. Watson and Robert J. Nash[†]

MolecularNature Limited, Plas Gogerddan, Aberystwyth, Cardiganshire SY23 3EB, U.K.

[†]Institute of Grassland and Environmental Research, Plas Gogerddan, Aberystwyth, Cardiganshire SY23 3EB, U.K.

1 INTRODUCTION

Most preparations used in traditional medicines are formulated in water. However, drug discovery programmes have typically used solvents such as methanol, chloroform or hexane for extraction. That such extraction methods can miss very biologically active metabolites is shown by the recent discovery of calystegine alkaloids in potato tubers and other solanaceous foods despite daily analysis for decades of these food plants in laboratories all over the world for toxic components.[1] It is becoming apparent that the water soluble fractions of medicinal plants and microbial cultures contain many interesting novel structures. Such chemicals include carbohydrate analogues with molecular weights usually less than 250 Daltons. The chemical and biological diversity in terms of structural information in small sugars is illustrated by the 2^{10} stereoisomers of sucrose alone - even glucose would have 2×2^5 isomers in the pyranose and furanose forms. This amazing diversity in such small molecules displays a remarkable economy in structural information in nature, completely surpassing in molecular weight terms anything achieved by the amino acids. One interesting class of polar natural products with a wide stereochemical diversity is the polyhydroxylated alkaloids found in plants and micro-organisms. Such compounds are increasingly being discovered as chemical fingerprinting and extraction methods for polar compounds are improved. Their potential as therapeutic agents is also arousing considerable interest and so this paper will review their distribution and therapeutic activities known to date.

The first natural polyhydroxylated alkaloid to be detected was the piperidine alkaloid nojirimycin (Figure 1a), which was isolated from a *Streptomyces* filtrate in 1966,[2] but most have been discovered since 1983. These alkaloids can be considered as analogues of monosaccharides in which the ring oxygen has been replaced by nitrogen. They are monocyclic and bicyclic polyhydroxylated derivatives of the following ring systems: pyrrolidine, piperidine, pyrrolizidine (two fused pyrrolidines), octahydroindolizine or indolizidine (fused piperidine and pyrrolidine) and *nor*tropane. Systematically these alkaloids have been described in the literature as derivatives of the parent heterocyclic compounds or sugars. 1-Deoxynojirimycin (2*S*-hydroxymethyl-3*R*,4*R*,5*S*-trihydroxy-piperidine or 1,5-dideoxy-1,5-imino-D-glucitol, Figure 1b) was originally synthesised by removing the anomeric hydroxyl group of nojirimycin[3] but it was later isolated from plant sources and bacterial cultures.[4,5] This derivative is more stable than nojirimycin and has become a model compound in this area of research, giving rise to a trivial nomenclature for 1-deoxy analogues of other alkaloids of this type. Thus, the 1-deoxy piperidine analogue of mannose has been given the trivial name of 1-deoxymannojirimycin (DMJ) (Figure 1c). The nomenclature used most frequently for particular compounds and their common abbreviations will be used in this review.

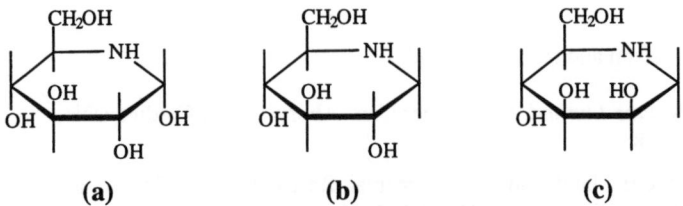

Figure 1. *Diagram illustrating the common nomenclature for polyhydroxylated piperidine alkaloids. (a) 5-Amino-5-deoxy-D-glucopyranose has the trivial name nojirimycin, (b) 1-deoxynojirimycin (DNJ) is the trivial name for 1,5-dideoxy-1,5-imino-D-glucitol and similarly (c) 1-deoxymannojirimycin (DMJ) is the common name for 1,5-dideoxy-1,5-imino-D-mannitol.*

2 THE DISTRIBUTION OF POLYHYDROXYLATED ALKALOIDS

Table 1 lists the microorganisms and plants reported to produce polyhydroxylated alkaloids which have a structural resemblance to carbohydrates. Some of the alkaloids such as 2R,5R-dihydroxymethyl-3R,4R-dihydroxypyrrolidine (DMDP) and 1,4-dideoxy-1,4-imino-D-arabinitol (D-AB1) (Figure 2) would appear to be fairly widespread secondary metabolites as they have been reported from species of both tropical and temperate plants from quite unrelated families and DMDP is also produced by a species of *Streptomyces*.[6] Others, such as the indolizidine alkaloid castanospermine (Figure 6), have only been found in two closely related legume genera. The polyhydroxylated pyrrolizidine alkaloids (Figure 5) appeared to be restricted to the Leguminosae until their discovery in 1994 in a member of the Casuarinaceae[7] and then in the closely related Myrtaceae[8] and recently they have also been found in abundance in members of the Hyacinthaceae.[9,10] The polyhydroxylated *nor*tropane alkaloids (Figure 7) seem to be largely limited to the closely related families Solanaceae[11,12] and Convolvulaceae,[13] where they co-occur with tropane alkaloids. However, they have also been found in species of *Morus* (Moraceae)[14,15] which is a family not noted for production of tropane alkaloids.

While there may be phylogenetic reasons for particular distributions of the polyhydroxylated alkaloids in plants, caution should nevertheless be exercised in using the presence of these compounds as taxonomic markers. One reason is that these alkaloids can be released into the soil by producer plants and micro-organisms from where some, such as DMDP and castanospermine can be readily taken up and accumulated in plant tissues of completely unrelated neighbouring species.[16] It may also be the case that micro-organisms (*Rhizobium*, other rhizosphere organisms, or endophytes) closely associated with specific plants may also produce polyhydroxylated alkaloids which could then be mistakenly considered of plant origin.

From Table 1 it can be seen that the majority of naturally-occurring polyhydroxylated alkaloids have been isolated from plants. Although the first alkaloid of this type to be discovered (nojirimycin) was isolated from the fermentation broth of a species of *Streptomyces*,[2] only a limited number of bacteria (principally actinomycetes) have subsequently been found that produce polyhydroxylated alkaloids. Also, only three of the compounds listed in Table 1 are known to be produced by fungi (nectrisine, swainsonine and 2-*epi*lentiginosine). However, the apparent restricted distribution of polyhydroxylated alkaloids amongst micro-organisms may be misleading as the techniques used to screen the cultures may frequently be inappropriate for detecting such highly polar compounds. There are actually quite a range of microbial products that can be described as carbohydrate analogues and some of these with therapeutic potential are discussed in Section 1.5.1.

The recent increase in the rate at which novel water-soluble alkaloids are reported would suggest that many more await discovery from diverse sources, including plants and microorganisms in many taxa not previously considered to be alkaloid producers. Several hundred related alkaloids have also been synthesised which have allowed investigations of structure-activity relationships, but it is now becoming apparent that many of these synthetic structures also occur naturally. For example, the 7-*O*-β-D-glucopyranosyl derivative of α-homonojirimycin was designed to be a transition state analogue of sucrose[17] before it and its aglycone were discovered as natural products.[18] Similarly, the synthetic piperidine alkaloids β-homonojirimycin,[19,20] α-homomanno-jirimycin[21] and β-homomannojirimycin[19] were all recently isolated from *Aglaonema treubii* (Araceae).[22]

3 POLYHYDROXYLATED ALKALOIDS AS GLYCOSIDASE INHIBITORS

Most of the polyhydroxylated alkaloids listed in Section 1.2. that have been studied in detail have been shown to inhibit glycosidases in a reversible and competitive manner.[89,90] Glycosidases are enzymes that catalyse the hydrolysis of the glycosidic bonds in complex carbohydrates and glycoconjugates. The wide variety of functions in which glycosidases are involved makes them essential for the survival and existence of all living organisms. For example, digestive glycosidases break down large sugar-containing molecules to release monosaccharides which can be more easily taken up and used metabolically by the organism; lysosomal glycosidases catabolise glycoconjugates intracellularly and a wide range of glycosidases are involved in the biosynthesis of the oligosaccharide portions of glycoproteins and glycolipids which play vital roles in mammalian cellular structure and function. For example, membrane glycoproteins include receptors for biologically important molecules such as hormones, low-density lipoprotein or acetylcholine whilst others are involved in cell-cell adhesion. The oligosaccharide chains play an important role in the correct functioning of these proteins by stabilising them and ensuring they have the correct conformation and they may also be involved in the targeting mechanism of certain proteins.[91]

Since the mode of action of glycosidases involves the cleavage of glycosidic bonds between sugar molecules, individual glycosidases show specificity for certain sugar molecules and for a specific anomeric configuration of that sugar. Polyhydroxylated alkaloids can be extremely potent and specific inhibitors of glycosidases by mimicking the pyranosyl or furanosyl moiety of their natural substrates. Therefore, the number, position and configuration of the hydroxyl groups of each alkaloid dictate the type of glycosidases which are inhibited. For example, the configurations of the hydroxyl substituents of the glucosidase inhibitor nojirimycin correspond to those of glucose in the pyranose configuration. Nojirimycin exists in aqueous solution in both α- and β-forms with an equilibrium of 60 % of the former and 40 % of the latter and each of these forms are responsible for the inhibition of α- and β-glucosidases respectively.[3]

Table 1. *The naturally-occurring polyhydroxylated alkaloids.*

Alkaloid	Source and Reference
Pyrroline (Figure 2)	
3,4-Dihydroxy-5-hydroxymethyl-1-pyrroline (Nectrisine or FR-900483)	*Nectria lucida* F-4490 (ATCC 20722) (Ascomycetes)[23]
Pyrrolidines (Figures 2 and 3)	
2*R*-Hydroxymethyl-3*S*-hydroxypyrrolidine (CYB-3)	*Castanospermum australe* (Leguminosae) seeds/leaves[24]
N-Hydroxyethyl-2-hydroxymethyl-3-hydroxypyrrolidine	*Castanospermum australe* (Leguminosae) seeds[25]
1,4-Dideoxy-1,4-imino-D-arabinitol (D-AB1)	*Angylocalyx* spp. (Leguminosae) seeds/leaves/bark[26] *Arachniodes standishii* (Polypodiaceae) leaves[26,27] *Morus bombycis* (Moraceae) leaves[14] *Eugenia* spp. (Myrtaceae) leaves/bark[16] *Hyacinthoides non-scripta* (Hyacinthaceae) bulb/leaves[28] *Scilla campanulata* (Hyacinthaceae) bulb[9] *Adenophora triphylla* var. *japonica* (Campanulaceae) whole plant[29]
1,4-Dideoxy-1,4-imino-(2-*O*-β-D-glucopyranosyl)-D-arabinitol	*Morus bombycis* (Moraceae) leaves[14] *Morus alba* (Moraceae) roots[15]
1,4-Dideoxy-1,4-imino-D-ribitol	*Morus alba* (Moraceae) roots[15]
2*R*,5*R*-Dihydroxymethyl-3*R*,4*R*-dihydroxy-pyrrolidine (DMDP)	*Derris elliptica* (Leguminosae) leaves[30] *Lonchocarpus* spp (Leguminosae) seeds/leaves[31] *Endospermum* sp. (Euphorbiaceae) leaves[32] *Omphalea diandra* (Euphorbiaceae) leaves[18] *Streptomyces* sp. KSC-5791 (Actinomycetes)[6] *Nephthytis poissoni* (Araceae) fruit/leaves[16,33] *Aglaonema* spp (Araceae) leaves[22,33] *Hyacinthoides non-scripta* (Hyacinthaceae) bulb/leaves[28] *Campanula rotundifolia* (Campanulaceae) leaves[34] *Hyacinthus orientalis* (Hyacinthaceae) bulb[35] *Scilla campanulata* (Hyacinthaceae) bulb[9] *Adenophora* spp. (Campanulaceae) roots[36]
6-DeoxyDMDP	*Angylocalyx* spp. (Leguminosae) seeds/leaves/bark[37]
6-*C*-butyl-DMDP	*Adenophora triphylla* var. *japonica* (Campanulaceae) whole plant[29]
6-Deoxy-6-*C*-(2,5-dihydroxyhexyl)-DMDP	*Hyacinthoides non-scripta* (Hyacinthaceae) fruits/stalks[9] *Scilla campanulata* (Hyacinthaceae) bulb[9]
2,5-Imino-2,5,6-trideoxy-D-*gulo*-heptitol	*Hyacinthus orientalis* (Hyacinthaceae) bulb[35]
2,5-Dideoxy-2,5-imino-DL-*glycero*-D-*manno*-heptitol (homoDMDP)	*Hyacinthoides non-scripta* (Hyacinthaceae)[28] *Hyacinthus orientalis* (Hyacinthaceae) bulb[35] *Scilla campanulata* (Hyacinthaceae) bulb[9] *Muscari armeniacum* (Hyacinthaceae) bulbs[10]
HomoDMDP-7-*O*-apioside	*Hyacinthoides non-scripta* (Hyacinthaceae) bulb/leaves[28] *Scilla campanulata* (Hyacinthaceae) bulb[9] *Muscari armeniacum* (Hyacinthaceae) bulbs[10]
HomoDMDP-7-*O*-β-D-xylopyranoside	*Hyacinthoides non-scripta* (Hyacinthaceae) fruits/stalks[9] *Scilla campanulata* (Hyacinthaceae) bulb[9] *Muscari armeniacum* (Hyacinthaceae) bulbs[10]
6-Deoxy-homoDMDP	*Hyacinthus orientalis* (Hyacinthaceae) bulb[35] *Muscari armeniacum* (Hyacinthaceae) bulbs[10]
Gualamycin	*Streptomyces* sp. NK11687[38]
Broussonetinines A and B, Broussonetines A, B, C, D, E, F, G, H, I, J, K and L	*Broussonetia kazinoki* (Moraceae) branches[39-44]
Piperidines (Figure 4)	
Nojirimycin	*Streptomyces roseochromogenes* R-468[2,3,45] *Streptomyces lavandulae* SF-425[2,3,45] *Streptomyces nojiriensis* SF-426[2,3,45]
1-Deoxynojirimycin (DNJ)	*Morus* sp. (Moraceae) roots[4] *Bacillus amyloliquefaciens, B. polymyxa, B. subtilis*[46] *Streptomyces lavandulae* subsp. *trehalostaticus* no. 2882[5] *Omphalea queenslandiae* (Euphorbiaceae) leaves[47] *Endospermum medullosum* (Euphorbiaceae) leaves[47] *Morus bombycis* (Moraceae) leaves[14] *Hyacinthus orientalis* (Hyacinthaceae) bulb[35] *Adenophora triphylla* var. *japonica* (Campanulaceae) whole plant[29]

Table 1. *Continued.*

Alkaloid	Source and Reference
Piperidines (Figure 4)	
1-Deoxynojirimycin-2-*O*-, 3-*O*-, 4-*O*-* α-D-glucopyranosides and 2-*O*-, 6-*O*-α-D-galactopyranoside and 2-*O*, 3-*O*-, 4-*O*-, 6-*O*-β-D-glucopyranoside	*Morus alba* (Moraceae) roots[15] *Streptomyces lavandulae* GC-148 [48]
N-Methyl-1-deoxynojirimycin	*Morus alba* (Moraceae) roots[15]
α-Homonojirimycin (HNJ)	*Omphalea diandra* (Euphorbiaceae) leaves[48] *Endospermum medullosum* (Euphorbiaceae) leaves[47] *Nephthytis poissoni* (Araceae) leaves[16] *Aglaonema treubii* (Araceae) leaves/ roots[22] *Hyacinthus orientalis* (Hyacinthaceae) bulb[35]
α-Homonojirimycin-7-*O*-β-D-glucopyranoside (MDL 25,637)	*Omphalea diandra* (Euphorbiaceae) leaves (tentative)[18] *Nephthytis poissoni* (Araceae) leaves[16] *Aglaonema treubii* (Araceae) leaves/ roots[22] *Hyacinthus orientalis* (Hyacinthaceae) bulb[35] *Lobelia sessilifolia* (Campanulaceae) whole plant[36]
α-Homonojirimycin-5-*O*-α-D-galactopyranoside	*Aglaonema treubii* (Araceae) leaves/ roots[22]
α-4-*Epi*homonojirimycin	*Aglaonema treubii* (Araceae) leaves/ roots[22,50]
α-1-Deoxy-1-*C*-methyl-homonojirimycin (Adenophorine)	*Adenophora* spp. (Campanulaceae) roots[36]
Adenophorine-1-*O*-β-D-glucopyranoside	*Adenophora* spp. (Campanulaceae) roots[36]
1-Deoxyadenophorine	*Adenophora* spp. (Campanulaceae) roots[36]
5-Deoxyadenophorine	*Adenophora* spp. (Campanulaceae) roots[36]
5-Deoxyadenophorine-1-*O*-β-D-glucopyranoside	*Adenophora* spp. (Campanulaceae) roots[36]
β-1-*C*-butyl-deoxygalactonojirimycin	*Adenophora* spp. (Campanulaceae) roots[36]
β-Homonojirimycin	*Aglaonema treubii* (Araceae) leaves/ roots[22] *Hyacinthus orientalis* (Hyacinthaceae) bulb[35]
β-4,5-Di*epi*homonojirimycin	*Aglaonema treubii* (Araceae) leaves/ roots[51]
Nojirimycin B (Mannojirimycin)	*Streptomyces lavandulae* SF-425 [52]
1-Deoxymannojirimycin (DMJ)	*Lonchocarpus sericeus* (Leguminosae) seeds/leaves[53] *Streptomyces lavandulae* GC-148 [54] *Omphalea diandra* (Euphorbiaceae) leaves[49] *Endospermum medullosum* (Euphorbiaceae) leaves[47] *Derris malaccensis* (Leguminosae)[55] *Angylocalyx* spp. (Leguminosae) seeds/leaves/bark[16] *Hyacinthus orientalis* (Hyacinthaceae) bulb[35] *Adenophora triphylla* var. *japonica* (Campanulaceae) whole plant[29]
α-Homomannojirimycin	*Aglaonema treubii* (Araceae) leaves/ roots[22] *Hyacinthus orientalis* (Hyacinthaceae) bulb[35]
β-Homomannojirimycin	*Aglaonema treubii* (Araceae) leaves/ roots[22] *Hyacinthus orientalis* (Hyacinthaceae) bulb[35]
Galactostatin	*Streptomyces lydicus* PA-5726 [56]
Fagomine	*Fagopyrum esculentum* (Polygonaceae) seeds[57] *Xanthocercis zambesiaca* (Leguminosae) seeds,[58] leaves/ roots[59] *Morus bombycis* (Moraceae) leaves[14] *Morus alba* (Moraceae) roots[15] *Lycium chinense* (Solanaceae) roots[60]
3-*Epi*fagomine	*Morus alba* (Moraceae) roots[15] *Xanthocercis zambesiaca* (Leguminosae) leaves/ roots[59]
3,4-Di*epi*fagomine	*Xanthocercis zambesiaca* (Leguminosae) leaves/ roots[59]
Fagomine-4-*O*-β-D-glucopyranoside	*Xanthocercis zambesiaca* (Leguminosae) seeds,[58] leaves/ roots[59]
Fagomine-3-*O*-β-D-glucopyranoside	*Xanthocercis zambesiaca* (Leguminosae) leaves/ roots[59]
6-Deoxy-fagomine	*Lycium chinense* (Solanaceae) roots[60]
α-1-*C*-ethyl-fagomine	*Adenophora triphylla* var. *japonica* (Campanulaceae) whole plant[29]
Pyrrolizidines (Figure 5)	
Alexine	*Alexa* spp. (Leguminosae) seeds/leaves[61]
3,7a-Di*epi*alexine	*Castanospermum australe* (Leguminosae) seeds/leaves[62]
7a-*Epi*alexine (Australine)	*Castanospermum australe* (Leguminosae) seeds/leaves[63]

Table 1. *Continued.*

Alkaloid	Source and Reference
Pyrrolizidines (Figure 5)	
1,7a-Di*epi*alexine	*Alexa* spp. and *Castanospermum australe* (Leguminosae) seeds/leaves[64]
7a-*Epi*alexaflorine	*Alexa grandiflora* (Leguminosae) leaves[65]
Casuarine	*Casuarina equisetifolia* (Casuarinaceae) bark[7] *Eugenia jambolana* (Myrtaceae) leaves[8]
Casuarine-6-*O*-α-D-glucopyranoside	*Casuarina equisetifolia* (Casuarinaceae) bark[7] *Eugenia jambolana* (Myrtaceae) leaves[8]
(1*S*, 2*R*, 3*R*, 7a*R*)-1,2-dihydroxy-3-hydroxymethylpyrrolizidine (Hyacinthacine A₁)	*Muscari armeniacum* (Hyacinthaceae) bulbs[10]
(1*R*, 2*R*, 3*R*, 7a*R*)-1,2-dihydroxy-3-hydroxymethylpyrrolizidine (Hyacinthacine A₂)	*Muscari armeniacum* (Hyacinthaceae) bulbs[10]
(1*R*, 2*R*, 3*R*, 5*R*, 7a*R*)- 1,2-dihydroxy-3-hydroxymethyl-5-methylpyrrolizidine (Hyacinthacine A₃)	*Muscari armeniacum* (Hyacinthaceae) bulbs[10]
(1*S*, 2*R*, 3*R*, 5*R*, 7a*R*)-1,2-dihydroxy-3,5-dihydroxy-methylpyrrolizidine (Hyacinthacine B₁)	*Hyacinthoides non-scripta* (Hyacinthaceae) fruits/stalks[9] *Scilla campanulata* (Hyacinthaceae) bulb[9]
(1*S*, 2*R*, 3*R*, 5*S*, 7a*R*)-1,2-dihydroxy-3,5-dihydroxy-methylpyrrolizidine (Hyacinthacine B₂)	*Scilla campanulata* (Hyacinthaceae) bulb[9]
(1*S*, 2*R*, 3*R*, 5*R*, 7*R*, 7a*R*)-3-hydroxymethyl-5-methyl-1,2,7-trihydroxypyrrolizidine (Hyacinthacine B₃)	*Muscari armeniacum* (Hyacinthaceae) bulbs[10]
(1*S*, 2*R*, 3*R*, 5*R*, 6*R*, 7*R*, 7a*R*)-3-hydroxymethyl-5-methyl-1,2,6,7-tetrahydroxypyrrolizidine (Hyacinthacine C₁)	*Hyacinthoides non-scripta* (Hyacinthaceae) fruits/stalks[9] *Muscari armeniacum* (Hyacinthaceae) bulbs[10]
(1*R*, 2*R*, 3*R*, 5*S*, 8*R*)-1,2-dihydroxy-3-hydroxymethyl-5-[(1*R*)-1,10-dihydroxy-6-oxo-decyl]-pyrrolizidine (Broussonetine N)	*Broussonetia kazinoki* (Moraceae) branches[44]
Indolizidines (Figure 6)	
Swainsonine	*Swainsona canescens* (Leguminosae) leaves[66] *Astragalus* spp. (Leguminosae) leaves/stems[67] *Oxytropis* spp. (Leguminosae) leaves/stems[67] *Rhizoctonia leguminicola* (Basidiomycetes)[68] *Metarhizium anisopliae* (Deuteromycetes)[69] *Ipomoea* sp. aff. *calobra* (Convolvulaceae) seeds[70] *Ipomoea carnea* (Convolvulaceae) leaves/stems[71]
Swainsonine *N*-oxide	*Astragalus lentiginosus* (Leguminosae)[67]
Lentiginosine	*Astragalus lentiginosus* (Leguminosae) leaves[72]
2-*Epi*lentiginosine	*Rhizoctonia leguminicola* (Basidiomycetes)[73] *Astragalus lentiginosus* (Leguminosae) leaves[72]
Castanospermine	*Castanospermum australe* (Leguminosae) seeds/leaves/bark[74] *Alexa* spp. (Leguminosae) seeds/leaves/bark[75]
6-*Epi*castanospermine	*Castanospermum australe* (Leguminosae) seeds/leaves/bark[76,77]
6,7-Di*epi*castanospermine	*Castanospermum australe* (Leguminosae) seeds[25]
7-Deoxy-6-*epi*castanospermine	*Castanospermum australe* (Leguminosae) seeds[78]

Table 1. *Continued.*

Alkaloid	Source and Reference
Nortropanes (Figure 7)	
Calystegine A$_3$ and Calystegine B$_2$	*Calystegia sepium* (Convolvulaceae) leaves/roots[79, 80] *Convolvulus arvensis* (Convolvulaceae) leaves/roots[79,81] *Atropa belladonna* (Solanaceae) leaves/roots[11,79,80] *Solanum* spp. (Solanaceae) tubers/leaves[1] *Ipomoea batatus* (Convolvulaceae) leaves/roots[16] *Datura wrightii* (Solanaceae) leaves[1] *Physalis alkekengi* var. *francheti* (Solanaceae) roots[82] *Hyoscyamus niger* (Solanaceae) leaves/roots[11] *Mandragora officinarum* (Solanaceae) leaves/roots/fruits[11] *Scopolia* spp. (Solanaceae) leaves/roots[11,83] *Ipomoea* sp. aff. *calobra* (Convolvulaceae) seeds[70] *Calystegia japonica* (Convolvulaceae) roots[16] *Lycium chinense* (Solanaceae) roots[60] B$_2$ only in *Morus alba* (Moraceae) fruits[84] *Ipomoea carnea* (Convolvulaceae) leaves/stems[71]
N-Methyl-calystegine B$_2$	*Lycium chinense* (Solanaceae) roots[60]
Calystegine A$_5$	*Physalis alkekengi* var. *francheti* (Solanaceae)[82] *Scopolia japonica* (Solanaceae) roots[83] *Hyoscyamus niger* (Solanaceae) leaves/roots[85] *Lycium chinense* (Solanaceae) roots[60]
Calystegine A$_6$	*Hyoscyamus niger* (Solanaceae) leaves/roots[85] *Lycium chinense* (Solanaceae) roots[60]
Calystegine A$_7$	*Lycium chinense* (Solanaceae) roots[60]
Calystegine B$_1$	*Convolvulus arvensis* (Convolvulaceae) leaves/roots[79,81] *Calystegia sepium* (Convolvulaceae) leaves/roots[80] *Physalis alkekengi* (Solanaceae) roots[82] *Hyoscyamus niger* (Solanaceae) leaves/roots[11] *Mandragora officinarum* (Solanaceae) leaves/roots/fruits[11] *Scopolia* spp. (Solanaceae) leaves/roots[11,83] *Ipomoea batatus* (Convolvulaceae) leaves/roots[16,84] *Duboisia leichhardtii* (Solanaceae) leaves[86] *Lycium chinense* (Solanaceae) roots[60]
Calystegine B$_1$-3-*O*-β-D-glucopyranoside	*Nicandra physalodes* (Solanaceae) fruits[87]
Calystegine B$_3$	*Physalis alkekengi* (Solanaceae) roots[82] *Scopolia japonica* (Solanaceae) roots[83] *Hyoscyamus niger* (Solanaceae) leaves/roots[85] *Lycium chinense* (Solanaceae) roots[60]
Calystegine B$_4$	*Scopolia japonica* (Solanaceae) roots[83] *Duboisia leichhardtii* (Solanaceae) leaves[86] *Lycium chinense* (Solanaceae) roots[60]
Calystegine B$_5$	*Lycium chinense* (Solanaceae) roots[60]
Calystegine C$_1$	*Morus alba* (Moraceae) roots[15] *Ipomoea batatus* (Convolvulaceae) roots[16] *Scopolia japonica* (Solanaceae) roots[83] *Duboisia leichhardtii* (Solanaceae) leaves[86] *Lycium chinense* (Solanaceae) roots[60] *Ipomoea carnea* (Convolvulaceae) leaves/stems[71]
N-Methyl-calystegine C$_1$	*Lycium chinense* (Solanaceae) roots[60]
Calystegine C$_2$	*Duboisia leichhardtii* (Solanaceae) leaves[86] *Lycium chinense* (Solanaceae) roots[60]
Calystegine N$_1$	*Hyoscyamus niger* (Solanaceae) leaves/roots[85] *Lycium chinense* (Solanaceae) roots[60]

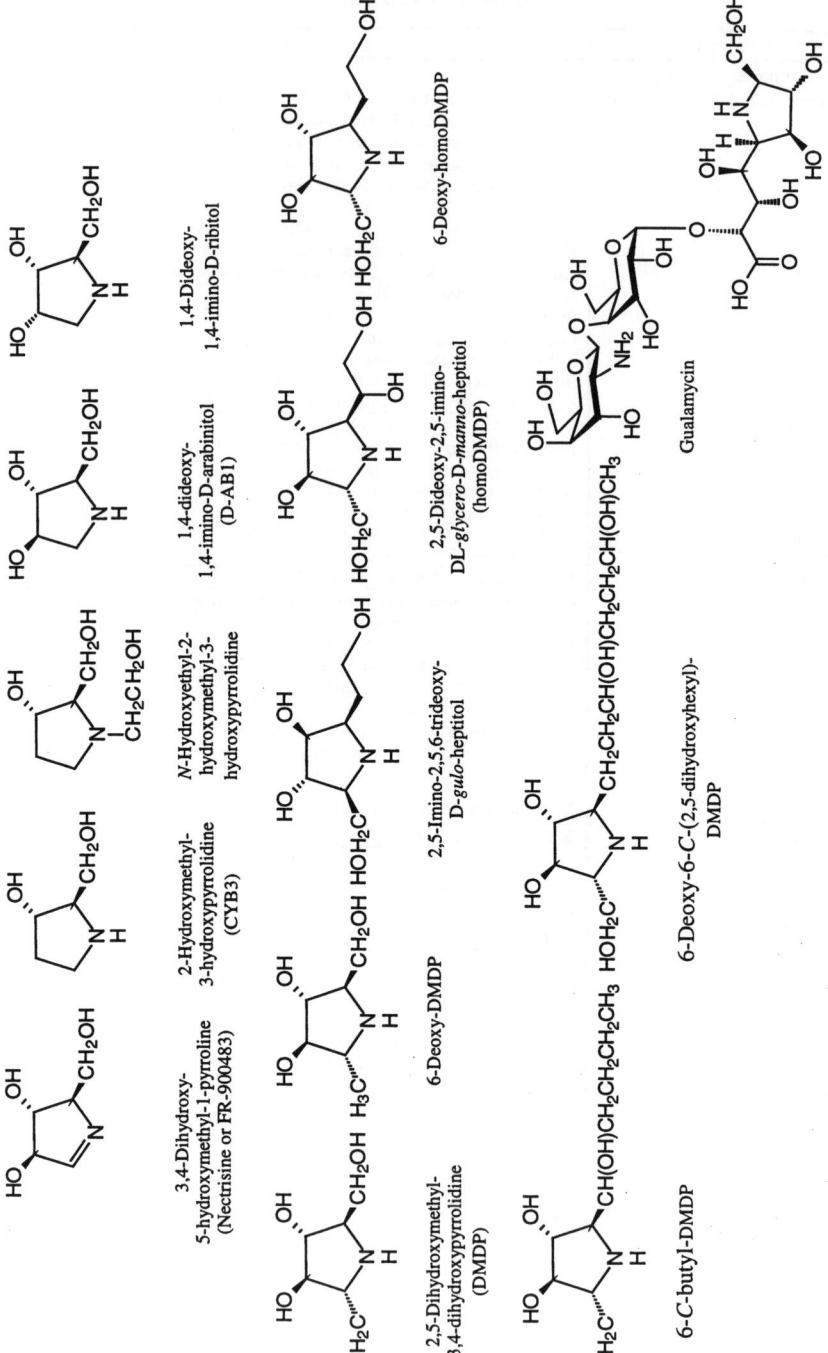

Figure 2. *Some naturally-occurring polyhydroxylated pyrroline and pyrrolidine alkaloids.*

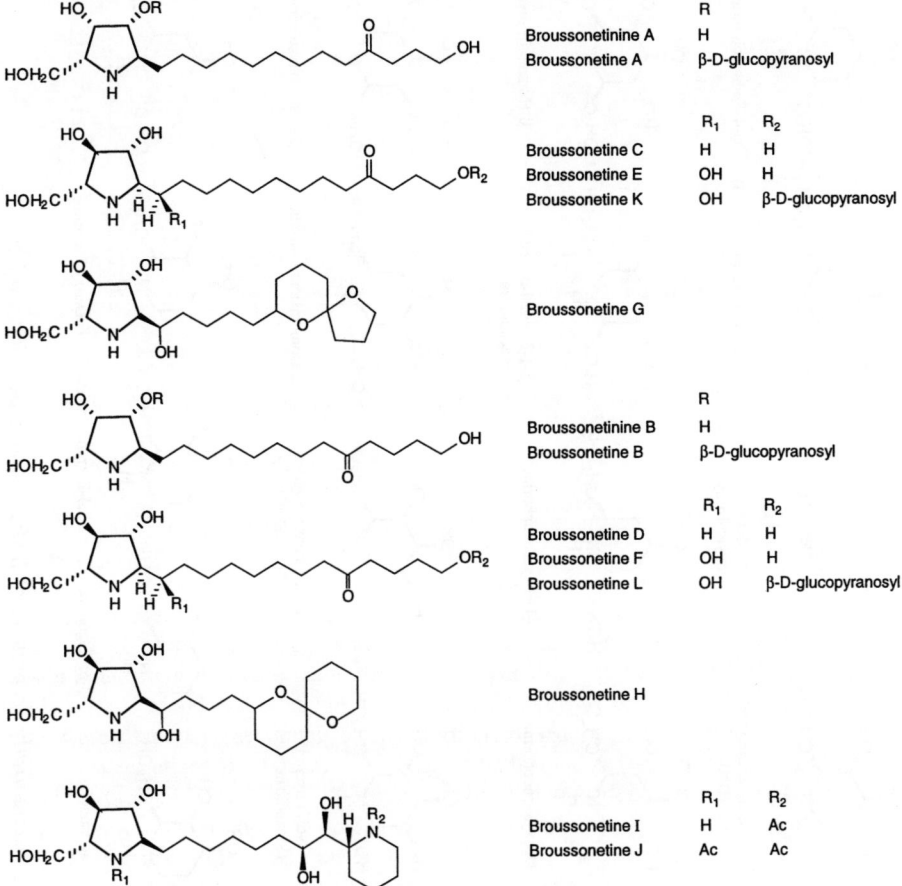

Figure 3. *Polyhydroxylated pyrrolidine alkaloids and derivatives isolated from* Broussonetia kazinoki.

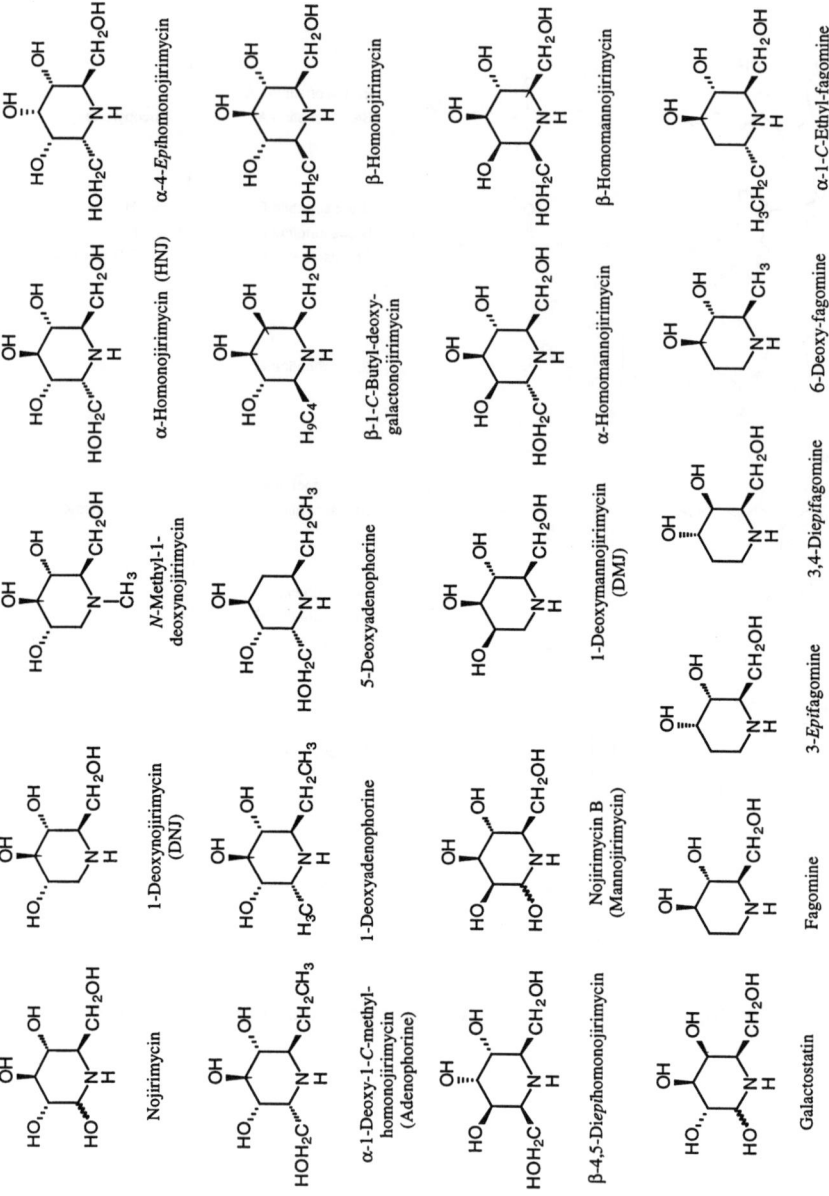

Figure 4. *Some naturally-occurring polyhydroxylated piperidine alkaloids.*

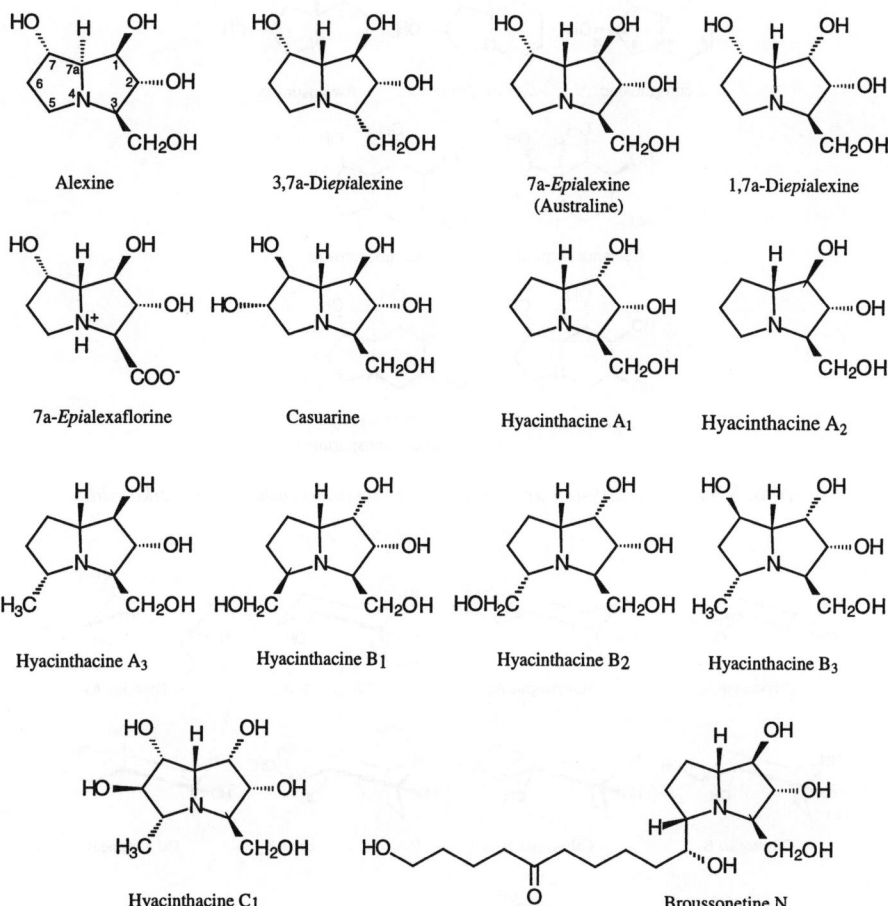

Figure 5. *Some naturally-occurring polyhydroxylated pyrrolizidine alkaloids[1].*

---

[1] Recently, Wormald and co-workers [88] reported that 7,7a-di*epi*alexine, originally identified as a natural product from *Castanospermum australe*,[63] was in fact incorrectly described in the original paper as identification was based on a comparison with erroneous NMR data published for 7a-*epi*alexine.[62] Therefore, 7,7a-di*epi*alexine can no longer be regarded as a natural product and so it is not depicted here.

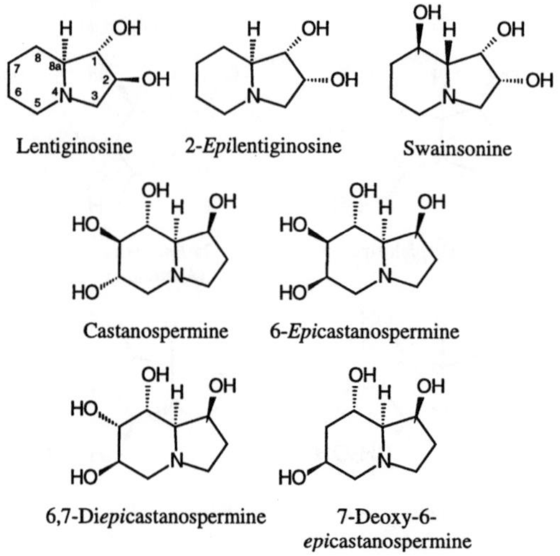

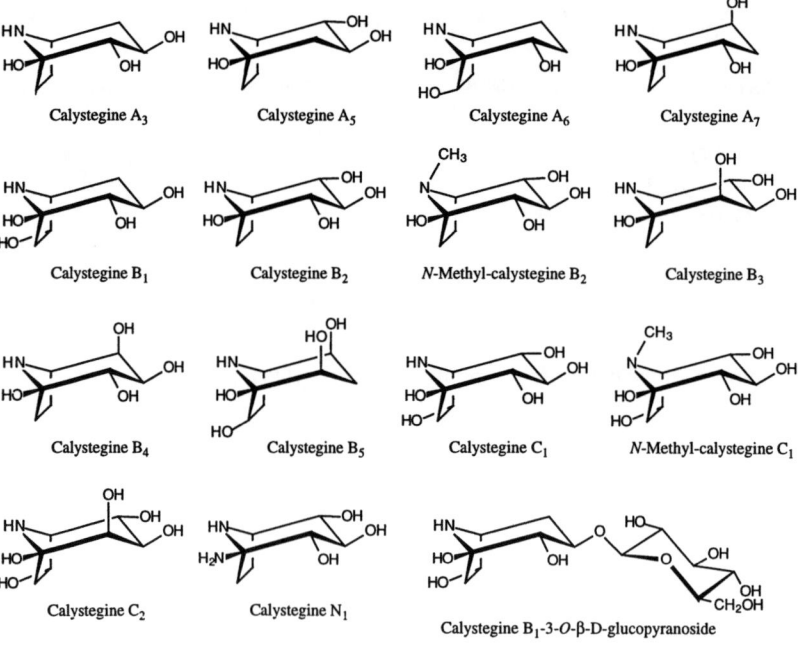

Figure 6. *Some naturally-occurring polyhydroxylated indolizidine alkaloids.*

Figure 7. *Some naturally-occurring polyhydroxylated nortropane alkaloids.*

Although the spatial arrangement of the hydroxyl groups of polyhydroxylated alkaloids serves as a means of recognition by specific glycosidases, it is the influence of the endocyclic nitrogen atom on the conformation and electrostatic properties of the molecule that is important for inhibition of enzyme activity. When a polyhydroxylated alkaloid binds to the active site of a glycosidase, it has been suggested that protonation of the compound leads to the formation of an ion pair between the inhibitor and a carboxylate anion in the active site of the enzyme.[89] The protonated inhibitor closely resembles the transition state of the natural substrate and hence the enzyme has a high affinity for the molecule. However, the strength of the binding and hence the effectiveness of the inhibition depends to a degree on the pKa of the inhibitor and the pH optimum of the enzyme.[92]

Although the nature of the glycosidases that will be inhibited by certain polyhydroxylated alkaloids can be predicted, to some extent, from the number, position and configuration of the hydroxyl groups, there can be marked differences in the inhibition of isoenzymes of a given glycosidase in different species[93,94] and even within the same cell. For example, in human liver cells there are multiple forms of α-mannosidase which are structurally, functionally and genetically quite distinct.[95] They also have different subcellular locations. The lysosomal α-mannosidase, which catabolises glycoconjugates, has an acidic pH optimum (around 4.0) whereas the cytosolic α-mannosidase functions best at near neutral pH (6.5). There are also two forms of α-mannosidase in the Golgi (α-mannosidase I and II, respectively) which are involved in glycoprotein processing, as is another α-mannosidase which is associated with the endoplasmic reticulum. The lysosomal, Golgi α-mannosidase II and neutral α-mannosidases are all inhibited by polyhydroxylated alkaloids with the same substituents and chirality as mannofuranose (e.g. swainsonine) whereas Golgi α-mannosidase I is inhibited by mannopyranose analogues such as DMJ.[96]

There is now a vast amount of literature on the inhibition of glycosidases by particular polyhydroxylated alkaloids and related compounds (both natural and synthetic) and an overview of this area is given in some reviews.[89,90,97] It should be noted, however, that the polyhydroxylated alkaloids discussed above are all inhibitors of exoglycosidases, which cleave the glycosidic linkage at the non-reducing terminus of the saccharide chain, liberating monosaccharide units. However, certain animal cells also possess endoglycosidases which act on glycosidic linkages within saccharide chains giving rise to smaller oligosaccharide units. Endoglycosidases are not generally inhibited by polyhydroxylated alkaloids. A large amount of research has been conducted on the inhibition of exoglycosidases involved in glycoprotein formation and catalysis which has been shown to be important in both the toxicity and the potential therapeutic uses of the polyhydroxylated alkaloids (Sections 1.4. and 1.5.). Table 2 lists some of the exoglycosidases that the polyhydroxylated alkaloids are reported to inhibit. This is not a comprehensive list, as not all naturally-occurring polyhydroxylated alkaloids have been tested on a wide range of glycosidases. It should also be remembered that due to the variations in specificity of isoenzymes, both between species and within the same cell, it is difficult to predict, for example, that a potent inhibitor of certain α-D-glucosidases will inhibit all α-D-glucosidases as strongly, if at all.

Table 2. *A list of some of the exoglycosidases that have been found to be inhibited by naturally-occurring polyhydroxylated alkaloids.*

Alkaloid	Glycosidases Inhibited
Pyrroline	
Nectrisine (FR-900483)	α-glucosidase, α-mannosidase[23,98]
Pyrrolidines	
2R-Hydroxymethyl-3S-hydroxypyrrolidine (CYB-3)	weak inhibitor of isomaltase[99]
1,4-Dideoxy-1,4-imino-D-arabinitol (D-AB1)	potent inhibitor of α-glucosidases (including maltase, isomaltase, sucrase and trehalase, Glucosidase II), α-D-arabinosidase, β-glucosidases (including lactase and cellobiase) and α-mannosidases (Mannosidase II and lysosomal)[90,99-101]
1,4-Dideoxy-1,4-imino-(2-O-β-D-glucopyranosyl)-D-arabinitol	weak inhibitor of α- and β-glucosidases[15]
1,4-Dideoxy-1,4-imino-D-ribitol	weak inhibitor of α-glucosidase[97]
2R,5R-Dihydroxymethyl-3R,4R-dihydroxy-pyrrolidine (DMDP)	β-glucosidases (including cellobiase) and α-glucosidases (including invertase, sucrase, isomaltase, trehalase), human lysosomal β-mannosidase, β-galactosidase (including lactase) and β-xylosidase[31,90,93,94,102]
6-DeoxyDMDP	β-mannosidase, β-galactosidase and α-fucosidase[37]
6-C-butyl-DMDP	good β-glucosidase inhibitor and poor inhibitor of β-galactosidase[29]
6-Deoxy-6-C-(2,5-dihydroxyhexyl)-DMDP	moderate inhibitor of amyloglucosidase and weak inhibitor of α-fucosidase[9]
2,5-Imino-2,5,6-trideoxy-D-*gulo*-heptitol	moderate inhibitor of α-fucosidase[35]
2,5-Dideoxy-2,5-imino-DL-*glycero*-D-*manno*-heptitol (homoDMDP)	potent inhibitor of β-glucosidases and β-galactosidases (including lactase) and trehalase, moderate inhibitor of other α-glucosidases (including maltase) and α- and β-mannosidases[9,28]
HomoDMDP-7-O-apioside	good inhibitor of β-galactosidases (including lactase) and amyloglucosidase[9]
HomoDMDP-7-O-β-D-xylopyranoside	potent inhibitor of β-glucosidases and β-galactosidases (including lactase) and moderate inhibitor of amyloglucosidase and β-mannosidase[9]
6-Deoxy-homoDMDP	potent inhibitor of α-glucosidases (including maltase)[35]
Broussonetines A and B	not inhibitory[40]
Broussonetinines A and B	potent inhibitors of β-galactosidase and α-mannosidase[40]
Broussonetines C and D	potent inhibitors of β-galactosidase and β-mannosidase[39]
Broussonetines E and F	potent inhibitors of β-glucosidase, β-galactosidase and β-mannosidase and good inhibitors of α-glucosidase[40]
Broussonetines G, H, K and L	potent inhibitors of β-glucosidase, β-galactosidase and β-mannosidase[41,43]
Piperidines	
Nojirimycin	β-glucosidases and α-glucosidases (including sucrase, maltase, isomaltase and amylase), β-galactosidases and N-acetyl-β-D-glucosaminidases[89,103]
1-Deoxynojirimycin (DNJ)	potent inhibitor of α-glucosidases (including trehalase, invertase, sucrase, maltase, isomaltase), more weakly inhibitory to Glucosidase I and II, and β-glucosidases, also α-mannosidases, α-fucosidase, α- and β-galactosidases (including lactase)[31,89,90,102]
1-Deoxynojirimycin-2-O-, 3-O-, 4-O-α-D-glucopyranosides and 2-O-, 6-O-α-D-galactopyranoside and 2-O-, 3-O-, 4-O-, 6-O-β-D-glucopyranoside	varying levels of inhibition of α-glucosidases[15]
N-Methyl-1-deoxynojirimycin	more potent inhibitor of α-glucosidases than DNJ[90]
α-Homonojirimycin (HNJ)	α-glucosidases (including maltase, sucrase and trehalase), β-glucosidases, lactase and α-galactosidases[18,49] and moderate inhibitor of Glucosidase I and II[104]
α-Homonojirimycin-7-O-β-D-glucopyranoside (MDL 25,637)	α-glucosidases (including Glucosidase II and trehalase)[105]
α-4-*Epi*homonojirimycin	Poor inhibitor of α- and β-galactosidases[50,51]
α-1-Deoxy-1-C-methyl-homonojirimycin (Adenophorine)	good α-galactosidase inhibitor and moderate to poor inhibitor of α-glucosidases[36]
Adenophorine-1-O-β-D-glucopyranoside	good trehalase inhibitor but moderate to poor inhibitor of other α-glucosidases[36]

Table 2. *Continued.*

Alkaloid	Glycosidases Inhibited
Piperidines	
1-Deoxyadenophorine	good α-galactosidase inhibitor and moderate to poor inhibitor of α-glucosidases[36]
5-Deoxyadenophorine	good α- and β-galactosidase inhibitor and poor trehalase inhibitor[36]
5-Deoxyadenophorine-1-*O*-β-D-glucopyranoside	Potent inhibitor of α-glucosidases and α-galactosidases[36]
β-1-*C*-butyl-deoxygalactonojirimycin	Potent α-galactosidase inhibitor only[36]
β-Homonojirimycin	weak inhibitor of β-glucosidase[22]
β-4,5-Di*epi*homonojirimycin	Potent inhibitor of α-glucosidases (including maltase and sucrase) and α-galactosidase, moderate inhibitor of α-fucosidase[51]
Nojirimycin B (Mannojirimycin)	α-mannosidases[52]
1-Deoxymannojirimycin (DMJ)	α-mannosidases (Mannosidase I) and α-fucosidase[31,45,106]
α-Homomannojirimycin	weak inhibitor of human liver α-mannosidases[22]
β-Homomannojirimycin	β-mannosidase[22]
Galactostatin	β-galactosidases[56]
Fagomine	weak inhibitor of α-glucosidases (including isomaltase and sucrase)[59]
3-E*pi*fagomine	Isomaltase, β-galactosidases and lactase[59]
3,4-Di*epi*fagomine	not inhibitory[59]
Fagomine-4-*O*-β-D-glucopyranoside	weak inhibitor of β-galactosidase[59,99]
Fagomine-3-*O*-β-D-glucopyranoside	not inhibitory[59]
α-1-*C*-ethyl-fagomine	good β-galactosidase inhibitor and poor inhibitor of α-glucosidases[29]
Pyrrolizidines	
Alexine	Disaccharidase-type α-glucosidases (trehalase, amyloglucosidase) and β-galactosidase[61,64]
3,7a-Di*epi*alexine	Disaccharidase-type α-glucosidases (amyloglucosidase and sucrase)[64]
7a-*Epi*alexine (Australine)	Disaccharidase-type α-glucosidases (amyloglucosidase, sucrase, maltase) and Glucosidase I, β-glucosidase, β-galactosidase[63,104]
1,7a-Di*epi*alexine	α-glucosidases (amyloglucosidase), weakly inhibitory to β-glucosidases[64]
7a-*Epi*alexaflorine	weak inhibitor of amyloglucosidase and sucrase[65]
Casuarine	α-glucosidases (including trehalase, amyloglucosidase and Glucosidase I)[107]
Hyacinthacine A$_1$	Potent inhibitor of lactase, moderate inhibitor of amyloglucosidase and α-fucosidase and a poor inhibitor of α- and β-glucosidases and β-mannosidase[10]
Hyacinthacine A$_2$	Potent inhibitor of amyloglucosidase, good inhibitor of lactase, moderate inhibitor of β-glucosidase and trehalase and weak inhibitor of β-mannosidase[10]
Hyacinthacine A$_3$	good inhibitor of amyloglucosidase and moderate inhibitor of lactase[10]
Hyacinthacine B$_1$	Moderate to poor inhibitor of β-glucosidases and β-galactosidases[9]
Hyacinthacine B$_2$	Moderate to good inhibitor of β-glucosidases and β-galactosidases (including potent inhibition of lactase)[9]
Hyacinthacine B$_3$	good inhibitor of lactase and amyloglucosidase and poor inhibitor of α-fucosidase[10]
Hyacinthacine C$_1$	moderate inhibitor of amyloglucosidase[9]
Broussonetine N	good inhibitor of β-glucosidase, β-galactosidase and β-mannosidase[44]
Indolizidines	
Swainsonine	potent inhibitor of α-mannosidases[108-110]
Lentiginosine	amyloglucosidase[72]
2-*Epi*lentiginosine	not inhibitory[72]
Castanospermine	α-glucosidases (including amyloglucosidase, sucrase, maltase, isomaltase, trehalase, amylase, Glucosidase I and II) and β-glucosidases (including lactase and cellobiase), β-glucocerebrosidase and β-xylosidase[93,94,111]
6-*Epi*castanospermine	amyloglucosidase, neutral α-mannosidase[76,112]
6,7-Di*epi*castanospermine	amyloglucosidase and fungal β-glucosidase[25]
7-Deoxy-6-*epi*-castanospermine	amyloglucosidase and yeast α-glucosidase[78]
Nortropanes	
Calystegine A$_3$	potent inhibitor of β-glucosidases, α- and β-galactosidases (including lactase) and trehalase[82]

Table 2. *Continued.*

Alkaloid	Glycosidases Inhibited
Nortropanes	
Calystegine A_5	not inhibitory[82]
Calystegine A_6	moderate inhibitor of trehalase only[85]
Calystegine A_7	moderate inhibitor of trehalase[60]
Calystegine B_1	potent inhibitor of β-glucosidases (including cellobiase) and β-galactosidase (including lactase) and a weak inhibitor of trehalase[82]
Calystegine B_1-3-O-β-D-glucopyranoside	potent inhibitor of rice α-glucosidase and weak inhibitor of β-glucosidases[113]
Calystegine B_2	potent inhibitor of β-glucosidases, disaccharidase-type α-glucosidases (including maltase, sucrase and trehalase) and α- and β-galactosidase (including lactase)[83]
N-Methyl-calystegine B_2	potent inhibitor of α-galactosidases and a moderate inhibitor of trehalase[60]
Calystegine B_3	weak inhibitor of β-glucosidases, α-galactosidases and trehalase[82,86]
Calystegine B_4	moderate inhibitor of α-galactosidases and trehalase, but weak inhibitor of β-glucosidases (including cellobiase) and lactase[83]
Calystegine C_1	potent inhibitor of disaccharidase-type α-glucosidases and β-glucosidases, moderate inhibitor of α- and β-galactosidases[82,83,86]
N-Methyl-calystegine C_1	weak inhibitor of some α-galactosidases[60]
Calystegine C_2	moderate inhibitor of β-glucosidases, α-mannosidases and trehalase[86]
Calystegine N_1	weak inhibitor of disaccharidase-type α-glucosidases and potent inhibitor of β-glucosidases (including cellobiase), lactase and α-galactosidase[85]

4 MAMMALIAN TOXICITY OF POLYHYDROXYLATED ALKALOIDS

As might be expected for a group of compounds that affect carbohydrate metabolism, the polyhydroxylated alkaloids have been reported to have a wide range of effects on a number of organisms and some cause serious livestock poisonings. In fact, it was the toxicity to livestock of the legumes *Swainsona canescens* and *Castanospermum australe* in Australia that first led to the isolation of the toxic principles swainsonine[66] and castanospermine.[74]

Cattle eating *Swainsona* species developed a syndrome called "pea struck" which is due to decreased α-mannosidase activity resulting in accumulation of mannose-rich oligosaccharides in lysosomes. This results in neuronal vacuolation, axonal dystrophy, loss of cellular function and ultimately death.[114] The disorder "locoism" in the Western United States was also found to be due to swainsonine in locoweeds (*Astragalus* and *Oxytropis* species).[67,115] The clinical symptoms of swainsonine poisoning in livestock on ingestion of plants containing this compound include depression, tremors, nervousness, emaciation, gastrointestinal malfunction and reproductive alterations such as abortion and birth defects.[116] Poisoning of animals in China by *Oxytropis ochrocephala* and *O. kansuensis* has also been shown to be due to the presence of swainsonine.[117]

The concentration of swainsonine detected in all plants implicated in poisonings is extremely low. For example, swainsonine was isolated from the vegetative parts of *A. lentiginosus* in a yield of 0.007 % (dry weight)[67] and the levels detected in *O. ochrocephala* and *O. kansuensis* were 0.012 % and 0.021 % (dry weights), respectively.[117] However, swainsonine can accumulate within the tissues of the body as it has the ability to permeate the plasma membrane freely, but once inside lysosomes it becomes protonated due to the low pH and so becomes concentrated.[118] Therefore, poisoning by swainsonine generally takes several weeks of ingestion before it becomes apparent. If affected animals are denied access to plants containing swainsonine early enough, lysosomal function will return to normal as the alkaloid is excreted from the body via the urine.[119] Although many of the lesions will disappear, some long-term effects may result in reduced animal performance.[116] A threshold of toxicity is, therefore, difficult to establish for swainsonine, but in 1994 Molyneux and his co-workers suggested that a concentration in the diet of as little as 0.001 % should be of concern.[120]

Even before swainsonine was identified as the toxic principle in *Swainsona* species, Dorling and co-workers[114] recognised that the lysosomal storage disorder induced in animals grazing these plants was biochemically and morphologically similar to the rare genetically-determined mannosidosis that occurs in Angus cattle[121] and humans.[122] This disease is characterised by accumulation in cells and excretion in the urine of mannose-rich oligosaccharides which results from a deficiency in all tissues of lysosomal α-D-mannosidase.[123] The condition is ultimately fatal. In infants the clinical syndrome may include mental retardation, central nervous system derangement, abnormal musculoskeletal development, locomotor incoordination, abnormal facial appearance and an increased susceptibility to recurrent infections.[122] However, swainsonine-induced mannosidosis is not an exact model of genetic mannosidosis since this compound inhibits α-mannosidase II in addition to the lysosomal α-mannosidase. Nevertheless, swainsonine has proved a valuable tool in the study of this lysosomal storage disease since it has provided the means of inducing a reversible phenocopy of the genetic disorder in animals and in tissue culture.[124] The use of such models could allow investigation of the viability of enzyme replacement therapy in the treatment of this disease.[125]

Comparison of the sequences of the major oligosaccharides which accumulate and are excreted in the urine of humans suffering from mannosidosis with those excreted in the swainsonine-induced disease has allowed the intracellular substrate specificity of lysosomal α-mannosidase to be defined.[126] In 1983 Cenci di Bello and co-workers first realised that the human lysosomal α-mannosidase actually consists of two enzymes which are specific for different mannose linkages.[124] The major enzyme present catalyses α-1,3-linkages but the other is specific for α-1,6-linkages.[127] In human mannosidosis the α-1,3 activity is absent but the α-1,6 activity is unaffected. Swainsonine inhibits the activity of both enzymes. However, cattle do not possess the α-1,6 form of lysosomal α-mannosidase and so the storage products in the bovine form of genetic mannosidosis bear a greater similarity to those seen in the form of the disease induced by swainsonine.[127]

Castanospermum australe contains the potent α- and β-glucosidase inhibitor castanospermine at a concentration of approximately 0.3 % of the dry weight of the seed.[74] Pigs, cattle and horses have been reported to be poisoned when they have consumed the seeds of *C. australe*. The chief symptom is gastroenteritis, owing to inhibition of the activity of intestinal brush border digestive disaccharidases (e.g. sucrase, maltase and trehalase).[128] The upset can be so severe that the animals can die.[129] Other symptoms include myocardial degeneration and nephrosis.[128] Several cases of human poisoning have also occurred.[130] Histologic changes observed in poisoned livestock and rodent feeding trials include degenerative vacuolation of hepatocytes and skeletal myocytes.[116,128,131] Castanospermine inhibits the activity of lysosomal α-glucosidase which leads to the accumulation of glycogen within the lysosomes.[132] This mimics the situation in the genetically-determined storage disorder glycogenosis type II (Pompe's disease) which is caused by a deficiency of lysosomal α-glucosidase.[133]

However, castanospermine also inhibits the activity of lysosomal β-glucosidase and this affects the catabolism of glycosphingolipids.[134]

Recently, swainsonine has been detected in species of the Convolvulaceae from Australia[70] and Africa.[71] These plants are reported to produce neurological disorders in livestock with clinical symptoms similar to those caused by the swainsonine-containing members of the Leguminosae. However, these species also contain the *nor*tropane alkaloids calystegines which inhibit α- and β-glucosidases and α-galactosidase and so there may be a combination of toxic effects. The calystegines are also common in the Solanaceae and there are several *Solanum* species that can poison livestock. Symptomatically, these cases resemble "locoism".[135] For example, *S. dimidiatum* causes "Crazy Cow Syndrome" in Texas[136] and *S. kwebense* causes "Maldronksiekte" in South Africa.[137] Both disorders are characterised by neurological signs of cerebellar dysfunction, including staggering and incoordination, with severe cellular vacuolation and degeneration of Purkinje cells in the brain. Both plants contain a range of calystegines, including calystegine B_2.[1] Therefore, it seems likely that these syndromes are also lysosomal storage disorders caused by glycosidase inhibition produced by the calystegines.[1]

5 THERAPEUTIC POTENTIAL OF POLYHYDROXYLATED ALKALOIDS

Very few polyhydroxylated alkaloids are available commercially. Those which are available (principally DNJ, DMJ, castanospermine and swainsonine) have become standard reagents used to investigate the potential therapeutic and biochemical applications of this class of glycosidase inhibitor. So most of the following consideration of the therapeutic applications of polyhydroxylated alkaloids is based on a limited number of compounds that have been tested thoroughly, simply because these are the only ones which are readily available. It should also be noted that the doses required for beneficial effects in human disease states are generally below those causing the toxicities described in Section 1.4.

5.1 Anti-Cancer Agents

Both catabolic and glycoprotein processing glycosidases are involved in the transformation of normal cells to cancer cells and in tumour cell invasion and migration. Many tumour cells display aberrant glycosylation due to an altered expression of glycosyltransferases[138] and it has been known for a long time that the levels of glycosidases are elevated in the sera of many patients with different tumours.[139] Secreted glycosidases may be involved in the degradation of the extracellular matrix in tumour cell invasion.[140] Furthermore, the lysosomal system is highly active in transformed cells, presumably reflecting enhanced turnover of glycoproteins and other macromolecules, and possibly increased exocytosis of lysosomal hydrolases.[141] The use of polyhydroxylated alkaloids to prevent the formation of aberrant asparagine-linked oligosaccharides during glycoprotein processing and to inhibit catabolic glycosidases is being actively pursued as a therapeutic strategy for cancer.

Although a number of the polyhydroxylated alkaloids have been reported to show anti-cancer activity, research has concentrated on developing swainsonine as a drug candidate for the management of human malignancies. Swainsonine has a complex mode of action in the whole animal. It inhibits the growth of tumour cells and prevents the dissemination of malignant cells from the primary tumour to secondary sites (a process known as metastasis). However, swainsonine also has a direct stimulatory effect on the immune system. There is a wealth of literature available on the anti-cancer properties of swainsonine, probing various aspects of its activity using a multitude of different rodent and human cell lines and animal models. There follows a summary of the essential findings from this work.

The action of swainsonine to prevent tumour cell proliferation is related to its inhibition of the processing of their asparagine-linked oligosaccharides. Neoplastic transformation in both human and rodent tumour cells is often accompanied by increased expression of β-1,6-branched oligosaccharides in complex-type N-linked glycans, which are apparently required for efficient tumour cell metastasis *in vivo*.[142,143] Swainsonine blocks expression of complex-type oligosaccharides in malignant cells. It would seem that loss or truncation of β-1,6-branched oligosaccharides in metastatic tumour cells has the effect of reducing cell motility by increasing cellular adhesion, it reduces their capacity to invade other tissues and it also reduces solid tumour growth, possibly by decreasing the cellular response to autocrine growth stimulation.[141,142] However, since all cellular glycoproteins which normally express complex-type oligosaccharides are affected by swainsonine treatment, investigations were carried out to determine whether the effects of swainsonine on cellular proliferation could also be widespread amongst non-transformed tissues in the body. These studies showed that the effects of this compound would appear to be cell-specific, since it does not affect the processing of all glycoproteins equally.[141,144,145]

There is considerable evidence that swainsonine enhances the natural anti-tumour defences of the body.[146,147] In animal models, it was observed that the reduction in metastasis of tumour cells induced by swainsonine administration continued for a number of days after the drug was withdrawn.[148] It was discovered that this results from the ability of swainsonine to activate immune effector cells such as natural killer cells (which are peripheral blood lymphocytes), T-lymphocytes and macrophages.[149,150] These phagocytic cells have diverse functions that include antigen presentation, cytokine production (e.g. interleukins), immune surveillance and cytocidal activity against tumour cells. However, many of the properties of these cells are apparent only after they have been activated.[150] Swainsonine stimulates proliferation of spleen cells, thereby increasing the numbers of natural killer cells. It also increases lymphocyte sensitivity to interleukin-2 (IL-2) and other cytokines.[151,152] Studies showed that the extent of activation by swainsonine of peritoneal macrophages to cytotoxicity against tumour cells was comparable to that obtained with known macrophage-activating agents such as interferon and bacterial lipopolysaccharide. It was associated with increased secretion of interleukin-1 (IL-1) by the macrophages, induction of protein kinase C activity and enhanced secretion of the major histocompatibilty antigen on the cell surfaces.[153] Swainsonine can also induce tumouricidal activity in resident tissue-specific macrophages of both the lung and spleen,[150] with activation being both time- and dose-dependent. This is relevant to the clinical management of metastatic diseases since visceral organs are common sites for metastasis formation.

Before swainsonine could be used clinically, the possible adverse effects of this agent had to be evaluated. As previously discussed, the systemic administration of high doses of swainsonine to sheep has been reported to induce a neural lysosomal mannoside storage disease.[154] However, no evidence of an overt toxic reaction was observed when swainsonine was administered orally to rodents[155] so it was considered that the mannosidosis induced by swainsonine could be a species- or tissue-specific phenomenon.[156] Swainsonine is both water and lipid soluble and therefore, it diffuses efficiently into tissues. In tissue culture swainsonine has been shown to attain concentrations inside cells similar to that in the culture medium within minutes.[118] In 1989 Mohla and co-workers compared the fates of swainsonine administered to mice by different routes (orally, intravenously, intraperitoneally and subcutaneously).[157] They found that in all cases the compound was cleared rapidly from the blood. Four years later the same team published a preliminary pharmacokinetic evaluation of swainsonine in mice after intravenous administration.[158] They injected one dose of 3 µg/ml which was rapidly cleared from the blood with a half-life of 31.6 minutes. The swainsonine rapidly entered various body organs and tissues from the blood. The highest levels were found in the bladder (owing to urinary excretion of the compound), kidney and thymus, but the lowest levels were found in the brain, which suggested that toxicity to the central nervous system may not be a problem at swainsonine levels which are sufficient to prevent metastasis. In a follow-up study swainsonine levels in various tissues were compared for up to three days after discontinuing oral administration of this compound to mice in their drinking water.[159] These researchers found that swainsonine was predominantly retained for at least 72 hours in lymphoid tissue (spleen and thymus) which was consistent with the sustained immunomodulatory and anti-metastatic properties of this compound.

The results from the tissue culture and animal models were so encouraging that the efficacy of swainsonine was examined in phase I clinical trials in humans with advanced malignancies.[143,160,161] In the first study the subjects were nineteen terminally ill cancer patients with either leukemia or breast, colon, lung, pancreatic, or head and neck cancers. Swainsonine was administered by continuous intravenous infusion over 5 days in the dose range 50-550 µg/kg body weight/day. The serum concentrations of this drug reached 3-11.8 mg/l which is 100-400 times greater than the 50 % inhibitory concentration for Golgi α-mannosidase II and lysosomal α-mannosidases.[143] The patients had elevated levels of the liver enzyme aspartate aminotransferase in the serum that was indicative of hepatocyte damage and they also suffered from pulmonary and peripheral oedema.[160] However, the patient with head and neck cancer had over 50 % tumour remission and two others showed symptomatic improvement. A second clinical trial was then undertaken by the same researchers to evaluate swainsonine by oral administration biweekly (dose range 50-600 µg/kg body weight) to sixteen patients with advanced malignancies.[161] Serum drug levels peaked 3-4 hours following a single oral dose. The maximum tolerated dose was found to be 150 µg/kg body weight/day due to increased levels of aspartate aminotransferase in the serum and oedema, similar to that seen in the previous trial. Other adverse effects included fatigue, anorexia, abdominal pain and neurological symptoms. It was also found in both studies that the immune system was depressed rather than stimulated and it was thought that this was due to the high doses administered. Interestingly, recent observations in cases of induced chronic locoweed poisoning in cattle similarly indicate that the immune response is depressed by prolonged exposure to high doses of swainsonine.[162] Further clinical trials were suggested by Goss and co-workers in 1997[161] to investigate alternative dosing schedules with lower starting doses in order to determine whether the toxic effects of swainsonine can be overcome. Indeed, swainsonine is currently undergoing phase II clinical trials in Canada with apparently very promising results (Dr. Jeremy Carver, CEO of GLYCODesign Inc., personal communication).

In 1993 Dennis and co-workers synthesised derivatives of swainsonine with lipophilic groups attached to either position 2 or 8.[163] The 2-carbonoyloxy esters of swainsonine were relatively poor inhibitors of α-mannosidases *in vitro* but entered cells at a rate comparable to swainsonine where they were converted to swainsonine by intracellular esterases. *In vivo*, the analogues were found to have comparable activities to swainsonine as stimulators of bone marrow cell proliferation. This has led to the suggestion that these or similar analogues could be useful as pro-drugs which could be preferentially hydrolysed to release swainsonine only once they were inside tumour and/or lymphoid cells. As such they could be expected to have improved pharmacological properties and reduced side-effects.[163]

In addition to swainsonine, some of the other naturally-occurring polyhydroxylated alkaloids listed in Table 1 also show various anti-cancer properties. However, none of these compounds have been exploited to nearly the same extent as swainsonine. For example, the potent α-glucosidase inhibitor castanospermine (also an indolizidine alkaloid) has been reported to inhibit experimental metastasis in mice.[164]

Various isoenzymes of β-*N*-acetylglucosaminidase have also been investigated as potential targets for cancer therapy due to their altered expression in a range of human cancer cell types and their potential involvement in the degradation of the extracellular matrix in tumour cell invasion.[165,166] β-*N*-acetylglucosaminidases release *N*-acetyl-D-glucosamine from glycoproteins and a number of inhibitors of these enzymes have been chemically synthesised, such as 2-acetamido-1,5-imino-1,2,5-trideoxy-D-glucitol[167] and a synthetic derivative of castanospermine (6-acetamido-6-deoxycastanospermine).[168] However, potent natural product inhibitors of β-*N*-acetylglucosaminidase have also been isolated from microbial sources. These include nagstatin from *Streptomyces amakusaensis*, strain MG846-fF3[169] which is a nitrogenous *N*-acetyl-D-glucosamine analogue fused with an imidazole ring. Synthetic derivatives of nagstatin have subsequently been produced with enhanced activity.[170] Other inhibitors of β-*N*-acetylglucosaminidase are pyrostatins A and B (4-hydroxy-2-imino-1-methylpyrrolidine-5-carboxylic acid and 2-imino-1-methylpyrrolidine-5-carboxylic acid, respectively) which were isolated from filtrates of *Streptomyces* sp., strain SA-3501.[171]

5.2 Immune Stimulants

The effect of swainsonine on bone marrow proliferation was initially investigated because its potential use in cancer therapy required that it have minimal toxicity to bone marrow and other normal tissues. In fact, a frequent limitation of the suitability of drug candidates as chemotherapeutic agents is their toxicity to normal tissues, especially bone marrow.[172,173] It was found that swainsonine stimulated cell proliferation of murine bone marrow following haematological injuries although it did not stimulate healthy bone marrow progenitor cells *in vivo* when administered alone.[141,149,174] The possibility that swainsonine could confer protection against the cytotoxic effects of both cell cycle-specific and -nonspecific cytotoxic anticancer agents was first examined in a murine model system.[175] The results indicated that the intraperitoneal administration of swainsonine decreased the lethality of methotrexate, 5-fluorouracil, cyclophosphamide and doxorubicin in non-tumour-bearing mice. However, these responses were critically dependent on the dose, sequence and timing of swainsonine administration. In 1999, Klein and co-workers[174] reported that swainsonine protected the bone marrow cells of mice from the toxic effects of cyclophosphamide without interfering with the drug's ability to inhibit tumour growth. Also, swainsonine did not stimulate proliferation of haematopoietic tumour cells which would suggest that this drug could still be used in patients with this type of cancer.

Klein *et al.*[174] also assessed the protective properties of swainsonine *in vivo* with the myelosuppressive agent 3'-azido-3'-deoxythymidine (known as Zidovudine or AZT), which is often used in therapy for the acquired immune deficiency syndrome (AIDS). Swainsonine administered by intraperitoneal injection increased both total bone marrow cellularity and the number of circulating white blood cells in mice treated with doses of AZT that typically lead to severe myelosuppression, which is the major dose-limiting feature in chemotherapeutic regimens for AIDS using this drug.[176] In addition, swainsonine protected human myeloid progenitor cells from AZT toxicity *in vitro*.[174]

Therefore, the possibility exists that swainsonine could be used to accelerate the recovery of bone marrow cellularity and competence following high-dose chemotherapy or autologous bone marrow transplantation, or used as an adjuvant during AZT treatment. Thus, the complications and increased risk of opportunistic infections associated with the prolonged immune-deficient state in these clinical conditions could be minimised and the cure rates of cytoreductive treatment could be significantly improved. This promising therapeutic application of swainsonine is currently being pursued in the USA in phase I and II clinical trials by the Canadian company GLYCODesign Inc., under the product code GD0039.

Finally, it is worth mentioning that the pyrroline alkaloid nectrisine (Figure 2), which inhibits α-glucosidases in addition to α-mannosidases, has been reported to restore the immune response of immunosuppressed mice.[98]

5.3 Anti-Diabetic Agents

It can be seen from Table 2 that a large number of the naturally-occurring polyhydroxylated alkaloids are potent inhibitors of the various α-glucosidase-specific disaccharidases involved in mammalian digestion (e.g. sucrase, maltase, isomaltase, etc.,). These enzymes are expressed at the surface of the epithelial cells of the brush border in the small intestine. In the late 1970s it was realised that inhibitors of these enzymes, such as DNJ, could be used therapeutically in the oral treatment of the non-insulin-dependent (type II) diabetes mellitus.[4,46]

It was found that the activity of 1-deoxynojirimycin (DNJ) *in vivo* against intestinal sucrase was lower than that seen *in vitro* and this initiated a synthetic programme to produce derivatives with enhanced activity. The *N*-alkyl derivatives were most effective and this led to the development of *N*-hydroxyethyl-deoxynojirimycin (known as Miglitol or BAY m-1099) as a drug candidate.[177] This derivative has improved retention in the small intestine which increases its potential *in vivo* antidiabetic activity. Miglitol does not have hypoglycemic activity like insulin and the sulfonylureas. Instead, by inhibiting the breakdown of complex carbohydrates in the gut it reduces glucose uptake and so reduces the postprandial rise in blood glucose which is characteristic of diabetes.[178] However, Miglitol is absorbed appreciably from the gut into the bloodstream and additional effects of this compound have been observed. It has been reported to reduce postprandial insulin secretion, lessen diabetic glucosuria and reduce the carbohydrate-driven synthesis of very low density lipoproteins.[179] The 7-*O*-β-D-glucopyranosyl derivative of α-homonojirimycin (MDL 25637) is another drug candidate which acts in the same manner as Miglitol.[17,180] This was synthesised prior to its discovery as a natural product.[18]

A naturally-occurring glycoside of DNJ (2-*O*-α-D-galactopyranosyl-deoxynojirimycin) and fagomine (which can be considered to be 2-deoxy-DNJ) have both recently been shown to have potent anti-hyperglycemic effects in isolated perfused pancreases from streptozotocin-induced diabetic mice.[181,182] Fagomine has the ability to potentiate glucose-induced insulin secretion from isolated rat pancreatic islets.[183] Although the exact mechanism of action of this compound has not yet been elucidated, it would appear that fagomine does not affect the enzymes in the citric acid cycle, but rather it accelerates the steps in the glycolytic pathway that occur after the formation of glyceraldehyde 3-phosphate.[183]

Castanospermine can be regarded as a bicyclic derivative of DNJ, with an ethylene bridge between the hydroxymethyl group and the ring nitrogen. However, owing to its toxicity, this compound was considered unsuitable for therapeutic use in diabetes mellitus. Unsuccessful attempts were made to synthesise derivatives of castanospermine that were better tolerated.[177]

5.4 Anti-Viral Agents

Another therapeutic application of polyhydroxylated alkaloids is as anti-viral agents. Inhibitors of processing α-glucosidases, such as castanospermine and DNJ, have been shown to decrease the infectivity of human immunodeficiency virus (HIV) *in vitro* at concentrations which are not cytotoxic to lymphocytes, whereas specific inhibitors of processing α-mannosidases (swainsonine and 1-deoxymannojirimycin) have no effect on HIV.[184-186] Castanospermine and DNJ also reduce the infectivity of other retroviruses such as the feline equivalent of HIV[187] and human cytomegalovirus (CMV) which is an opportunistic pathogen in AIDS which is caused by HIV infection.[188]

HIV primarily infects cells of the immune system. Essential to infection is the interaction between the heavily glycosylated viral envelope glycoproteins gp120 and gp41 with the CD4 receptor which is a membrane glycoprotein found on the surface of T-lymphocytes and other cells of the immune system. In the presence of castanospermine and DNJ the glycosylation patterns of the viral coat glycoproteins are altered. Although this does not prevent formation of viral particles, they no longer have the ability to interact correctly with the CD4 receptor and so they are non-infectious.[186,189]

Trials were conducted to try to define the structural features necessary in polyhydroxylated alkaloids for antiviral activity.[185,190] It was discovered that *N*-substituted derivatives of DNJ were more potent *in vivo* than the natural product. In particular *N*-butyl-DNJ (SC-48334) had enhanced anti-HIV activity (possibly because the aliphatic chain increased uptake by the cells) and this compound was developed as a drug candidate and was evaluated in phase II clinical trials. In combination with dideoxynucleoside derivatives which target the viral reverse transcriptase activity (e.g. AZT), *N*-butyl-DNJ caused diarrhoea, abdominal pain and weight loss in human patients after oral administration.[191] Chemical modifications of *N*-butyl-DNJ were undertaken to try to eliminate the gastrointestinal toxicity. A perbutyrylated ester (SC-49483) was reported in 1995 to be undergoing a phase II clinical trial.[178] This is effectively a pro-drug as it was shown to be absorbed from the gut and metabolised in the gastrointestinal mucosa to release the active agent *N*-butyl-DNJ into the plasma. Although the anti-viral efficacy of this compound in humans has not been reported, it apparently did not cause diarrhoea in Rhesus monkeys.[178]

The mechanism of action of *N*-butyl-DNJ at the molecular level appears to involve the inhibition of the host cell enzymes α-glucosidases I and II. This arrests *N*-linked glycan processing such that immature glucosylated *N*-glycans are present on the HIV envelope glycoproteins gp120 and gp41. Conformational changes occur in the V1 and V2 loops of gp120 when it folds in the presence of these glucosylated *N*-glycans. As a consequence, the virus can still bind to its cellular receptor CD4 but gp120 is unable to undergo the post-CD4-binding conformational change required to expose the fusogenic peptide of gp41 that normally facilitates entry of the virus into the host cell.[189,192,193] However, because glucosidases I and II are located in the lumen of the endoplasmic reticulum, it was found that very high extracellular concentrations of *N*-butyl-DNJ were required to achieve enzyme inhibition. It was reported that although *N*-butyl-DNJ had a K_i of 0.2 μM against purified glucosidase I *in vitro*, extracellular concentrations of 0.5 mM were required to achieve inhibition of glucosidase I in intact cells in tissue culture.[194] When *N*-butyl-DNJ was evaluated in HIV patients, the serum levels achieved were only in the range of 10-50 μM which could explain the lower than anticipated antiviral efficacy of the drug that was observed during the clinical trials.[191,195] Recently, the pharmacokinetic behaviour of a more lipophilic derivative of DNJ (*N*-benzyl-DNJ) has been studied in rats.[196] The plasma half life and tissue residency of this derivative were reported to be greater than those of the *N*-alkyl derivatives.

Similarly, synthetic modification of castanospermine showed that the lipophilic 6-*O*-acyl derivatives were more potent inhibitors of HIV than the natural product.[197,198] In particular, the 6-*O*-butanoyl derivative (MDL 28,574) was approximately twenty times more active than castanospermine and fifty times more active than *N*-butyl-DNJ.[186] However, once inside cells, 6-*O*-butanoylcastanospermine appeared to be hydrolysed to release castanospermine and hence this compound can also be considered to be a pro-drug.[199,200] Recently, this compound was reported to be tolerated well by patients during a phase II clinical trial.[201] Studies *in vitro* have shown synergistic activity against HIV type 1 and 2 replication when castanospermine (and its 6-*O*-butanoyl derivative) are combined with AZT and other similar dideoxynucleoside drugs.[202,203]

Nectrisine (an inhibitor of both α-glucosidases and α-mannosidases) has also been reported to have anti-viral activity. It has been shown to inhibit the retrovirus Friend leukaemia virus *in vivo* in mice and it too can potentiate the activity of AZT.[204]

5.5 Treatment of Glycosphingolipid Lysosomal Storage Diseases

The glycosphingolipid (GLS) lysosomal storage diseases result from mutations in the genes that encode the enzymes required for GLS catabolism within lysosomes. They are a relatively rare group of human disorders that can give rise to progressive neurodegeneration due to lysosomal storage within cells of the central nervous system. In the most severe forms of these diseases, death occurs in early infancy.[195] An enzyme replacement therapy is currently available for type 1 Gaucher disease (caused by a defect in β-glucocerebrosidase)[205] but there are diseases associated with virtually every enzyme in the glycosphingolipid degradation pathway. However, a relatively new approach to the generic treatment of this family of disorders relies on the fact that the mutations giving rise to the enzyme defects generally do not totally eliminate the catalytic activity of the enzymes. Therefore, drugs could be used to regulate the rate of biosynthesis of the glycosphingolipids so that the amount of substrate the defective enzyme has to catabolise is reduced to a level that matches the residual enzyme activity, thus balancing synthesis with degradation and preventing storage. This type of treatment is termed "substrate deprivation".

The first step in GLS biosynthesis is the transfer of glucose from the sugar nucleotide uridine diphosphoglucose (UDP-glucose) to ceramide which is catalysed by the enzyme UDP-glucose: *N*-acetylsphingosine D-glucosyltransferase [EC 2.4.1.80]. In 1994 Platt and co-workers[206] discovered that this enzyme is inhibited by the α-glucosidase inhibitor *N*-butyl-DNJ (K_i 7.4 μM). Subsequently, a galactose analogue of *N*-butyl-DNJ (*N*-butyl-deoxy-galactonojirimycin or *N*-butyl-DGJ) was found to be a more selective inhibitor of this process *in vitro*.[207] These compounds act as competitive inhibitors for ceramide in the glucosyltransferase assay and as non-competitive inhibitors for UDP-glucose, indicating that inhibitory activity is by ceramide mimicry. Therefore, the potential GSL lysosomal storage diseases that could be targets for treatment using *N*-butyl-DNJ and *N*-butyl-DGJ are Gaucher's, Fabry, Tay-Sachs, Sandhoff, ceramide lactoside lipidosis and fucosidosis.[195] In 1997 Platt *et al.*[208] reported successfully preventing GSL storage in a mouse model of Tay-Sachs disease using *N*-butyl-DNJ.

When considering the potential clinical application of *N*-butyl-DNJ for the treatment of GLS lysosomal storage diseases, it is worth considering the findings of the clinical trials of this compound as an anti-HIV agent. In Section 5.4. it was reported that the serum levels of the compound attained after oral dosing were insufficient for inhibition of Glucosidase I because the drug could not reach its target due to the location of the catalytic site of the enzyme within the lumen of the endoplasmic reticulum.[194] However, the glucosyltransferase inhibited by *N*-butyl-DNJ is known to have its catalytic site orientated towards the cytosolic face of an early Golgi compartment which would make this enzyme target more accessible to the drug.[209] Also, Platt and Butters[195] considered that the serum levels reported for *N*-butyl-DNJ from the clinical trials would be sufficient to inhibit GLS biosynthesis. Since *N*-butyl-DNJ was reasonably well tolerated during the HIV clinical trials, the clinical evaluation of this compound for the treatment of GLS lysosomal storage diseases is planned.[195]

An alternative approach involving the treatment of Fabry disease with 1-deoxy-galactonojirimycin (DGJ) itself has recently been proposed.[210] Fabry disease is characterised by a deficiency in lysosomal α-galactosidase A. In some forms of the disease this deficiency arises from a genetic mutation that affects the folding of the enzyme, although the catalytic site is unaffected. Incomplete folding is recognised by the "quality control" system that operates in the endoplasmic reticulum resulting in abortive maturation of the mutant enzyme.[211] DGJ is a potent competitive inhibitor of α-galactosidase A. However, when lymphoblasts from Fabry patients were exposed to this compound at concentrations below those normally required for intracellular inhibition of α-galactosidase A, DGJ seemed to accelerate maturation of the mutant enzyme which was successfully transported from the endoplasmic reticulum to the Golgi apparatus and then correctly targeted to the lysosome, resulting in an elevated expression of enzyme activity. This phenomenon was also seen after oral administration of DGJ to transgenic mice that over-expressed the same mutant form of α-galactosidase A. Fan and co-workers[210] proposed that this may be caused by DGJ occupying the catalytic site of the defective enzyme, thereby stabilising its conformation which might, in turn, control the flexibility of folding of the protein which would allow resumption of the glycoprotein processing and lead to maturation of the enzyme. Therefore, DGJ acts as a "chemical chaperon" to force the mutant enzyme to assume the correct conformation in order to complete its biosynthesis. Fan *et al.*[210] also suggested that the binding of DGJ to α-galactosidase A could be pH-dependent such that the formation of the enzyme-inhibitor complex is favoured by the near neutral pH of the endoplasmic reticulum and dissociation occurs in the acidic environment of the lysosome. These researchers concluded that it may be possible to use other competitive inhibitors as "chemical chaperons" for the defective enzymes involved in other types of GSL lysosomal storage disorders.[210]

5.6 Other Therapeutic Applications

Castanospermine has been investigated as a potential therapeutic agent in the treatment of autoimmune diseases such as multiple sclerosis and arthritis. Castanospermine not only inhibited the development of arthritis in mice but it also inhibited the progression of the disease when treatment was commenced after the onset of symptoms.[212] It prevented the passage of leucocytes through vascular subendothelial basement membranes to sites of inflammation by interfering with the biosynthesis of *N*-linked oligosaccharide hydrolases and their subsequent targeting to lysosomes. These enzymes are expressed at the surface of leucocytes and their activity is necessary for cell migration.[213]

Castanospermine may also have a role, one day, in transplant surgery, as it can inhibit the rejection of transplanted tissues. Kidney, pancreas and heart allograft survival have all been shown to be prolonged in a dose-dependent manner in the presence of castanospermine.[214-216] The compound appears to modify the expression of adhesion molecules and other cell surface glycoprotein receptors that may be involved in the alloreactive response in allograft recipients. It downregulates the expression of genes encoding the glycoproteins CD54 and CD11a, which are both considered to have unique and vital roles in the rejection process. This downregulation may act to modulate the rejection process by influencing the interaction between endothelial cells and graft-infiltrating cells, reducing cell migration into the allograft.[217,218]

6 FUTURE PROSPECTS

One promising new area of research is the potential use of synthetic derivatives of polyhydroxylated alkaloids to specifically target the enzymes involved in the biosynthesis of the cell walls of human-pathogenic microorganisms. For example, recent studies on the structure of mycobacterial cell walls have identified a disaccharide linker between the arabinogalactan polysaccharide and the peptidoglycan that contains L-rhamnopyranose.[219] This is a common feature amongst the actinomycetes.[220] Rhamnose has no role in mammalian metabolism so compounds which interfere specifically with rhamnose metabolism should not have any deleterious effects on the host animals and this could provide a new approach to the treatment of diseases induced by mycobacteria, such as tuberculosis and leprosy, as well as diseases caused by other pathogens which also utilise rhamnose as part of their cell wall structure. The L-rhamnopyranose unit is introduced into the cell wall via the sugar nucleotide deoxythymidine diphosphorhamnose (dTDP-L-rhamnose) and it is possible that a chemotherapeutic approach to the treatment of these diseases could be to find compounds which inhibit either the microbial biosynthesis of dTDP-L-rhamnose from D-glucose-1-phosphate (a multi-enzyme conversion)[221] or its subsequent incorporation into the cell wall via rhamnosyl transferases. It is reasonable to suggest that the types of structures which might inhibit these biochemical steps should contain a rhamnose fragment or a rhamnose analogue. Recently, a wide range of synthetic iminosugar derivatives have been evaluated *in vitro* against the L-rhamnosidase naringinase from *Penicillium decumbens*[222,223] and some have also been shown to inhibit the conversion of dTDP-D-glucose to dTDP-L-rhamnose in cell-free extracts of *Mycobacterium smegmatis*.[223] Initial findings indicate the most promising candidate compounds are piperidine analogues of L-rhamnopyranose bearing amide substituents.

A similar line of investigation involves alkaloidal mimics of galactofuranose. The arabinogalactan polysaccharide in the cell walls of actinomycetes is composed of the furanose forms of both arabinose and galactose.[220] Uridine diphosphogalactofuranose (UDP-Gal*f*) is formed by the contraction of UDP-galactopyranose in a reaction catalysed by UDP-galactosyl mutase and the UDP-Gal*f* is then incorporated into the cell wall via a galactosyltransferase. Since galactofuranose has no role in mammalian metabolism, specific inhibition of either UDP-Gal mutase or the UDP-Gal*f* transferase by galactofuranose mimics may well be achieved without any harm to the mammalian hosts. *In vitro* studies have recently identified pyrrolidine analogues of galactofuranose that inhibit the biosynthesis of mycobacterial cell walls, probably through their action on the UDP-galactosyl mutase.[224]

There is no doubt that there are many more natural biologically active polar glycosidase inhibitors waiting to be discovered including glycosides and other derivatives of those known already. Some of these, or semi-synthetic derivatives, are likely also to be inhibitors of endoglycosidases as well as previously untested glycosyltransferases, which opens up a wider range of potential specific applications.

References

1. R.J. Nash, M. Rothschild, E.A. Porter, A.A. Watson, R.D. Waigh and P.G. Waterman, *Phytochemistry*, 1993, **34**, 1281.
2. S. Inouye, T. Tsurouka and T. Niida, *J. Antibiot. Ser. A*, 1966, **19**, 288.
3. S. Inouye, T. Tsurouka, T. Ito and T. Niida, *Tetrahedron*, 1968, **24**, 2125.
4. M. Yagi, T. Kouno, Y. Aoyagi and H. Murai, *Nippon Nogeikagaku Kaishi*, 1976, **50**, 571.
5. S. Murao and S. Miyata, *Agric. Biol. Chem.*, 1980, **44**, 219.
6. S. Watanabe, H. Kato, K. Nagayama and H. Abe, *Biosci. Biotechnol. Biochem.*, 1995, **59**, 936.
7. R.J. Nash, P.I. Thomas, R.D. Waigh, G.W.J. Fleet, M.R. Wormald, P.M. de Q. Lilley and D.J. Watkin, *Tetrahedron Lett.*, 1994, **35**, 7849.
8. M.R. Wormald, R.J. Nash, A.A. Watson, B.K. Bhadoria, R. Langford, M. Sims and G.W.J. Fleet, *Carbohydr. Lett.*, 1996, **2**, 169.
9. A. Kato, I. Adachi, M. Miyauchi, K. Ikeda, T. Komae, H. Kizu, Y. Kameda, A.A. Watson, R.J. Nash, M.R. Wormald, G.W.J. Fleet and N. Asano, *Carbohydr. Res.*, 1999, **316**, 95.
10. N. Asano, H. Kuroi, K. Ikeda, H. Kizu, Y. Kameda, A. Kato, I. Adachi, A.A. Watson, R.J. Nash and G.W.J Fleet, *Tetrahedron Asymm.*, 2000, **11**, 1.
11. B. Dräger, A. van Almsick and G. Mrachatz, *Planta Med.*, 1995, **61**, 577.
12. R.C. Griffiths. PhD Thesis, "Polyhydroxylated Alkaloids and their Ability to Inhibit Glycosidases", 1998, University of Wales, U.K.
13. T. Schimming, B. Tofern, P. Mann, A. Richter, K. Jenett-Siems, B. Dräger, N. Asano, M.P. Gupta, M.D. Correa and E. Eich, *Phytochemistry*, 1998, **49**, 1989.

14. N. Asano, E. Tomioka, H. Kizu and K. Matsui, *Carbohydr. Res.*, 1994, **253**, 235.

15. N. Asano, K. Oseki, E. Tomioka, H. Kizu and K. Matsui, *Carbohydr. Res.*, 1994, **259**, 243.

16. R.J. Nash, A.A. Watson and N. Asano. "Alkaloids: Chemical and Biological Perspectives", Ed. S.W. Pelletier, Elsevier Science Ltd., Oxford, U.K. 1996, Vol. 11, Chapter 5, p. 345.

17. P.S. Liu, *J. Org. Chem.*, 1987, **52**, 4717.

18. G.C. Kite, J.M. Horn, J.T. Romeo, L.E. Fellows, D.C. Lees, A.M. Scofield and N.G. Smith, *Phytochemistry*, 1990, **29**, 103.

19. K.E. Holt, F.J. Leeper and S. Handa, *J. Chem. Soc. Perkin Trans.*, 1994, **1**, 231.

20. O.R. Martin and O.M. Saavedra, *Tetrahedron Lett.*, 1995, **36**, 799.

21. I. Bruce, G.W.J. Fleet, I. Cenci di Bello and B. Winchester, *Tetrahedron*, 1992, **48**, 10191.

22. N. Asano, M. Nishida, H. Kizu, K. Matsui, A.A. Watson and R.J. Nash, *J. Nat. Prod.*, 1997, **60**, 98.

23. T. Shibata, O. Nakayama, Y. Tsurumi, M. Okuhara, H. Terano and M. Kohsaka, *J. Antibiot.*, 1988, **41**, 296.

24. R.J. Nash, E.A. Bell, G.W.J. Fleet, R.H. Jones and J.M. Williams, *J. Chem. Soc., Chem. Commun.*, 1985, **11**, 738.

25. R.J. Molyneux, Y.T. Pan, J.E. Tropea, M. Benson, G.P. Kaushal and A.D. Elbein, *Biochemistry*, 1991, **30**, 9981.

26. D.W.C. Jones, R.J. Nash, E.A. Bell and J.M. Williams, *Tetrahedron Lett.*, 1985, **26**, 3125.

27. J. Furukawa, S. Okuda, K. Saito and S.I-. Hatanaka, *Phytochemistry*, 1985, **24**, 593.

28. A.A. Watson, R.J. Nash, M.R. Wormald, D.J. Harvey, S. Dealler, E. Lees, N. Asano, H. Kizu, A. Kato, R.C. Griffiths, A.J. Cairns and G.W.J. Fleet, *Phytochemistry*, 1997, **46**, 255.

29. N. Asano, M. Nishida, M. Miyauchi, K. Ikeda, M. Yamamoto, H. Kizu, Y. Kameda, A.A. Watson, R.J. Nash and G.W.J. Fleet, *Phytochemistry*, 2000, **53**, 379.

30. A. Welter, J. Jadot, G. Dardenne, M. Marlier, and J. Casimir, *Phytochemistry*, 1976, **15**, 747.

31. S.V. Evans, L.E. Fellows, T.K.M. Shing and G.W.J. Fleet, *Phytochemistry*, 1985, **24**, 1953.

32. J.M. Horn, D.C. Lees, N.G. Smith, R.J. Nash, L.E. Fellows and E.A. Bell. "6th International Symposium on Insect - Plant Relationships", Eds. V. Labeyrie, G. Fabres and D. Lachaise, Dr. W. Junk Publishers, Dordrecht, Netherlands. 1987, p. 394.

33. J.V. Dring, G.C. Kite, R.J. Nash and T. Reynolds, *Bot. J. Linn. Soc.*, 1995, **117**, 1.

34. R.J. Nash, A.A. Watson, A.L. Winters, G.W.J. Fleet, M.R. Wormald, S. Dealler, E. Lees, N. Asano and H. Kizu. "Phytochemical Diversity: A Source of New Industrial Products" Eds. S. Wrigley, M. Hayes, R. Thomas and E. Chrystal, Royal Society of Chemistry, Cambridge, U.K. 1997, p. 106.

35. N. Asano, A. Kato, M. Miyauchi, H. Kizu, Y. Kameda, A.A. Watson, R.J. Nash and G.W.J. Fleet, *J. Nat. Prod.*, 1998, **61**, 625.

36. K. Ikeda, M. Takahashi, M. Nishida, M. Miyauchi, H. Kizu, Y. Kameda, M. Arisawa, A.A. Watson, R.J. Nash, G.W.J. Fleet and N. Asano, *Carbohydr. Res.*, 2000, **323**, 73.

37. R.J. Molyneux, Y.T. Pan, J.E. Tropea, A.D. Elbein, C.H. Lawyer, D.J. Hughes, and G.W.J. Fleet, *J. Nat. Prod.*, 1993, **56**, 1356.

38. K. Tsuchiya, S. Kobayashi, T. Harada, T. Kurokawa, T. Nakagawa, N. Shimada and K. Kobayashi, *J. Antibiot.*, 1995, **48**, 626.

39. M. Shibano, S. Kitagawa and G. Kusano, *Chem. Pharm. Bull.*, 1997, **45**, 505.

40. M. Shibano, S. Kitagawa, S. Nakamura, N. Akazawa and G. Kusano, *Chem. Pharm. Bull.*, 1997, **45**, 700.

41. M. Shibano, S. Nakamura, N. Akazawa and G. Kusano, *Chem. Pharm. Bull.*, 1998, **46**, 1048.

42. M. Shibano, S. Nakamura, M. Kubori, K. Minoura and G. Kusano, *Chem. Pharm. Bull.*, 1998, **46**, 1416.

43. M. Shibano, S. Nakamura, N. Motoya and G. Kusano, *Chem. Pharm. Bull.*, 1999, **47**, 472.

44. M. Shibano, D. Tsukamoto and G. Kusano, *Chem. Pharm. Bull.*, 1999, **47**, 907.

45. N. Ishida, K. Kumagai, T. Niida, T. Tsuruoka and H. Yumoto, *J. Antibiot. Ser. A*, 1967, **20**, 66.

46. D.D. Schmidt, W. Frommer, L. Müller and E. Truscheit, *Naturwissenschaften*, 1979, **66**, 584.

47. G.C. Kite, L.E. Fellows, D.C. Lees, D. Kitchen and G.B. Monteith, *Biochem. Syst. Ecol.*, 1991, **19**, 441.

48. Y. Ezure, *Agric. Biol. Chem.*, 1985, **49**, 2159.

49. G.C. Kite, L.E. Fellows, G.W.J. Fleet, P.S. Liu, A.M. Scofield and N.G. Smith, *Tetrahedron Lett.*, 1988, **29**, 6483.

50. O.R. Martin, P. Compain, H. Kizu and N. Asano, *Bioorg. Med. Chem. Lett.*, 1999, **9**, 3171.

51. N. Asano, M. Nishida, A. Kato, H. Kizu, K. Matsui, Y. Shimada, T. Itoh, M. Baba, A.A. Watson, R.J. Nash, P.M. de Q. Lilley, D.J. Watkin and G.W.J. Fleet, *J. Med. Chem.*, 1998, **41**, 2565.

52. T. Niwa, T. Tsuruoka, H. Goi, Y. Kodama, J. Itoh, S. Inouye, Y. Yamada, T. Niida, M. Nobe and Y. Ogawa, *J. Antibiot.*, 1984, **37**, 1579.

53. L.E. Fellows, A. Bell, D.G. Lynn, F. Pilkiewicz, I. Miura and K. Nakanishi, *J. Chem. Soc. Chem. Commun.*, 1979, **574**, 977.

54. Y. Ezure, N. Ojima, K. Konno, K. Miyazaki, N. Yamada, M. Sugiyama, M. Itoh and T. Nakamura, *J. Antibiot.*, 1988, **41**, 1142.

55. N. Asano, K. Oseki, H. Kizu and K. Matsui, *J. Med. Chem.*, 1994, **37**, 3701.

56. Y. Miyake and M. Ebata, *Agric. Biol. Chem.*, 1988, **52**, 661.

57. M. Koyama and S. Sakamura *Agric. Biol. Chem.*, 1974, **38**, 1111.

58. S.V. Evans, A.R. Hayman, L.E. Fellows, T.K.M. Shing, A.E. Derome and G.W.J. Fleet, *Tetrahedron Lett.*, 1985, **26**, 1465.

59. A. Kato, N. Asano, H. Kizu, K. Matsui, A.A. Watson and R.J. Nash, *J. Nat. Prod.*, 1997, **60**, 312.

60. N. Asano, A. Kato, M. Miyauchi, H. Kizu, T. Tomimori, K. Matsui, R.J. Nash and R.J. Molyneux, *Eur. J. Biochem.*, 1997, **248**, 296.

61. R.J. Nash, L.E. Fellows, J.V. Dring, G.W.J. Fleet, A.E. Derome, T.A. Hamor, A.M. Scofield, and D.J. Watkin, *Tetrahedron Lett.*, 1988, **29**, 2487.

62. R.J. Nash, L.E. Fellows, A.C. Plant, G.W.J. Fleet, A.E. Derome, P.D. Baird, M.P. Hegarty and A.M. Scofield, *Tetrahedron*, 1988, **44**, 5959.

63. R.J. Molyneux, M. Benson, R.Y. Wong, J.E. Tropea and A.D. Elbein, *J. Nat. Prod.*, 1988, **51**, 1198.

64. R.J. Nash, L.E. Fellows, J.V. Dring, G.W.J. Fleet, A. Girdhar, N.G. Ramsden, J.M. Peach, M.P. Hegarty and A.M. Scofield, *Phytochemistry*, 1990, **29**, 111.

65. A.C. De S. Pereira, M.A.C. Kaplan, J.G.S. Maia, O.R. Gottlieb, R.J. Nash, G.W.J. Fleet, L. Pearce, D.J. Watkin and A.M. Scofield, *Tetrahedron*, 1991, **47**, 5637.

66. S.M. Colegate, P.R. Dorling and C.R. Huxtable, *Aust. J. Chem.*, 1979, **32**, 2257.

67. R.J. Molyneux and L.F. James, *Science*, 1982, **216**, 190.

68. M.J. Schneider, F.S. Ungemach, H.P. Broquist and T.M. Harris, *Tetrahedron*, 1983, **39**, 29.

69. M. Hino, O. Nakayama, Y. Tsurumi, K. Adachi, T. Shibata, H. Terano, M. Kohsaka, H. Aoki and H. Imanaka, *J. Antibiot.*, 1985, **38**, 926.

70. R.J. Molyneux, R.A. McKenzie B.M. O'Sullivan and A.D. Elbein, *J. Nat. Prod.*, 1995, **58**, 878.

71. K.K.I.M. De Balogh, A.P. Dimande, J.J. Van der Lugt, R.J. Molyneux, T.W. Naudé and W.G. Welman. "Toxic Plants and Other Natural Toxicants", Eds. T. Garland and A.C. Barr, CAB International, Wallingford, Oxon, U.K. 1998, Chapter 84, p. 428.

72. I. Pastuszak, R.J. Molyneux, L.F. James and A.D. Elbein, *Biochemistry*, 1990, **29**, 1886.

73. T.M. Harris, C.M. Harris, J.E. Hill and F.S. Ungemach, *J. Org. Chem.*, 1987, **52**, 3094.

74. L.D. Hohenschutz, E.A. Bell, P.J. Jewess, D.P. Leworthy, R.J. Pryce, E. Arnold and J. Clardy, *Phytochemistry*, 1981, **20**, 811.

75. R.J. Nash, L.E. Fellows, J.V. Dring, C.H. Stirton, D. Carter, M.P. Hegarty and E.A. Bell, *Phytochemistry*, 1988, **27**, 1403.

76. R.J. Molyneux, J.N. Roitman, G. Dunnheim, T. Szumilo and A.D. Elbein, *Arch. Biochem. Biophys.*, 1986, **251**, 450.

77. R.J. Nash, L.E. Fellows, A. Girdhar, G.W.J. Fleet, J.M. Peach, D.J. Watkin and M.P. Hegarty, *Phytochemistry*, 1990, **29**, 1356.

78. R.J. Molyneux, J.E. Tropea and A.D. Elbein, *J. Nat. Prod.*, 1990, **53**, 609.

79. D. Tepfer, A. Goldmann, N. Pamboukdjian, M. Maille, A. Lépingle, D. Chevalier, J. Dénarié and C. Rosenberg, *J. Bacteriol.*, 1988, **170**, 1153.

80. A. Goldmann, M.L. Milat, P.H. Ducrot, J.Y. Lallemand, M. Maille, A. Lépingle, I. Charpin and D. Tepfer, *Phytochemistry*, 1990, **29**, 2125.

81. R.J. Molyneux, Y.T. Pan, A. Goldmann, D.A. Tepfer and A.D. Elbein, *Arch. Biochem. Biophys.*, 1993, **304**, 81.

82. N. Asano, A. Kato, K. Oseki, H. Kizu and K. Matsui, *Eur. J. Biochem.*, 1995, **229**, 369.

83. N. Asano, A. Kato, H. Kizu, K. Matsui, A.A. Watson and R.J. Nash, *Carbohydr. Res.*, 1996, **293**, 195.

84. N. Asano, A. Kato, K. Matsui, A.A. Watson, R.J. Nash, R.J. Molyneux, L. Hackett, J. Topping, and B. Winchester, *Glycobiology*, 1997, **7**, 1085.

85. N. Asano, A. Kato, Y. Yokoyama, M. Miyauchi, M. Yamamoto, H. Kizu and K. Matsui, *Carbohydr. Res.*, 1996, **284**, 169.

86. A. Kato, N. Asano, H. Kizu, K. Matsui, S. Suzuki and M. Arisawa, *Phytochemistry*, 1997, **45**, 425.

87. R.C. Griffiths, A.A. Watson, H. Kizu, N. Asano, H.J. Sharp, M.G. Jones, M.R. Wormald, G.W.J. Fleet and R.J. Nash, *Tetrahedron Lett.*, 1996, **37**, 3207.

88. M.R. Wormald, R.J. Nash, P. Hrnciar, J.D. White, R.J. Molyneux and G.W.J. Fleet, *Tetrahedron Asymmetry*, 1998, **9**, 2549.

89. G. Legler, *Adv. Carbohydr. Chem. Biochem.*, 1990, **48**, 319.

90. N. Asano, H. Kizu, K. Oseki, E. Tomioka, K. Matsui, M. Okamoto and M. Baba, *J. Med. Chem.*, 1995, **38**, 2349.

91. A.D. Elbein. "Swainsonine and Related Glycosidase Inhibitors", Eds. L.F. James, A.D. Elbein, R.J. Molyneux and C.D. Warren, Iowa State University Press, Ames, Iowa, U.S.A. 1989, Chapter 14, p. 155.

92. P. Lalegerie, G. Legler and J.M. Yon, *Biochimie*, 1982, **64**, 977.

93. A.M. Scofield, P. Witham, R.J. Nash, G.C. Kite and L.E. Fellows, *Comp. Biochem. Physiol.*, 1995, **112A**, 187.

94. A.M. Scofield, P. Witham, R.J. Nash, G.C. Kite and L.E. Fellows, *Comp. Biochem. Physiol.*, 1995, **112A**, 197.

95. B. Winchester, *Biochem. Soc. Trans.*, 1984, **12**, 522.

96. B. Winchester, S. Al Daher, N.C. Carpenter I. Cenci di Bello, S.S Choi, A.J. Fairbanks and G.W.J. Fleet, *Biochem. J.*, 1993, **290**, 743.

97. B. Winchester and G.W.J. Fleet, *Glycobiology*, 1992, **2**, 199.

98. H. Kayakiri, K. Nakamura, S. Takase, H. Setoi, I. Uchida, H. Terano, M. Hashimoto, T. Tada and S. Koda, *Chem. Pharm. Bull.*, 1991, **39**, 2807.

99. A.M. Scofield, L.E. Fellows, R.J. Nash and G.W.J. Fleet, *Life Sci.*, 1986, **39**, 645.

100. G.W.J. Fleet, S.J. Nicholas, P.W. Smith, S.V. Evans, L.E. Fellows and R.J. Nash, *Tetrahedron Lett.*, 1985, **26**, 3127.

101. M.T.H. Axamatwaty, G.W.J. Fleet, K.A. Hannah, S.K. Namgoong and M.L. Sinnott, *Biochem. J.*, 1990, **266**, 245.

102. L.E. Fellows, S.V. Evans, R.J. Nash and E.A. Bell, *ACS Symp. Ser.*, 1986, **296**, 72.

103. U. Fuhrmann, E. Bause and H. Ploegh, *Biochim. Biophys. Acta*, 1985, **825**, 95.

104. Y. Zeng, Y.T. Pan, N. Asano, R.J. Nash and A.D. Elbein, *Glycobiology*, 1997, **7**, 297.

105. S.V. Kyosseva, Z.N. Kyossev and A.D. Elbein, *Arch. Biochem. Biophys.*, 1995, **316**, 821.

106. J. Bischoff and R. Kornfeld, *Biochem. Biophys. Res. Commun.*, 1984, **125**, 324.

107. A.A. Bell, L. Pickering, A.A. Watson, R.J. Nash, Y.T. Pan, A.D. Elbein and G.W.J. Fleet, *Tetrahedron Lett.*, 1997, **38**, 5869.

108. D.R.P. Tulsiani, T.M. Harris and O. Touster, *J. Biol. Chem.*, 1982, **257**, 7936.

109. I. Cenci di Bello, G. Fleet, J.C. Son, K.-I.. Tadano and B. Winchester. "Swainsonine and Related Glycosidase Inhibitors", Eds. L.F.James, A.D. Elbein, R.J. Molyneux and C.D. Warren, Iowa State University Press, Ames, Iowa, U.S.A. 1989, Chapter 27, p. 367.

110. I. Cenci di Bello, G.W.J. Fleet, K. Namgoong, K.-I. Tadano and B. Winchester, *Biochem. J.*, 1989, **259**, 855.

111. R. Saul, J.P. Chambers, R.J. Molyneux and A.D. Elbein, *Arch. Biochem. Biophys.*, 1983, **221**, 593.

112. B.G. Winchester, I. Cenci di Bello, A.C. Richardson, R.J. Nash, L.E. Fellows, N.G. Ramsden and G.W.J. Fleet, *Biochem. J.*, 1990, **269**, 227.

113. N. Asano, A. Kato, H. Kizu, K. Matsui, R.C. Griffiths, M.G. Jones, A.A. Watson and R.J. Nash, *Carbohydr. Res.*, 1997, **304**, 173.

114. P.R. Dorling, C.R. Huxtable and P. Vogel, *Neuropathol. Appl. Neurobiol.*, 1978, **4**, 285.

115. R.J. Molyneux, L.F. James and K.E. Panter. "Plant Toxicology", Eds. A.A. Seawright, M.P. Hegarty, L.F. James and R.F. Keeler, Queensland Poisonous Plants Committee, Yeerongpilly, QLD, Australia. 1985, p. 266.

116. B.L. Stegelmeier, L.F. James, K.E. Panter and R.J. Molyneux, *Am. J. Vet. Res.*, 1995, **56**, 149.

117. G.R. Cao, S.J. Li, D.X. Duan, R.J. Molyneux, L.F. James, K. Wang and C. Tong. "Poisonous Plants", Eds. L.F. James, R.F. Keeler, E.M., Bailey, P.R. Cheeke and M.P. Hegarty, Iowa State University Press, Ames, Iowa, U.S.A. 1992, p. 117.

118. K. Chotai, C. Jennings, B. Winchester and P. Dorling, *J. Cell Biochem.*, 1983, **21**, 107.

119. P.F. Daniel, C.D. Warren and L.F. James, *Biochem. J.*, 1984, **221**, 601.

120. R.J. Molyneux, L.F. James, M.H. Ralphs, J.A. Pfister, K.P. Panter and R.J. Nash. "Plant Associated Toxins -Agricultural, Phytochemical and Ecological Aspects", Eds. S.M. Colegate and P.R. Dorling, CAB International, Wallingford, Oxon, U.K. 1994, Chapter 21, p. 107.

121. J.D. Hocking, R.D. Jolly and R.D. Batt, (1972) *Biochem. J.*, 1972, **128**, 69.

122. P.A. Ockerman. "Lysosomes and Storage Diseases", Eds. H.G. Hers and F. van Hoof, Academic Press, London, U.K. 1973, p. 292.

123. R.D. Jolly, B.G. Winchester, J. Gehler, P.R. Dorling and G. Dawson, *J. Appl. Biochem.*, 1981, **3**, 273.

124. I. Cenci di Bello, P. Dorling and B. Winchester, *Biochem. J.*, 1983, **215**, 693.

125. B. Winchester *Biochem. Soc. Trans.*, 1992, **20**, 699.

126. P.F. Daniel, J.E. Evans, R. De Gasperi, B. Winchester and C.D. Warren, *Glycobiology*, 1992, **2**, 327.

127. R. De Gasperi, P.F. Daniel and C.D. Warren, *J. Biol. Chem.*, 1992, **267**, 9706.

128. R.A. McKenzie, K.G. Reichmann, C.K. Dimmock, P.J. Dunster and J.O. Twist, *Aust. Vet. J.*, 1988, **65**, 165.

129. Y.T. Pan, J. Ghidoni and A.D. Elbein, *Arch. Biochem. Biophys.*, 1993, **303**, 134.

130. S.L. Everist. "Poisonous Plants of Australia", Angus and Robertson, Sydney, Australia. 1974, p. 282.

131. B.L. Stegelmeier, R.J. Molyneux, A.D. Elbein and L.F. James, *Vet. Pathol.*, 1995, **32**, 289.

132. R. Saul, J.J. Ghidoni, R.J. Molyneux and A.D. Elbein, *Proc. Natl. Acad. Sci. U.S.A.*, 1985, **82**, 93.

133. R.H. Glew, A. Basu, E.M. Prence and A.T. Remaley, *Lab. Invest.*, 1985, **53**, 250.

134. I. Cenci di Bello, D. Mann, R.J. Nash and B. Winchester. "Lipid Storage Disorders", Eds. R. Salvayre, L. Douste-Blazy and S. Gatt, Plenum Publishing Corp., New York, U.S.A. 1988, p. 635.

135. R.J. Molyneux, R.J. Nash and N. Asano. "Alkaloids: Chemical and Biological Perspectives", Ed. S.W. Pelletier, Elsevier Science Ltd., Oxford, U.K. 1996, Vol. 11, Chapter 4, p. 303.

136. J.S. Menzies, C.H. Bridges and E.M. Bailey Jr., *Southwest. Vet.*, 1979, **32**, 45.

137. J.G. Pienaar, T.S. Kellerman, P.A. Basson, W.L. Jenkins and J. Vahrmeijer, *Onderstepoort J. Vet. Res.*, 1976, **43**, 67.

138. S. Hakomori, *Cancer Res.*, 1985, **45**, 2405.

139. J.W. Woollen and P. Turner, *Clin. Chim. Acta.*, 1965, **12**, 671.

140. R.J. Bernacki, M.J. Niedbala and W. Korytnyk, *Cancer Metastasis Rev.*, 1985, **4**, 81.

141. K. Olden, P. Breton, K. Grzegorzewski, Y. Yasuda, B.L. Gause, O.A. Oredipe, S.A. Newtown and S.L. White, *Pharmacol. Ther.*, 1991, **50**, 285.

142. J.W. Dennis, K. Koch, S. Yousefi and I. VanderElst, *Cancer Res.*, 1990, **50**, 1867.

143. J.A. Baptista, P. Goss, M. Nghiem, J.J. Krepinsky, M. Baker and J.W. Dennis, *Clin. Chem.*, 1994, **40**, 426.

144. M.A. Spearman, J.E. Damen, T. Kolodka, A.H. Greenberg, J.C. Jamieson and J.A. Wright, *Cancer Lett.*, 1991, **57**, 7.

145. B. Korczak and J.W. Dennis, *Int. J. Cancer*, 1993, **53**, 634.

146. T. Kino, N. Inamura, K. Nakahara, S. Kiyoto, T. Goto, H. Terano, M. Kohsaka, H. Aoki and H. Imanaka, *J. Antibiot.*, 1985, **38**, 936.

147. M.J. Humphries, K. Matsumoto, S.L. White, R.J. Molyneux and K. Olden, *Cancer Res.*, 1988, **48**, 1410.

148. K. Olden, S.L. White, S. Mohla, S.A. Newton, Y. Yasuda, D. Bowen and M.J. Humphries, *Oncology*, 1989, **3**, 83.

149. S.L. White, T. Nagai, S.K. Akiyama, E.J. Reeves, K. Grzegorzewski and K. Olden, *Cancer Commun.*, 1991, **3**, 83.

150. P.C. Das, J.D. Roberts, S.L. White and K. Olden, *Oncol. Res.*, 1995, **7**, 425.

151. M. Yagita and E. Saksela, *Scand. J. Immunol.*, 1990, **31**, 275.

152. M.P. Colombo, A. Modesti, G. Parmiani and G. Forni, *Cancer Res.*, 1992, **52**, 4853.

153. K. Grzegorzewski, S.A. Newton, S.K. Akiyama, S. Sharrow, K. Olden and S.L. White, *Cancer Commun.*, 1989, **1**, 373.

154. C.R. Huxtable and P.R. Dorling, *Am. Assoc. Pathol.*, 1982, **107**, 124.

155. C.R. Huxtable and P.R. Dorling, *Acta Neuropathol.*, 1985, **68**, 65.

156. D.R.P. Tulsiani and O. Touster, *J. Biol. Chem.*, 1987, **262**, 6506.

157. S. Mohla, M.J. Humphries, S.L. White, K. Matsumoto, S.A. Newton, C.C. Sampson, D. Bowen and K. Olden, *J. Natl. Med. Assoc.*, 1989, **81**, 1049.

158. D. Bowen, J. Adir, S.L. White, C.D. Bowen, K. Matsumoto and K. Olden, *Anticancer Res.*, 1993, **13**, 841.

159. D. Bowen, W.M. Southerland, C.D. Bowen and D.E. Hughes, *Anticancer Res.*, 1997, **17**, 4345.

160. P.E. Goss, J. Baptiste, B. Fernandes, M. Baker and J.W. Dennis, *Cancer Res.*, 1994, **54**, 1450.

161. P.E. Goss, C.L. Reid, D. Bailey and J.W. Dennis, *Clin. Cancer Res.*, 1997, **3**, 1077.

162. B.L. Stegelmeier, P.W. Snyder, L.F. James, K.E. Panter, R.J. Molyneux, D.R. Gardner, M.H. Ralphs and J.A. Pfister. "Toxic Plants and Other Natural Toxicants", Eds. T. Garland and A.C. Barr, CAB International, Wallingford, Oxon, U.K. 1998, Chapter 57, p. 285.

163. J.W. Dennis, S.L. White, A.M. Freer and D. Dime, *Biochem. Pharmacol.*, 1993, **46**, 1459.

164. G.K. Ostrander, N.K. Scribner and L.R. Rohrschneider, *Cancer Res.*, 1988, **48**, 1091.

165. B. Woynarowska, H. Wikiel, M. Sharma, N. Carpenter, G.W.J. Fleet and R.J. Bernacki, *Anticancer Res.*, 1992, **12**, 161.

166. S. Martino C. Emiliani A. Tabilio F. Falzetti, J.L. Stirling and A. Orlacchio, *Biochem. Biophys. Acta,* 1997, **1335**, 5.

167. G.W.J. Fleet, P.W. Smith, R.J. Nash, L.E. Fellows, R.B. Parekh and T.W. Rademacher, *Chem. Lett.*, 1986, 1051.

168. P.S. Liu, M.S. Kang and P.S. Sunkara, *Tetrahedron Lett.*, 1991, **32**, 719.

169. T. Aoyagi, H. Suda, K. Uotani, F. Kojima, K. Aoyama, K. Horiguchi, M. Hamada and T. Takeuchi, *J. Antibiot.*, 1992, **45**, 1404.

170. K. Tatsuta, S. Miura, S. Ohta and H. Gunji, *J. Antibiot.*, 1995, **48**, 286.

171. T. Aoyama, F. Kojima, C. Imada, Y. Muraoka, H. Naganawa, Y. Okami, T. Takeuchi and T. Aoyagi, *J. Enzym. Inhib.*, 1995, **8**, 223.

172. E. Frei and G.P. Cannellos, *Am. J. Med.*, 1980, **69**, 585.

173. W. Hryniuk and M.N. Levine, *J. Clin. Oncol.*, 1986, **4**, 1162.

174. J-L.D. Klein, J.D. Roberts, M.D. George, J. Kurtzberg, P. Breton, J-C. Chermann, and K. Olden, *Br. J. Cancer*, 1999, **80**, 87.

175. O.A. Oredipe, S.L. White, K. Grzegorzewski, B.L. Gause, J.K. Cha, V.A. Miles and K. Olden, *J. Natl. Cancer Inst.*, 1991, **83**, 1149.

176. G.X. McLeod and S.M. Hammer, *Ann. Int. Med.*, 1992, **117**, 487.

177. B. Junge, M. Matzke and J. Stoltefuss. "Handbook of Experimental Pharmacology – Oral Antidiabetics", Eds. J. Kuhlmann and W. Puls, Springer-Verlag, Berlin & Heidelberg, Germany. 1996, Vol. 119, p. 411.

178. G.S. Jacob, *Curr. Opin. Struct. Biol.*, 1995, **5**, 605.

179. L. Müller. "Novel Microbial Products for Medicine and Agriculture", Eds. A.L. Demain, G.A. Somkuti, J.C. Hunter-Cervera and H.W. Rossmoore, Society for Industrial Microbiology, Fairfax, Virginia, U.S.A. 1989, Chapter 13, p. 109.

180. B.L. Rhinehart, K.M. Robinson, P.S. Liu, A.J. Payne, M.E. Wheatley and S.R. Wagner, *J. Pharmacol. Exp. Ther.*, 1987, **241**, 915.

181. M. Kimura, F.-J. Chen, N. Nakashima, I. Kimura, N. Asano and S. Koya, *J. Trad. Med.*, 1995, **12**, 214.

182. H. Nojima, I. Kimura, F.-J. Chen, Y. Sugihara, M. Haruno, A. Kato and N. Asano, *J. Nat. Prod.*, 1998, **61**, 397.

183. S. Taniguchi, N. Asano, F. Tomino and I. Miwa, *Horm. Metab. Res.*, 1998, **30**, 679.

184. A.S. Tyms, E.M. Berrie, T.A. Ryder, R.J. Nash, M.P. Hegarty, D.L. Taylor, M.A. Mobberley, J.M. Davis, E.A. Bell, D.J. Jeffries, D. Taylor-Robinson and L.E. Fellows, *Lancet*, 1987, **ii**, 1025.

185. G.W.J. Fleet, A. Karpas, R.A. Dwek, L.E. Fellows, A.S. Tyms, S. Petursson, S.K. Namgoong, N.G. Ramsden, P.W. Smith, J.C. Son, F. Wilson, D.R. Witty, G.S. Jacob and T.W. Rademacher, *FEBS Lett.*, 1988, **237**, 128.

186. D.L. Taylor, P. Sunkara, P.S. Liu, M.S. Kang, T.L. Bowlin and A.S. Tyms, *AIDS*, 1991, **5**, 693.

187. E.B. Stephens, E. Monck, K. Reppas and E.J. Butfiloski, *J. Virol.*, 1991, **65**, 1114.

188. D.L. Taylor, L.E. Fellows, G.H. Farrar, R.J. Nash, D. Taylor-Robinson, M.A. Mobberley, T.A. Ryder, D.J. Jeffries and A.S. Tyms, *Antivir. Res.*, 1988, **10**, 11.

189. P.B. Fischer, M. Collin, G.B. Karlsson, W. James, T.D. Butters, S.J. Davis, S. Gordon, R.A. Dwek and F.M. Platt, *J. Virol.*, 1995, **69**, 5791.

190. A. Karpas, G.W.J. Fleet, R.A. Dwek, S. Petursson, S.K. Namgoong, N.G. Ramsden, G.S. Jacob and T.W. Rademacher, *Proc. Natl. Acad. Sci. U.S.A.*, 1988, **85**, 9229.

191. M.A. Fischl, L. Resnick, R. Coombs, A.B. Kremer, J.C. Pottage, R.J. Fass, K.H. Fife, W.G. Powderly, A.C. Collier, R.L. Aspinall, S.L. Smith, K.G. Kowalski and C.B. Wallemark, *J. AIDS*, 1994, **7**, 139.

192. P.B. Fischer, G.B. Karlsson, T.D. Butters, R.A. Dwek and F.M. Platt, *J. Virol.*, 1996, **70**, 7143.

193. P.B. Fischer, G.B. Karlsson, R.A. Dwek and F.M. Platt, *J. Virol.*, 1996, **70**, 7153.

194. G.B. Karlsson, T.D. Butters, R.A. Dwek and F.M. Platt, *J. Biol. Chem.*, 1993, **268**, 570.

195. F.M. Platt and T.D. Butters, *Biochem. Pharmacol.*, 1998, **56**, 421.

196. E.D. Faber, L.A.G.M. Van den Broek, E.E.Z. Oosterhuis, B.P. Stok and D.K.F. Meijer, *Drug Deliv.*, 1998, **5**, 3.

197. P.S. Sunkara, D.L. Taylor, M.S. Kang, T.L. Bowlin, P.S. Liu, A.S. Tyms and A. Sjoerdsma, *Lancet*, 1989, **i**, 1206.

198. W.K. Anderson, R.A. Coburn, A. Gopalsamy and T.J. Howe, *Tetrahedron Lett.*, 1990, **31**, 169.

199. C.G. Bridges, S.P. Ahmed, M.S. Kang, R.J. Nash, E.A. Porter and A.S. Tyms, *Glycobiology*, 1995, **5**, 249.

200. M.S. Kang, *Glycobiology*, 1996, **6**, 209.

201. G.J. Richmond, P. Zolnouni, J. Stall, M. McPherson, P. Hamedani, V. Cross, E. Sidarous M. Stoltz, *Proc. Int. Conf. AIDS*, 1996, **11**, 80.

202. V.A. Johnson, B.D. Walker, M.A. Barlow, T.J. Paradis, T.C. Chou and M.S. Hirsch, *Antimicrob. Agents Chemother.*, 1989, **33**, 53.

203. D.L. Taylor, T.M. Brennan, C.G. Bridges, M.S. Kang and A.S. Tyms, *Antivir. Chem. Chemother.*, 1995, **6**, 143.

204. K. Tatatsuki, T. Hattori, T. Kaizu, M. Okamoto, Y. Yokato, K. Nakamura and H. Kayakiri. Antiretroviral pyrroline and pyrrolidine sulfonic acid derivatives. 1990. European Patent Application EP O 407 701 A2.

205. T.M. Cox, *Trends Exp. Clin. Med.*, 1994, **4**, 144.

206. F.M. Platt, G.R. Neises, R.A. Dwek and T.D.Butters, *J. Biol. Chem.*, 1994, **269**, 8362.

207. F.M. Platt, G.R. Neises, G.B. Karlsson, R.A. Dwek and T.D. Butters, *J. Biol. Chem.*, 1994, **269**, 27108.

208. F.M. Platt, G.R. Neises, G. Reinkensmeier, M.J. Townsend, V.H. Perry, R.L. Proia, B. Winchester, R.A. Dwek and T.D. Butters, *Science*, 1997, **276**, 428.

209. F.M. Platt and T.D. Butters, *Trends Glycosci. Glycotechnol.*, 1995, **7**, 495.

210. J-Q. Fan, S.Ishii, N. Asano and Y. Suzuki, *Nat. Med.*, 1999, **5**, 112.

211. S. Ishii, R. Kase, T. Okumiya, H. Sakuraba and Y. Suzuki, *Biochem. Biophys. Res. Comm.*, 1996, **220**, 812.

212. D.O. Willenborg, C.R. Parish and W.B. Cowden, *Immunol. Cell Biol.*, 1992, **70**, 369.

213. M.R. Bartlett, W.B. Cowden and C.R. Parish, *J. Leukocyte Biol.*, 1995, **57**, 207.

214. P.M. Grochowicz, K.M. Bowen, A.D. Hibberd, D.A. Clark, W.B. Cowden and D.O. Willenborg, *Transplant. Proc.*, 1992, **24**, 2295.

215. P.M. Grochowicz, Y.C. Smart, K.M. Bowen, A.D. Hibberd, D.A. Clark, W.B. Cowden and D.O. Willenborg, *Transplant. Proc.*, 1993, **25**, 2900.

216. A.D. Hibberd, P.M. Grochowicz, Y.C. Smart, K.M. Bowen, D.A. Clark and B. Purdon, *Transplant. Proc.*, 1995, **27**, 448.

217. P.M. Grochowicz, A.D. Hibberd, K.M. Bowen, D.A. Clark, G. Pang, W.B. Cowden, T.C. Chou, L.K. Grochowicz and Y.C. Smart, *Transplant. Proc.*, 1997, **29**, 1259.

218. A.D. Hibberd, P.M. Grochowicz, Y.C. Smart, K.M. Bowen, D.A. Clark, W.B. Cowden and D.O. Willenborg, *Transplant. Proc.*, 1997, **29**, 1257.

219. G.D. Besra, K.-H. Khoo, M.R. McNeil, A. Dell, H.R. Morris and P.J. Brennan, *Biochemistry*, 1995, **34**, 4257.

220. R.E. Lee, P.J. Brennan and G.S. Besra, *Curr. Top. in Microbiol. Immunol.*, 1996, **215**, 1.

221. Y. Tsukioka, Y. Yamashita, T. Oho, Y. Nakano and T. Koga, *J. Bacteriol.*, 1997, **179**, 1126.

222. B.G. Davis, T.W. Brandstetter, L. Hackett, B.G. Winchester, R.J. Nash, A.A. Watson, R.C. Griffiths, C. Smith and G.W.J. Fleet, *Tetrahedron*, 1999, **55**, 4489.

223. J.P. Shilvock, J.R. Wheatley, R.J. Nash, A.A. Watson, R.C. Griffiths, T.D. Butters, M. Müller, D.J. Watkin, D.A. Winkler and G.W.J. Fleet, *J. Chem. Soc., Perkin Trans.*, 1999, **1**, 1.
224. R.E. Lee, M.D. Smith, R.J. Nash, R.C. Griffiths, M. McNeil, R.K. Grewal, W.X. Yan, G.S. Besra, P.J. Brennan and G.W.J. Fleet, *Tetrahedron Lett.*, 1997, **38**, 6733.

'LESSONS FROM NATURE': CAN ECOLOGY PROVIDE NEW LEADS IN THE SEARCH FOR NOVEL BIOACTIVE CHEMICALS FROM TROPICAL RAINFORESTS?

Paul Reddell and Victoria Gordon

CSIRO Tropical Forest Research Centre, PO Box 780, Atherton, Queensland, 4883, AUSTRALIA

A role for ecology in providing a powerful, knowledge-based approach for more effectively targeting the search for new bioactive chemicals from tropical rainforests is discussed. To illustrate this, we outline some of our current research on chemical defences of fruits and seeds of rainforest plants and show how this ecological knowledge can be applied directly to target groups of plants for screening for chemicals with specific biostatic and biocidal properties.

Fruits and seeds of certain groups of rainforest plants which have significant resources invested in individual propagules and/ or persist for extended periods in the warm, moist environment on the forest floor appear to be a particularly rich source of bioactive defence chemicals (i.e. toxins and feeding deterrents). Analysis of broad groups of secondary metabolites, together with bioassays on crude extracts from these fruits and seeds show highly localized occurrence, and often very potent, biocidal activities within specific layers of these propagules. These findings are consistent with our understanding of seed ecology from which we would predict such localized activity because of the biological 'trade-offs' inherent in packaging, protection and dispersal of propagules. Broad groups of secondary metabolites also occur more frequently, and often at higher concentrations in fruits and seeds than in other tissues (e.g. leaves and bark) from these plants. Similarly, much stronger biocidal activities are demonstrated from crude extracts from fruits and seed layers than from other plant parts. Again this is consistent with their ecology, which suggests that there would be a proportionally greater investment (energy per unit mass) in defence of seeds than other plant parts.

Further significant opportunities exist to use tropical forest ecology to target promising natural sources of chemicals with specific bioactivities. 'Symbioses' between forest ecologists and natural products chemists would hasten this discovery process. Such collaboration would also contribute to our understanding of the chemical mechanisms that underlie many of the fundamental ecological interactions that drive and maintain tropical forests.

1 TROPICAL FORESTS AS POTENTIAL SOURCES OF NEW AGRICULTURAL AND INDUSTRIAL CHEMICALS

Tropical rainforests are the most biologically diverse terrestrial ecosystems on earth. They cover less than 7% of the land's surface but they are home to more than 50% of all species. For example, they are estimated conservatively to contain more than 125,000 species of higher plants, 5 million species of insects and 1.5 million species of fungi.[1]

Tropical rainforests are structurally complex ecosystems; they contain many different ecological niches, often requiring highly specialized survival strategies. They are also frequently resource limited and highly competitive environments. Competition and resource limitations in such complex ecosystems confer strong evolutionary pressures on organisms to improve their survival 'fitness'. This is likely to result in significant and on-going chemical 'innovation' by organisms, and as a consequence, tropical rainforests are presumably repositories of high chemical diversity, both in terms of molecular structures and biological activity. This presumption of high chemical diversity based on ecology is borne out by the limited comparative chemical data that are available. For example, in surveys for alkaloids, the flora of tropical rainforests consistently yields a higher proportion of alkaloid bearing species than do those of other ecosystems.[2]

Traditionally, tropical rainforests have been an important source of natural medicines and agricultural chemicals used in subsistence societies,[3] with up to 70% of the world's population still relying on such traditional and natural medicines in their primary health care.[4] More recently, bio-prospecting in these forests has provided industrialized societies with the source, or lead structure, for more than 50 commercial drugs (e.g. the vinca alkaloids, pilocarpine, curare, quinine, reserpine and diosgenin) and a number of agricultural and industrial chemicals such as the carbamate and rotenoid insecticides.[4-6] However, it is estimated that less than 10% of plant species from these forests have been investigated chemically (with less than 1% assessed in any detail), and only a miniscule number of tropical forest microbes, insects and vertebrates have been examined.[7] Based on the likely high chemical diversity of these forests, there remains a very significant opportunity for the discovery of new bioactive chemicals (and novel lead structures) with potential applications in medicine, agriculture and industry.

2 ECOLOGY AS A 'KNOWLEDGE-BASED' SOURCE OF LEADS IN THE SEARCH FOR NEW BIOACTIVE CHEMICALS

One limitation in the process of discovery of novel bioactive chemicals from tropical forests has been the immensity of the screening task (i.e. the number of species, samples types and extracts involved) using current approaches. Current approaches to discovery fall loosely into three broad groups that reflect an increasing degree of structuring or 'targeting' (and presumably efficiency) of the collection and assessment procedures. These groups are:

1. Serendipity, or the isolation and characterization of new chemicals with no specific approach to targeting collections from particular environments or taxonomic groups;
2. Surveys which involve structured and systematic collection of biological material in particular locations and environments, or from particular taxonomic groupings;
3. 'Knowledge-based' targeting of collections. To date these have principally relied on either ethnobotanical knowledge of usage in traditional societies or on information on taxonomic relatedness to species already known to contain specific groups of novel or interesting chemical structures.

All three approaches have been used in tropical forests.

One potentially very powerful, knowledge-based strategy that appears rarely used involves applying an understanding of forest ecology. Ecology is in essence the study of interactions between organisms, and between organisms and their physical environment. Most of these interactions are chemically mediated, and the chemical entities (secondary metabolites) involved have effectively been developed, field tested and refined by on-going evolutionary processes and selection pressures i.e. they are nature's often sophisticated solutions to specific ecological problems.[8-10] Consequently, where adequate and appropriate ecological knowledge exists, it can provide very strong clues as to where to expect the most frequent occurrences and highest diversities of chemicals that have specific bioactive properties.

For example, chemical defences produced by organisms against their predators and competitors are widespread in nature and include a diverse range of compounds with fungicidal, antibiotic and insecticidal properties together with those that are toxic or act as feeding deterrents to vertebrates.[11, 12] However, these chemical defences are particularly well developed in certain ecosystems, environments and/or ecological niches where the 'co-evolutionary arms race' is most active. They may also be localized within certain parts of the organisms or induced or best expressed at certain stages of lifecycles or certain times of year to coincide with particular pressures from predators or the requirement to protect specific vulnerable resources. With appropriate ecological information, environmental 'hotspots' (i.e. places, niches), target groups and/ or tissues in which these chemicals are most prevalent can be defined and used to focus the search for new agrochemicals with specific biocidal actions (e.g. as fungicides, antibiotics, insecticides, nematicides and vertebrate feeding deterrents). In addition, many of these defence chemicals (especially those directed at mammalian herbivores) are likely to have properties associated with their toxicity or deterrent action that also make them good candidates for screening as lead molecules in development and discovery of drugs for certain purposes (e.g. cell toxins in cancer treatment; deterrents as muscle relaxants, tranquilizers and sedatives). In this instance, although it is not the primary purpose for which the molecules were 'designed' or used in nature, understanding ecology can provide clues as to where to look for particular types of biological activity that are likely to require particular modes of action or receptor affinities.

3 'SEEDS OF DISCOVERY': AN EXAMPLE OF APPLYING ECOLOGY TO MORE EFFECTIVELY TARGET SOURCES OF NEW BIOACTIVE CHEMICALS

A simple example from some of our current work in tropical rainforests in north Queensland demonstrates how ecology can be used to target organisms for screening for specific types of bioactivity, in this case, biostatic and biocidal properties.

Seeds are the primary regenerative propagules of most tropical rainforest plants and we are studying their ecology primarily to understand how forests will respond to future disturbances and perturbations. Simply put, we are interested in the ecology of seeds (and seedlings) because they represent the future of the forest. However, our specific interests in one aspect of their ecology, how seeds are packaged and defended against a wide range of predators and pathogens, also have another practical application in providing a knowledge-based rationale on which to focus the search for new bioactive chemicals from these forests. Three aspects of our current thinking on seed ecology allow us to reduce the range of potential target plants (and sample types) to a smaller set of knowledge-based 'best-bets', with obvious implications in improving the potential efficiency of the search process and the likelihood of a bioactivity 'hit'.

Firstly, we postulated (based on knowledge of the timing of cycles of flowering and fruiting, levels of seed production, fruit anatomy, dispersal modes, germination characteristics and degree of exposure to potential predators) that certain types of seeds should have strongly developed chemical defence strategies against specific groups of predators. In particular, two seed ecological types (species with large-seeds and those that comprise the soil seed bank) would have especially effective and well-developed chemical defences against a wide range of threatening organisms (see Table 1 for some key attributes of these two seed ecological types). Secondly, we suspected that in species where seeds were protected primarily by chemical defences, these defences would be either at relatively higher concentrations (and hence easier to detect) or molecules of much greater potency than those in other plant parts (such as the leaves and bark). Our rationale in this case being that there would be greater 'investment'

Table 1: *Comparison of key ecological attributes of two groups of rainforest plants with predicted requirement for effective chemical defence of seeds*

	Large seeded species	**Seed bank species**
Successional status	Generally late successional	Pioneer/ early successional
Resource investment in individual seeds	High; large seeds containing high concentrations of nutrients and energy-rich compounds (lipids, oils)	Moderate; small seeds with moderate to high concentrations of nutrients ± energy-rich compounds
Dispersal syndrome	Vertebrate or no specialized dispersal mode	Wind, water or invertebrate (for arillate species)
Fruiting structure	Often fleshy, indehiscent, multi-layered fruit containing a single seed	Generally dry, dehiscent fruits containing a number of dry seeds with dark testa
Potential seed longevity	Moderate; weeks to months (years in some exceptional cases where seeds have woody outer cases)	Long; years in the upper few cm of soil profile in warm & moist conditions conducive to invertebrate & microbial activity
Dormancy mechanisms	Generally none, but mechanical dormancy due to woody seed cases in some species	Obligate dormancy; either physiological (e.g. phytochrome responses) and/or mechanical (e.g. impermeable testas)
Germination environment	On litter or soil surface in low light under closed forest canopy	From the soil in high light environments following canopy opening
Time for germination	Moderate to slow (many weeks to months)	Rapid; days to weeks once dormancy is broken
Exposure to potential predators	High	High
Presumed defences & deterrents against	Mammalian, invertebrate and microbial predators	Soil dwelling arthropods, nematodes, fungi and bacteria
Phenology of fruit production	Variable; some species have 'masting' behaviour to satiate predators	Regular; usually annual with large seed crops

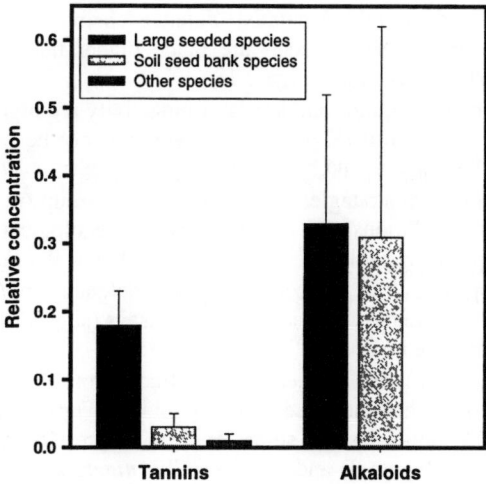

Figure 1 *Comparative concentrations of tannins and alkaloids in fruits from three seed ecological groups. Data are the mean derived from all species examined. Tannin concentrations are in g g^{-1} freeze dried weight, alkaloid concentrations are based on a semi-quantitative precipitation test.*

(energy per unit mass) in the defence of structures which (a) have relatively high concentrations of nutrients and energy-rich compounds (e.g. lipids, carbohydrates) and (b) affect the potential for representation of genes in the next generation in the forest. Finally, due to size constraints, fruits and seeds often represent trade-offs in conflicting biological roles (e.g. the requirement to attract dispersal agents and the need to ensure that the seed is not consumed or destroyed by these agents). On this basis we hypothesized that defence chemicals would often need to be highly localized and, in many cases, because of the limited amount that could be included in the 'package', particularly potent in their action. Localization within the fruit suggests that it would be easy to miss the active chemicals if whole fruit were extracted and assayed, whilst high potency of action at low concentrations or in low total amounts are both potentially desirable characteristics of bioactive molecules for industry.

With the above in mind we are currently surveying representatives from two seed ecological types (large seeded species, seed bank species) in north Queensland rainforests. Our objectives are to establish what putative defence chemicals their fruits and seeds contain, where these chemicals are localized and how effective they are as deterrents against specific groups of predators. We are also comparing the occurrence (and relative concentrations) of these defence chemicals with (1) those in fruits and seeds of species from other types where from their ecology we would predict that chemical defences are less important and (2) other plant parts (leaves, stems, bark, roots) from the same species. To date our approach has been based on fractionating the fruits into layers, developing profiles of the broad groups of secondary metabolites present and undertaking bioassays of activity of crude extracts against a range of potential predator organisms.

Preliminary results from our studies support much of which we hypothesized and clearly show the power of using ecology to target the search for bioactive chemicals. For example, our studies of crude extracts from the large-seeded species show:

- the concentrations of broad groups of putative defence chemicals (e.g. tannins, alkaloids, saponins) present in these species are much higher than in any other ecological group that we have examined (Figure 1);
- groups of putative defence compounds and levels of bioactivity are highly localized within particular layers of the fruit (Figure 2) and would likely be missed if the whole fruit was used in the analyses and bioassays;
- a diversity of different chemical strategies is used within this group (i.e. there are a large range of chemical 'solutions' to a similar set of ecological problems), indicating they are very promising source of potential new chemical structures;
- in some instances, apparent low concentrations of defence compounds are associated with very strong biocidal activities in the bioassays (suggesting that some very potent compounds may be involved);
- microbial growth is inhibited more frequently by extracts from the large seeded species than by extracts from the other seed types examined. For example, 67% of the large seeded species examined show some level of suppression of growth of *Staphylococcus aureus* (e.g. Figure 3) and/or *Aspergillus niger*, whereas only 17% of seed bank species and none from the other ecology groups showed any anti-microbial activity in their seeds or fruits;
- extracts from fruits and seeds consistently have much higher levels of anti-microbial activity than extracts from other parts of the same species. For example, none of the leaf samples from the large seeded species showed any significant inhibition of growth of either *S. aureus* or *A. niger* in our bioassays.

We are currently collecting and screening further species and fractionating our crude extracts as a first step in attempting to identify some of the specific bioactive components.

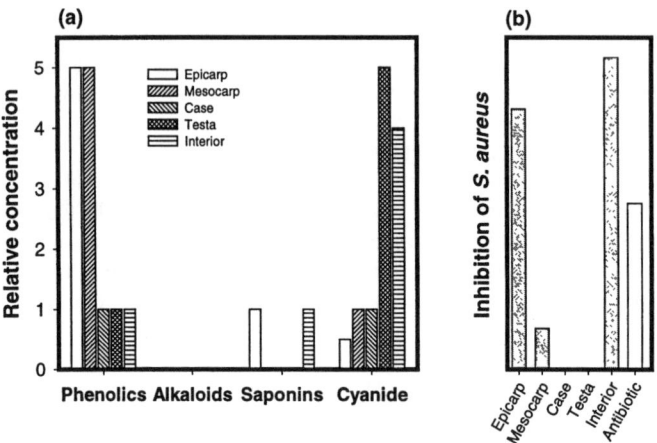

Figure 2   *Distribution of (a) four broad groups of secondary metabolites and (b) bioactivity against* Staphylococcus aureus, *within the fruit and seed layers of a large seeded species from the Proteaceae.*

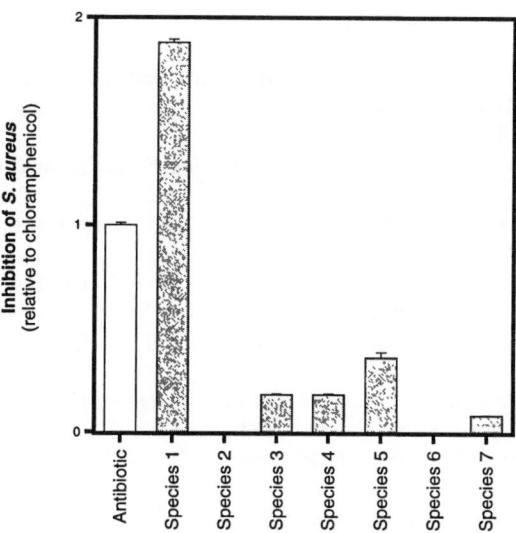

Figure 3 *Comparison of growth inhibition of* Staphylococcus aureus *by crude extracts (50% acetone) of seed interiors from a typical subset of large seeded species in our studies.*

4   SYMBIOSIS: A FUTURE FOR COLLABORATION BETWEEN ECOLOGISTS AND CHEMISTS?

There are many other instances where existing knowledge of forest ecology could be used to more effectively target natural sources (environments, niches, groups of organisms) of chemicals with particular types of bioactivity (including antibiotics, fungicides, insecticides, nematicides, vertebrate feeding deterrents, mammalian cell toxins, seed germination inhibitors, antioxidants and 'energy-dissipating' molecules, essential oils, and scents and fragrances). However, surprisingly this approach has not been used more widely in the past.

Symbiosis, mutually beneficial association between dissimilar organisms, is a powerful driving force in nature and a strategy used frequently to deal with resource limitations in complex and difficult environments. There are good reasons why we should learn from nature and look for opportunities for 'symbioses' between ecologists and natural product chemists in the study of bioactive chemicals from tropical forests. Such interactions will result in:

- the more rapid and efficient discovery of new bioactive chemical entities with potential for specific applications in agriculture, medicine and industry, and;
- better conservation and management of these forests through a more mechanistic understanding of their ecology and an enhanced appreciation of their potential contribution to our current and future welfare.

Nature has many lessons, if only we care to take note.

References

1. A. Gentry. *'Human Medicinal Agents from Plants'*, American Chemical Society, Washington, 1993, Chapter 2, p. 13.
2. D. A. Levin. *Am. Nat.*, 1976, **110**, 261
3. R. E. Schultes and R. F. Raffauf. *'The Healing Forest'*, Dioscorides Press, Oregon, 1990.
4. N. R. Farnsworth. *'Biodiversity'*, National Academy Press, Washington, 1988, Chapter 9, p. 83
5. *'Medicinal Resources of the Tropical Forest'*, M. J. Balick, E. Elisabetsky and S. A. Laird (Editors), Columbia University Press, New York, 1996.
6. M. F. Balandrin, J. A. Klocke, E. S. Wurtele and W. H. Bollinger. *Science*, 1985, **228**, 1154
7. N. Myer. *'The Primary Source: Tropical Forests and Our Future'*, W. W. Norton and Company, New York, 2nd edn., 1992.
8. D. H. Williams, M. J. Stone, P. R. Hauck and S. K. Rahman. *J. Nat. Prod.*, 1989, **52**, 1189
9. *'Herbivores: Their Interactions with Secondary Plant Metabolites'*, G. A. Rosenthal and M. R. Berenbaum (Editors), Academic Press, London, 1992.
10. T. Cavalier-Smith. *'Secondary Metabolites: Their Function and Ecology'*, John Wiley and Sons, Chichester, 1992, p. 64.
11. J. B. Harborne. *'Introduction to Ecological Biochemistry'*, Academic Press, London, 4th edn., 1993.
12. P. G. Waterman. *'Secondary Metabolites: Their Function and Ecology"*, John Wiley and Sons, Chichester, 1992, p. 255.

BRAZILIAN BIODIVERSITY: A SOURCE OF PHYTOMEDICINES, NATURAL DRUGS AND LEADS FOR THE PHARMACEUTICAL AND AGROCHEMICAL INDUSTRIES

Benjamin Gilbert

Oswaldo Cruz Foundation, Pharmaceutical Technology Institute – FarManguinhos, 21041-250 Rio de Janeiro, Brazil; gilbert @far.fiocruz.br

The immense biodiversity of Brazil is a major world source of new pharmaceutical and agrochemical leads. Among these are a number of plants whose crude extracts have been used traditionally with success against diseases that are common to the tropical and temperate zones of the earth and others that are repellent to insects or other arthropods. Some of these crude extracts yield readily purified active principles that are already known to science, but are not economically exploited at their source in Brazil. In many cases these also provide leads for synthetic or semi-synthetic bio-active compounds. Other extracts exhibit biological activity as crude mixtures at a higher level or in a manner distinct to that observed with components isolated to date. Of particular interest are a rich diversity of alkaloids, especially indole and benzyliso-quinoline derived groups, bitter principles, especially quassinoids, hydroxylated and modified steroids and furanoid di- and tri-terpenoids, quinones, among them naphtho- and anthra-quinones and modified structures derived from these, as well as at least one quinone methide, and lignans. There is no doubt that the investigation of macromolecules, particularly proteins, glycoproteins and polysaccharides will yield other leads.

1 INTRODUCTION

Brazil has the largest biodiversity in the world contained within a single political unit. This results from the presence of a number of quite different habitats which include two distinct rain forest biotopes – the Amazon and the Atlantic forests – a very large porous soil savannah known as the "cerrado", a dry semi-desert and a temperate well-watered south. All of these areas comprise a million square kilometres or more. In recent times the biodiversity of those areas that are suitable for agriculture and cattle raising has suffered considerable destruction even though many of the plants and micro-organisms present are of undoubted economic value. The rational use of these can be engineered to protect the remaining natural reserves by providing a higher source of income than is obtainable from destructive practices such as logging and conventional agriculture. A summary of some of the unexplored opportunities is provided.

2 PHARMACEUTICAL LEADS

These may be divided into two sections: those plants, or occasionally animal materials, that have been used traditionally for healing disease - certainly over a thousand species, and those that have no or very little recorded history of medicinal use – the vast majority of the Brazilian biodiversity.   Looking at the first group one notices how few of these species are used outside the region. The recently published "Physicians' Desk Reference – PDR for Herbal Medicines" lists 559 recommended medicinal plants.[1] Of these only some 30 are tropical American species and 15 are native to Brazil. This is a very small proportion for such a rich biodiversity. In Table 1 are listed just one or two examples of plants used for each of a diversity of ailments.

Table 1 *Some medicinal plants used in Brazil which, with few exceptions, are not on the external market and which offer leads for the pharmaceutical industry. Many of these plants have been examined both chemically and pharmacologically.*

Disease or Condition	Species	Part used
Blood Anemia, post-malarial	*Bertholletia excelsa* H.B.K. (Lecythidaceae)	outer fruit case
Bronchial and pulmonary Including asthma (bronchodilator), bronchitis (wheeze), cough	*Mikania glomerata* Spreng. (Compositae) *Paraharncornia amapa* Ducke (Apocynaceae) *Cissampelos sympodialis* Eichl. (Menispermaceae) *Justicia pectoralis* var. *stenophylla* Leonard (Acanthaceae)	leaves latex root aerial part
	Lippia microphylla Cham. (Verbenaceae) *Aspidosperma carapanauba* Pichon., *A. desmanthum* Benth., *A. nitidum* Benth. (Apocynaceae) *Copaifera reticulata* Ducke, *C. multijuga* Hayne and other spp (Leguminosae, Caesalpinaceae) *Hymenaea courbaril* L (Leg., Caesalpinaceae) *Siparuna erythrocarpa* DC., *S. apiosyce* DC. (Monimiaceae)	leaves inner bark trunk oil trunk sap leaves
'lung diseases', restoration of lung lesion	*Kalanchoe brasiliensis* Camb. (Crassulaceae) *Caesalpinia ferrea* Mart. (Leguminosae)	leaf sap dry pod
Dental Anti-caries	*Stevia rebaudiana* Bert. (Compositae) *Zizyphus joazeiro* Mart. (Rhamnaceae)	leaves inner bark, leaves
Diabetes	*Bauhinia forficata* Link., *B. cheilantha* Steud., *B. macrostachya* Wall, *B. ungulata* L. (Leguminosae, Faboidae) *Myrcia multiflora* (Lam.) D.C. (*M. sphaerocarpa* D.C.); other *Myrcia* spp (Myrtaceae) *Cissus sycioides* L. (Vitaceae) *Chrysobalanus icaco* Drude et Engl. (Chrysobalanaceae)	leaves leaves leaves leaves
Fever **of indeterminate origin**	*Potomorphe umbellata* (L.) Miq. (Piperaceae) *Sambucus australis* Cham. & Schlt. (Caprifoliaceae)	aerial parts flower heads

Table 1 *Continued*

Disease or Condition	Species	Part used
Gastro-intestinal Amebiasis, giardiasis	*Licania macrophylla* Benth. (Chrysobalanaceae)	inner bark
constipation	*Carica papaya* L. (Caricaceae)	fruit, leaves, seeds
diarrhea, salmon- ellosis, shigellosis	*Psidium guajava* L. var. *pomifera* L. (Myrtaceae) *Simarouba amara* Aubl. (Simaroubaceae)	younger leaves bark, root bark
dyspepsia, indigestion	*Baccharis trimera* (Less.) DC, *B.* *genistelloides* Baker (Compositae)	aerial parts
hang-over	*Vernonia condensata* Baker (Compositae)	leaves
infection	*Quassia amara* L. (Simaroubaceae)	leaves
stomach ache	*Piper callosum* Ruiz et Pav. (Piperaceae)	leaves
gastritis and stomach ulcer	*Maytenus ilicifolia* Mart.,*M. acquifolia* (Celastraceae) *Myracroduon urundeuva* Fr. All. (Anacardiaceae)	leaves inner bark
haemorrhoids	*Scoparia dulcis* L. (Scrophulariaceae) *Eleutherine plicata* Hert. (Iridaceae)	whole plant rhizome
worms, intestinal	*Ficus insipida* Willd. (*F. anthelmintica* Mart.) (Moraceae)	latex
Geriatric Arteriosclerosis	*Cuphea carthagenensis* (Jacq.) Macbr. [*C.* *balsamona* Cham. & Schlecht] (Lythraceae) *Pfaffia paniculata* (Mart.) O. Kuntze (Amaranthaceae)	whole plant rhizoma
Gynaecological Amenorrhea	 *Aeolanthus suaveolens* G.Don (Labiatae)	leaves, flowers, seeds
cholic, menstrual	*Guatteria ouregou* (Aubl.) Dunal. (Annonaceae)	seeds
uterine infection., vaginitis (healing agent, astringent)	*Stryphnodendron adstringens* (Mart.) Cov. [= *S. barbatiman* Mart.] (Leguminosae) *Lippia gracillis* H.B.K., *L.* aff *alnifolia* Mart. et Schauer, *L. sidoides* Cham. (Verbenaceae) *Ouratea hexasperma* (St. Hil.) Bail. (Ochnaceae) *Dalbergia subcymosa* Ducke (Leguminosae)	bark leaves inner bark inner bark
Heart and Circulatory Angina pectoris, hypertension and heart problems	 *Cecropia hololeuca* Miq.,*C. peltata* Vell.,*C.* *glazioui* Sneth (*C. palmata*) (Moraceae) *Cereus grandiflorus* Mill. (Cactaceae)	 young leaves and buds stem
hyper-cholesteremia	*Croton cajucara* Benth. (Euphorbiaceae)	leaves, bark
Immune system Immune deficiency (immune stimulant)	*Tabebuia avellanedeae* Lor. ex Griseb and other species (Bignoniaceae) *Uncaria tomentosa* D.C., *U. guianensis* J.F. Gmel. (Rubiaceae)	inner bark inner bark

Table 1   *Continued*

Disease or Condition	Species	Part used
Infections, microbial		
General (antibiotic)	*Copaifera langsdorffii* Desf. and other spp (Leguminosae, Caesalpinaceae)	trunk oil
	Alternanthera brasiliana (L.) Kuntze and other sp. (Amaranthaceae)	aerial part
	Lippia gracillis H.B.K., *L. microphylla* Cham., *L.* aff. *alnifolia* Mart. et Schauer, *L. sidoides* Cham. (Verbenaceae)	leaves
	Capraria biflora L. (Scrophulariaceae)	leaves, flowers
mouth and throat	*Vismia cayennensis* Choisy, *V. guianensis* D.C. (Guttiferae)	resin from trunk, leaves, fruits
Inflammation		
Bladder and prostate	*Hymenaea courbaril* L. and other *Hymenaea* spp. (Leguminosae)	trunk sap
general	*Hyptis crenata* Pohl ex Benth (Labiatae)	aerial parts
arthritis and rheumatism	*Echinodorus macrophyllus* (Kunth.) Mich. (Alismataceae)	leaves
	Brosimum (Brosimopsis) acutifolium Huber (Moraceae)	bark
	Bauhinia rutilans Spruce ex Benth. (Leguminosae)	bark, stem
Liver		
Hepatitis	*Phyllanthus amarus* L., *P. niruri* Schum., *P. tenellus* Roxb. (Euphorbiaceae)	aerial parts
malfunction	*Solanum paniculatum* L. (Solanaceae)	root
Mental and Nervous Diseases		
Anxiety, nervousness (tranquilliser)	*Erythrina mulungu* Mart. (Leguminosae, Faboideae)	bark
spasms(antispasmodic)	*Achyrocline satureioides* D.C. (Compositae)	flower heads
nervous exhaustion	*Paullinia cupana* H.B.K. (Sapindaceae)	seeds
sleeplessness (sedative)	*Lippia alba* N.E.Br., *L. citriodora* L. (Verbenaceae)	leaves
Mutagenesis and Neoplasias		
Inhibitors of mutagenesis or of cancer	*Tabebuia* spp. (see Immune stimulation above)	inner bark
	Copaifera spp (see Skin diseases below)	trunk oil
	Himatanthus sucuuba (Spruce) Woodson (Apocynaceae)	latex
Pain		
General	*Ageratum conyzoides* L. (Compositae)	leaves
	Piper callosum Ruiz et Pav. (Piperaceae)	leaves
muscular	*Carapa guianensis* Aubl. (Meliaceae)	seed oil

Table 1   *Continued*

Disease or Condition	Species	Part used
Parasitic diseases		
Helminths	*Mammea americana* L. Guttiferae	bark, fruit case
leishmaniasis	*Kalanchoe brasiliensis* Camb. (Crassulaceae)	leaf sap
malaria	*Bidens pilosa* L.; *B. binnatus* L. (Compositae)	aerial parts
	Cassia occidentalis L. (Leg.Caesalpinaceae)	seeds
Skin		
General skin diseases	*Copaifera* spp. (Leguminosae)	trunk oil
	Virola surinamensis Warb., *V. sebifera* Aubl.	leaves, resin
	(Myristicaceae)	and bark
	Arrabidaea chica (H.B.K.) Bur. (Bignoniaceae)	leaves
itch, irritation,	*Caraipa grandifolia* Mart.; *C. densifolia* Mart.	bark
pityriasis	(Guttiferae)	
leprosy	*Carpotroche brasiliensis* Endl. (Flacourtiaceae)	seed oil
micoses	*Lippia sidoides* Cham., *L. gracilis* H.B.K.,	leaves
	L. aff. alnifolia Mart. et Schauer (Verbenaceae)	
	Ocotea barcellensis (Meiss.) Mez (Lauraceae)	trunk oil
Snake bites	*Eclipta alba* Hassk. syn. *E. prostrata* L.	aerial parts
	(Compositae)	
	Pentaclethra filamentosa Benth. (Leguminosae)	bark
Urinary system		
Diuretic	*Costus spicatus* Swartz. and other *Costus* spp.	aerial parts
	(Zingiberaceae)	
lithiasis, kidney	*Phyllanthus* spp. (see hepatitis above)	aerial parts
stones		
Viral diseases		
AIDS, HIV 1 & 2	*Eugenia florida* DC (Myrtaceae)	leaves
	Uncaria spp. (see Immune system, above)	bark
	Alternanthera spp. (see Infections, above)	aerial parts
Herpes simplex 1 or 2	*Copaifera* spp (see skin diseases above)	trunk oil
	Spondias mombin Jacq. (Anacardiaceae)	leaves, twigs
	Caraipa minor Huber, *C. densifolia* Mart.	bark
	(Guttiferae)	
measles	*Sambucus australis* Cham. et Schlecht	flowers
	(Caprifoliaceae)	

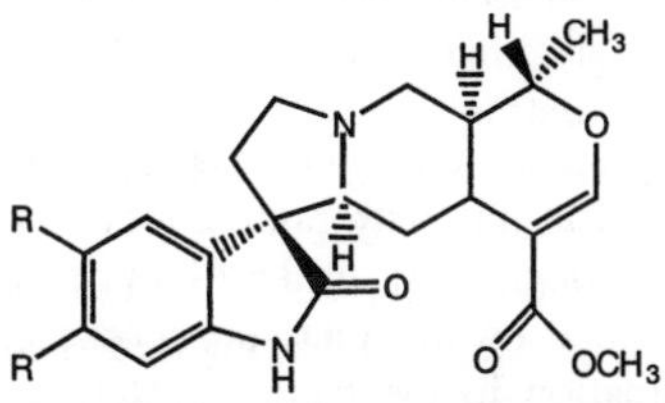

Figure 1   *The structure and stereochemical similarity of carapanaubine (R=OMe) from* Asidosperma carapanauba *Pichon,* Apocynaceae,[25] *and isopteropodine (R=H), the main immune stimulant of* Uncaria tomentosa *D.C.* Rubiaceae,[26] *suggest the involvement of the former alkaloid in the multiple medicinal uses of the bark of the "carapanaúba"(Nitidae) group of* Aspidoderma *species.*[21,27,28]

The widespread and continued use of these plants and acceptance by many phytomedicine experts is strong evidence that they are effective for the purposes described. In many cases compounds that contribute to the observed pharmacological activity have been identified,[2] but it should be remembered that plant medicines often act effectively as crude extracts which lose activity when fractionated in the attempt to isolate a single active component. This suggests that the activity of these plant medicines is due to a synergy, which may involve multiple water-soluble components that are often lost or intentionally discarded during conventional work-up procedures. Some of these components have been shown to be immune stimulants at very low concentrations and this type of activity probably makes up one of the effects in a combination of actions.[3] The many *Tabebuia* species employed in medicine, for example, contain both tannins and naphthoquinones, and the latter have been shown to stimulate the immune system.[4] The naphthoquinone, lapachol, which occurs in the heartwood of *Tabebuia* species, can be used as an anticancer agent[5,6] but not at the levels which would occur in bark teas recommended by herbalists.[7] At a recommended oral dosage of 20 mg/kg body weight in humans, the activity of lapachol may be related to interference with a redox mechanism or, by way of its metabolite β-lapachone, with DNA repair and direct action on topoisomerase-I.[8,9,10] However, there is less than 0.001% lapachol and 0.000004% β-lapachone in the bark, which is insufficient to cause anticancer activity in bark teas by the above mechanisms.[11] So, in dealing with traditional medicinal plants, we must look at two distinct markets – that for the isolated substance and that for the herbal drug as a synergistic mixture of compounds.

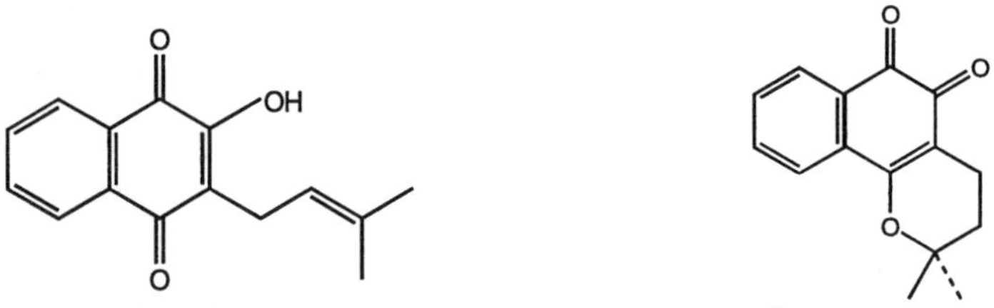

lapachol   β-lapachone

Figure 2 *Lapachol occurs in the heartwood of many* Tabebuia *species. Its interfence in oxidative processes has been documented.[8,9] Its metabolite β-lapachone inhibits enzymes associated with DNA replication.[10,30] However the use of* Tabebuia *bark (pau d'arco) as a cancer preventative can hardly be attributed to these activities.*

In order to access these drugs, with their potential for world sales, the would-be commercial organisation has to fall in line with international and national legislation. It is really only possible to do this through existing Brazilian bodies that can deal with the governmental regulatory aspects and ensure that the people of the region where the plant occurs have a real participation in the benefits. Here two more considerations come into mind: the commercial product is normally not just a crude extract or an isolated substance but is usually a formulated product. When derived from a pure chemical entity, it is usually not the original natural substance, but a derivative that exhibits higher stability or has preferred physico-chemical and pharmacological properties. Alternatively, it may be a purely synthetic substance that incorporates the idea derived from the original plant. Brazilian legislators are concerned that such modifications do not deprive the local population of its right to benefit by diluting royalties to an insignificant level. Thus, there is much to be said for installing production facilities in the region (as for example Merck of Germany's subsidiary Vegetex, in the northern state of Maranhão) and in working through a Brazilian organisation such as the Oswaldo Cruz Foundation's Institute of Pharmaceutical Technology or the State Agricultural Company EMBRAPA. There are also private intermediaries such as EXTRACTA Moléculas Naturais Ltda.,[12] which works with the Universities of Rio de Janeiro and Pará as well as with some environmental NGOs such as PRONATURA.[13] Local manufacture produces greater benefit than royalties and stimulates the development of the biotechnology of genetic improvement and propagation in the region where the impacts of industrial production on the environment can be adjusted to protect rather than just exploit natural reserves.

How can benefits be brought to a regional population and not just to a handful of technical employees in an already economically developed city or to their modern agro-organisation in the interior? This can be achieved through the cooperatives and commercial branches of the Brazilian government's Agricultural Reform Programme, which can also provide access to the most diverse habitats where substantial residues of original flora still exist. This assures direct health benefits to settlers in the outback and assists the recovery of the natural vegetation destroyed by loggers and cattle ranchers whose abandoned and degraded areas have fallen to the reform agency INCRA (National Institute of Colonisation and Agricultural Reform) to restore. In this way the spread of natural and anthropogenic fires is inhibited by the establishment of economically viable cultivation of native plants.

An important consideration in the agricultural production of plants containing medicinally active components is that they are sometimes produced by, or in consequence of, microbial agents present in or on the plant. These also constitute part of the biodiversity regulated by law, may be pathogenic to other economic crops and are not always easily cultured. Production in the natural habitat raises no problems. As an example one may notice that the maytansinoids, ansa-macrolides, which form part of the chemical make-up of *Maytenus ilicifolia* and *M. aquifolia,* Celastraceae, the two species most highly considered for stomach ulcer treatment, have been shown to be produced by bacteria associated with the plants.[14]

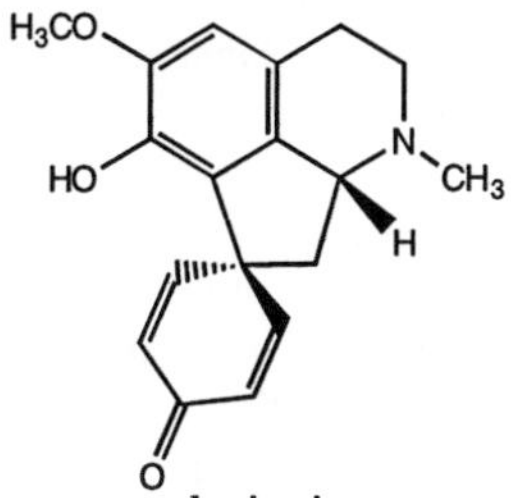

glaziovine 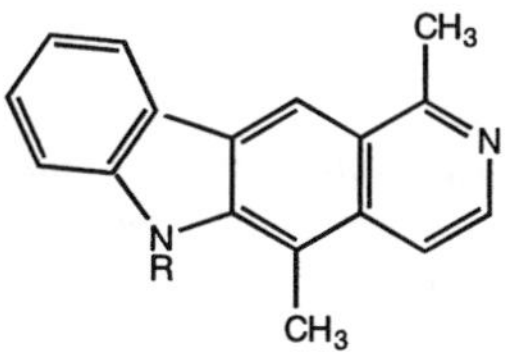   olivacine

Figure 3 *Glaziovine, an anti-depressant from Ocotea glaziovii,*[15,16] *and olivacine, found in Aspidoderma olivaceum and A. nigricans are examples of potentially valuable medicines derived from plants without apparent traditional use in Brazil.*[31,32] *The anti-leukaemic activity of olivacine is considerably enhanced by quaternization of the N_b nitrogen.*[33]

The majority of Brazilian plants have no known traditional medicinal use. Also, the diseases that affect modern society are not the same as those that led to the development of traditional medicine. Therefore the screening of "non-medicinal" plants is a valid route to new pharmaceutical leads. High throughput screening (HTS) is generally applied to this task. However it has been pointed out that the nature of the extraction procedure and particularly of the "clean-up" procedure which is used before submitting to HTS may well discard proteins, tannins or pro-tannins, polysaccharides and the like that have been shown in several cases to be responsible for, or contributory to, medicinal action. There are, of course, other restrictions to HTS, but it remains the only practical procedure for dealing with thousands of samples and it is the method used by EXTRACTA in Brazil. Other methods were formerly used and resulted, for example, in the discovery that glaziovine, a proaporphine from *Ocotea glaziovii* Mez, Lauraceae, had antidepressant activity.[15,16] In the formulation of pharmaceuticals a number of additives are used which include colouring matters and rheological agents. Of the former, two have been commercialised and, as they are used by indigenous peoples in body paint, are probably safe. These are "chica red", or carajurone, a flavonoid quinone methide which rapidly develops by enzymatic action in harvested leaves of the easily cultivated climber, *Arrabidaea chica* L., Bignoniaceae,[17] and the stable deep blue pigment formed from genipin from the fruit of *Genipa americana* L., Rubiaceae, on reaction with primary amines which is indelible on the skin.[18,19] Of the rheological agents, polysaccharides which are known to and used by the rural population are numerous. Among commercial possibilities are those from the very long pods of the Leguminous *Inga edulis* Mart. (small fast growing shade tree) and *Parkia pendula* Benth. (forest giant).

Carajurone/carajuretin   Genepin monomer

Figure 4   *Carajurone (R=CH₃) and the related carajuretin (R=H) result by the very rapid enzymatic aerial oxidation of flavans present in the leaves of Arrabidaea chica[17]. The plant is one of the most used Amazonian medicinal plants and the anti-microbial and other activities may be related to the presence of these red pigments[27]. Some indigenous tribes use them as face paints and carajurone has been commercialised in the past as "chica" red. Genipa americana fruit juice, by the action of an endogenous glucosidase followed by aerial oxidation, provides another pigment. The juice has multiple medicinal uses probably associated with its antimicrobial action[27]. Here the dye is blue-black and is believed to be an ethylene-bis-"ψ-aza-azulene", derived by condensation of the genipin monomer with primary amines[19].*

3 AGENTS ACTIVE AGAINST ARTHROPODS

Up till about 1950 Brazil produced a large amount of a rotenone containing insecticide, known commercially as "Derris" and derived mainly from the root bark of *Lonchocarpus (Derris) urucu* Killip., Leguminosae-Faboideae. This woody climber was planted along the northern margins of the Amazon and near some major tributaries and constituted one of the major economic products of the region along with Brazil nuts, chiclé and copaiba oil. The plantations, although 50 years without use, are, in many cases, still there and could enter into production at short notice. The species, being native, has not been ousted by the forest. There may also have been a smaller scale production of insecticide-containing extracts from *Ryania speciosa* Vahl., Flacourtiaceae, and *Quassia amara* L., Simaroubaceae, but residual plantations have not been located by the author. *Ryania*, which contains the highly toxic ryanodine would probably not be acceptable by present day standards, but quassia is used as an internal medicine for amoebiasis and malaria and has been used as a bitter agent in drinks. Both derris and quassia are economically viable insecticidal agents when formulated as ground plant material. As noted above for medicinal agents, derris is more active unrefined than its rotenoid content (about 25% on dried root bark) would seem to justify so that chemical processing, if done at all, is limited to straight extraction and concentration. *Quassia amara* is a member of a large number of insect repellent Simaroubaceae and Meliaceae species which occur in the Amazon and other tropical habitats of Brazil. The French ORSTOM group comment on the insecticidal activity of the twig bark of *Simarouba amara* Aubl.,[20] a species widespread in northern and central-western Brazil where its traditional use is as a mosquito repellent,[21] and, like quassia, as an internal medicine for intestinal problems.

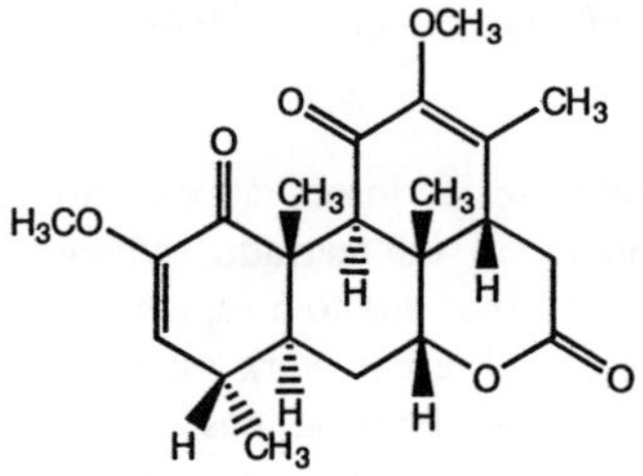

quassin

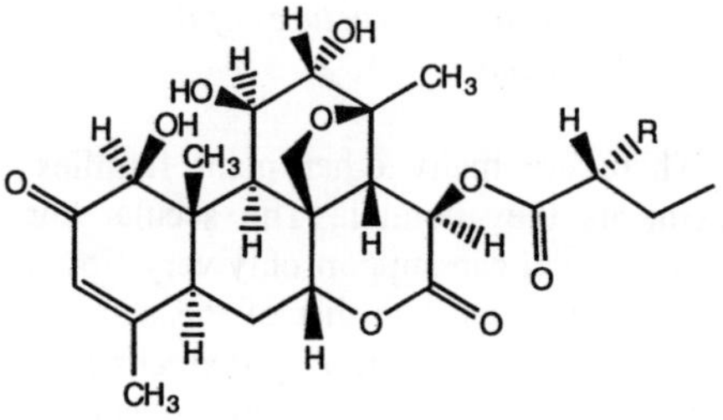

quassimarin (R=OAc)
simalikalactone-D (R=CH₃)

Figure 5 *Biological activities associated with quassinoids, from Simaroubaceae species, may not be linked to a common mechanism of action.[34] Different testing procedures have indicated insecticidal activity for quassin,[34] anticancer activity for quassimarin[35] (also present in* Quassia amara*) and antimalarial activity for simalikalactone-D.[36,37]*

Many of the bitter tasting oxygenated triterpenoids with rearranged or partly degraded structures that occur in the two plant families repel insects and this property is taken advantage of by local populations in areas where such agents are a necessity for life. However, in spite of their low toxicity and biodegradability, little attention seems to have been given to this group by industry with the sole exception of neem, *Azadirachta indica* A. Juss., Meliaceae, an Asian species. From the fruit of one very abundant tree species, *Carapa guianensis* Aubl., Meliaceae, we have developed an odourless insect repellent candle and this technology has now been licensed to a number of manufacturers[22].

The demand for some hundreds of tons of fruit from *Carapa guianensis*, expected shortly to exceed the natural supply, is stimulating replanting of this species in the states of Amapá and Pará where it had been decimated by the timber trade as a substitute for mahogany. Propagation studies already made with this species and with *Quassia amara* and *Lonchocarpus urucu* (by EMBRAPA) would make industrial production of these natural insecticides and repellents a short term undertaking.

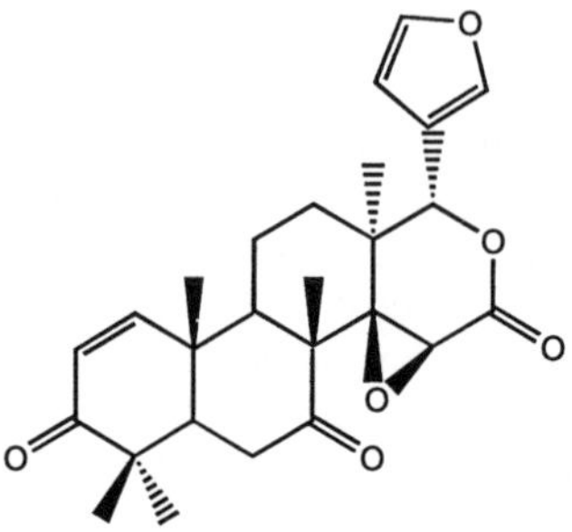

Figure 6 *7-Deacetyl-7-oxogedunin is one of a number of β-furanoid tetranor-triterpenoid bitter principles found in* Carapa guianensis *Aubl., Meliaceae.[29] Insect antifeedant or insecticidal activities have been demonstrated for bitter furanoid structures of the same or rearranged types from other Meliaceae species.[38]*

There are many other plant families that naturally resist insect attack through chemicals they contain. The secular survival of plants in the cerrado, where the untreated soil can support only very slow growth, is sometimes due to a repulsion even more efficacious than that of the above mentioned group. Present work on *Pterodon emarginatus* Vog. (syn. *P. pubescens* Benth.) Leguminosae-Faboideae, has shown that formulations of crude seed oil are as effective at the same concentration on skin as is N,N-diethyl-*m*-toluamide (DEET) against *Aedes aegyptii* mosquitoes. This activity is probably associated with the same components that were earlier shown to be lethal to schistosome larvae[23]. The long duration of palm fronds as roofing materials in ambients infested by termites and other cellulose feeding arthropods may also be associated with insect repellency. The popular use of palm material for keeping away mosquitoes is indeed recorded, and has been confirmed experimentally by our group in the traditional usage (pyrolysis)[24]. Many palm trees, including the species, *Attalea excelsa* Mart., in question, occur in immense stands sometimes covering hundreds of square kilometres. These are only a few examples of ecologically based agro-chemicals which have stood the test of time, have been used through centuries by man, are biodegradable and, if manufactured in the region, will aid to preserve one of the world's important reserves of biodiversity.

References and Footnotes

1. J. Gruenwald, T. Brendler and C. Jaenicke, scient. editors, *'Physicians' Desk Reference – PDR for Herbal Medicines'*, Medical Economics Company, Montvale, New Jersey, 1999.
2. A. R. M. de Souza Brito and A. A. Souza Brito in M. J. Balick, E. Elisabetsky and S. A. Laird, eds., *'Medicinal Resources of the Tropical Forest'*, Columbia Univ. Press, New York, 1996, chap. 28, p. 386.
3. H. Wagner and A. Proksch, *"Immunostimulatory Drugs of Fungi and Higher Plants"* in *'Economic and Medicinal Plant Research'*, Acad. Press, London, 1985, Vol. 1, Chapter 4; H. K. M. Wagner, *"Immunostimulants and Adaptogens from Plants"* in J. T. Arnason et al., *'Phytochemistry of Medicinal Plants'* Plenum Press, New York, 1995, Chap. 1; H. Wagner and M. Wiesenauer, *'Phytotherapie'*, Gustav Fischer, Stuttgart, 1995, Chapter 1, pp. 14.
4. H. Wagner, B. Kreher and K. Jurcic, *Arzneimittelforschung*, 1988, **38**, 273.
5. K. V. Rao, T. J. McBride and J. J. Oleson, *Cancer Research*, 1968, **28**, 1952.
6. C. F. de Santana, O. G. de Lima, I. L. d'Albuquerque, A. L. Lacerda and D. G. Martins, *Rev. Inst. Antibioticos*, 1968, **8**, 89.
7. K. Jones, *'Pau d'Arco, Immune Power from the Rain Forest'*, Healing Arts Press, Rochester, VT, 1995, pp. 13-14.
8. E. G. Ball, C. B. Anfinson and O. Cooper, *J.Biol. Chem.*, 1947, **168**, 257.
9. M. Gonçalves, R. Garcia Carnero, M. Blanco, E.C. Gurucharri-Lloyd, *Cancer Treatment Reports*, 1976, **60** (1), 1-8.
10. D. A. Boothman, D. K. Trask and A. B. Pardee, *Cancer Res.*, 1989, **49**, 605.
11. V. L. M. Q. Britto, A. B. de Oliveira, J. A. Lombardi and F. C. Braga, 2nd *IUPAC Internat. Conf. on Biodiversity*, 11-15 July 1999, Belo Horizonte, Brazil, Abstracts, p.105.
12. EXTRACTA, www.biorio.org.br/extracta.
13. PRONATURA, instpronatura@openlink.com.br.
14. J. O. Pereira and S. C. França, personal communication, 1999.
15. B. Gilbert, M. E. A. Gilbert, O. Ribeiro, M. M. de Oliveira, E. Wenkert, B. Wickberg, V. Hollstein and H. Rapoport, *J. Amer. Chem. Soc.*,1964, **86**, 694.
16. J. Buckingham, exec. editor, *'Dictionary of Natural Products"*, Chapman and Hall, 1994, vol.3, p.2558 , references cited therein.
17. E. Chapman, A.G. Perkin and R. Robinson, *J. Chem. Soc.*, 1927, 3015.
18. C. D. Mell, *Textile Colorist*, 1928, **50**, 602-4; *Chem. Abs.*, 1929, **23**, 708.
19. H. Inouye, Y. Takeda, K. Inoue, I. Kawamura, M. Yatsuzuka, R. Tonyama, T. Ikumoto, T. Shingu, T. Yokoi, *Tennen Yuki Kagobutsu Toronkai Koen Yoshishu*, 1983 (26), 577; *Chem. Abstr.*, 1984, **100**, 99895p.
20. P. Grenand, C. Moretti and H. Jacquemin, *'Pharmacopées traditionelles en Guyane'* ORSTOM, Paris, 1987, pp. 396-405.
21. J. E. C. Martins, *'Plantas Medicinais de Uso na Amazônia'*, CEJUP, Belém, 1989, pp. 82-83.
22. Fundação Oswaldo Cruz, Brazil Patent, PI 9800437-9, 22 Jan. 1998.
23. W. B. Mors, M. F. dos Santos F., H. J. Monteiro, B. Gilbert and J. Pellegrino, *Science*, 1967, **157**, 950.
24. W. B. Mors, *'Useful Plants of Brazil'*, Holden-Day, San Francisco, 1966, p. 21.
25. B. Gilbert, J.A. Brissolese, N. Finch, W. I. Taylor, H. Budzekiewicz, M. N. Wilson and C. Djerassi, *J. Amer. Chem. Soc.*, 1963, **85**, 1523.

26. H. Wagner, B. Kreuzcamp and K. Jurcic, *Planta Medica*, 1985, **51**, 419.
27. J. A. Duke and R. Vasquez, *'Amazonian Ethnobotanical Dictionary'*, CRC Press, Boca Raton, 1994, p. 25.
28. M. A. R. de Tenório, M. E. van den Berg, O. F. de Menezes and P. Salles in D. Bouchillet, ed. *'Medicinas Tradicionais e Medicina Ocidental na Amazônia'*, CEJUP, Belém, p. 443.
29. W. D. Ollis, A. D. Ward, M. M. de Oliveira and R. Zelnik, *Tetrahedron*, 1970, **26**, 1637.
30. A. R. Schuerch and W. Werbi, *Eur. J. Biochem.*, 1978, **84**, 197.
31. J. Schmutz and F. Hunzicker, Pharm. *Acta Helv.*, 1958, **33**, 344.
32. B. Gilbert, A. P. Duarte, Y. Nakagawa, J. A. Joule, S. E. Flores, J. Aguayo Brissolese, J. Campello, E. P. Carrazzoni,, R. J. Owellen, E. C. Blossey, K. S. Brown Jr. and C. Djerassi, *Tetrahedron*, 1965, **21**, 1141.
33. I. Koseki, K. Paulilo M. M. de Oliveira, I. T. Nakamura, C. E. M. P. D. M. Cappellaro and M. H. B. Catroxo, *Arq. Inst. Biol.*, São Paulo, 1989, **56**, 19.
34. J. Polonsky, *Fortschritte der Chemie Organischer Naturstoffe*, 1985, **47**, 221.
35. S. M. Kupchan and D. R. Streelman, *J. Org. Chem.*, 1976, **41**, 3481.
36. W. Trager and J. Polonsky, *Am. J. Trop. Med. Hyg.*, 1981, **30**, 531.
37. J. A. Cabral, J. D. McChesney and W. K. Milhous, *J. Nat. Prod.*, 1993, **56**, 1954.
38. K. L. Mikolajczak, D. Weisleder, L. Parkanyi and J. Clardy, *J. Nat. Prod.*, 1988, **51**, 606.

A MODERN PERSPECTIVE TO THE TRADITIONAL USE OF PLANTS IN THE HIGHLANDS OF SCOTLAND

Richard Constadouros

Agros Associates, Croc na Boull, Muir of Ord, Ross-shire, IV6 7TW, Scotland

1 INTRODUCTION

This paper demonstrates how local initiatives to protect and enhance biodiversity in the Highlands and Islands of Scotland are re-examining remedies based on indigenous plants developed over 1000 years ago in the region and which have almost died out. In many cases, in doing so, the basis for their traditional use is being verified on sound scientific fact. In addition, as part of this rediscovery process, additional uses are being found for many of these plants.

The scope of this paper explores the reasoning behind this new interest and then examines the historical context and highlights some of the major plant species traditionally used. It then moves on to the current initiatives based on a study that was commissioned by the Development Agency in the region, which are resulting in the introduction of a range of unique materials as ingredients for the pharmaceutical, cosmetic, chemical, and food and beverage industries.

2 BACKGROUND

Over the past decade or so there has been an increasing interest in the use of plants as feedstock for industry in the Highlands and indeed Scotland as a whole. This has often been led by small, cosmetic and specialist food companies, wishing to cash in on Scotland's current image of quality and purity. Many of these companies started by collecting their own plant raw materials but as their businesses grew so this task became unsustainable and they turned to ingredient suppliers. However, as it turned out their products were sourced almost totally from Europe.

At the same time there are two major push factors at influence in the Highlands. The first is the move to increase the biodiversity of the area and to move away from extensive plantations of single species conifers of alien species. This is occurring through the regeneration of the traditional mixed native woodlands of Scots pine, (*Pinus sylvestris*), birch, (*Betula alba* and *pendula*) and other native trees and an under story of species such as juniper (*Juniper communis*), wild cranberries, (*Vaccinium macrocarpa*) and blaeberries, (*Vaccinium myrtillus*).

The second is as a result of Common Agricultural Policy and changes in the traditional forestry sector. Both are resulting in losses of income to farmers, landowners and crofters who are all taking a closer look at their resources to identify new sources of income from their land.

3 HISTORICAL CONTEXT

This section draws heavily on Mary Beith's book on traditional medicines of the Highlands and Islands entitled Healing Threads.

The development of traditional medicine in the Highlands is attributed to the Celts although earlier inhabitants were using plants as medicines. In Celtic medicine the healing tradition was a mixture of common sense and mysticism in a vigorous blend. Yet as Mary Beith says, to date there has been a curious neglect of a wide view of the whole body of that tradition. However in spite of this, that tradition has survived until quite recently. Celtic medicine drew from various sources such as Greek and Arabic as well as an intelligent observation of nature and understanding of the human mind.

The core philosophy seems to have been the importance of preventative medicine and attention to the overall health of the body through cleanliness and hygiene. The remedies were all well documented. This healing tradition was part of the Celtic perspective on life, which once pervaded the whole of Britain and much of mainland Europe.

Gaelic traditional medicine, which developed a strong association with the official Highland physicians of the Middle Ages, had its foundation in Celtic society, which in turn had absorbed even earlier traditions. Two parallel traditions thus developed. The monastic and university trained physicians treating the wealthy and urban population and the folk healers working in the rural areas. Both very often used the same remedies. In the rural areas the clan chieftains extended their patronage to doctors and a hereditary system developed with a particular family such as the Beatons supplying the clan doctor. This tradition is now projected as a mix of nostalgia for a fairy other-world and a darker side of witches and demons.

So the question should now be asked: is there a future in the past? The answer lies in two areas. The first is in the general context of today's trend. There is still an interest in traditional medicines in the Highlands by doctors and pharmacists as evidenced by the existence of the Scottish Society of Historical Medicine. There is also the impact of plant based medicines and self medication in Eastern Europe and Germany , which continues the Celtic tradition of preventative medicine. In addition, there is the role of Chinese and Ayurvedic medicines and traditions from the rain forests of South America. All of this is fuelled by the interest of consumers in self administered, health and body care using plant based materials.

The second area is more specific and is based on an examination of the plant species used as food additives and in traditional medicines in the Highlands. A list below gives the ones considered by traditional practitioners as being highly valued.

4 MAJOR PLANT SPECIES USED BY CELTIC AND LATER HEALERS

Common Name	Latin Name	Usage
All Heal	*Prunella vulgaris*	With golden rod in the treatment of green wounds
Ash	*Fraximus excelsior*	Leaves used in poultices
Betony	*Betonica officinalis*	With others in wound treatment
Blaeberry	*Vaccinium myrtillus*	Astringent drink, stomach treatment
Bog myrtle	*Myrica gale*	Food flavouring, insect repellent
Bogbean	*Menyanthus trifoliata*	Stomach pains, skin poultices, tonic
Carrot	*Dacus carota*	Muscle pains and cancerous sores
Celandine	*Ranunculus ficaria*	Treatment of skin growths
Centaury	*Centaurium erythraea*	Tonic, aperitif, appetite promoter
Chickweed	*Stellaria media*	Use as a poultice
Dandelion	*Taraxicum leontoda*	Treatment of ulcers
Elder	*Sambucus nigra*	Skin tonic and in face cream
Eyebright	*Euphrasia officinalis*	Eyewash
Ferns		Treatments for burns, ointments
Heather	*Calluna vulgaris*	Flavouring for beverages, anti stress
Herb Robert	*Geranium robertum*	Skin cancer used up to 1922
Juniper	*Juniperus communis*	Lotions, ointments, beverage flavour
Lovage	*Ligusticum scoticum*	Food flavouring
Periwinkle	*Vinca minor*	Bruises
Plantain	*Plantago major*	Treatment of wounds and abrasion
St Johns Wort	*Hypericum perforatum*	With others, wound treatments
Seaweeds		Skin disorders
Vetch	*Lathyrus montanus*	Throat infections, beverages
Violet	*Viola oderata*	Cosmetic astringent

5 CURRENT INITIATIVES

In order to evaluate the above list in relation to the demands of today it is necessary to examine current initiatives in the Highlands.

In 1994 the development agency, Highlands and Islands Enterprise, commissioned a study to investigate the potential for the use of plants indigenous to the Highlands and Islands for the 4 industry sectors, food, pharmaceuticals, cosmetics and chemicals. The rationale for this was that:

- Scottish cosmetic companies were sourcing their ingredients from Europe;
- consumer demand for products with natural ingredients could provide commercial opportunities;
- there was a need to find alternative land uses.

The study looked at 120 plant species and found 17 with short and medium term potential prospects as well as 4 species that could be imported and cultivated. Also 15 of the species on the above list were found to be in trade today. These were:

Berries

Juniper	*Juniper communis*
Rowan	*Sorbus acuparia*
Sloe	*Prunus spinosa*
Bilberry	*Vaccinium myrtillus*
Cranberry	*Vaccinium macrocarpon*
Hawthorn	*Crataegus laevigata*
Sea Buckthorn	*Hippophae rhamnoides*

Other Plants

Meadowsweet	*Filipendula ulmaria*
Mountain arnica	*Arnica montana*
Bladderwrack	*Fucus vesiculosus*
Lovage	*Ligusticum scoticum*
Bog myrtle	*Myrica gale*
Bearberry	*Arctostaphylos urva-ursi*
Bitter vetch	*Lathyrus montanus*

Tree Products

Silver birch	*Betula alba* and *pendula*
Scots pine	*Pinus sylvestris*

These species were selected according to a number of criteria including their availability in quantity, whether currently in trade or considered by end users to have potential. In addition, Eyebright and St Johns Wort were among others also felt to have potential. It is possible to see the common species on this latter list and on the list of important plants in Celtic medicine.

The study report recommended that a vertically integrated commercial venture be established to market ingredients based on essential oils, extracts and chemical fractions derived from the species identified to UK, mainland Europe, US and possibly Japan. These ingredients would be processed using where appropriate, the latest technology such as CO_2 based extraction.

The raw plant material would initially be purchased from landowners from wild harvested sources but should long term potential be established, then landowners would be encouraged to grow these plants under some form of cultivation using improved planting material.

The proposals have raised a number of issues that are being considered by the management of the company, which has been established to develop this potential. The first of these relates to the harvesting of the plant material. Much of the Highlands is under one or more conservation designations which have the potential to restrict harvesting activities. The conservation organisations while supporting the venture in principal, as it supports the concept of increasing the biodiversity of the Highlands are concerned at the potential loss of biodiversity with uncontrolled harvesting of wild grown plants. In addition, the cost of harvesting is high with hand harvesting so that there is a need to mechanise this, which is where the concept of cultivation comes in. This in turn makes the selection of cultivars of certain of the target species with a high level of the active ingredient an important medium to long-term objective.

There is an increasing interest in developing new technologies for the extraction of materials from plants to improve quality and the efficiency of extraction. The venture aims to take advantage of these new technologies but this must take into account the requirements of the end users, which is where a partnership needs to develop. For many of these species, little is known of the compounds that may exist in them and which may be responsible for the perceived benefits identified by Celtic healers.

Finally, throughout the study and subsequent formation of the commercial venture there has been the recognition of the more advanced state of this industry in mainland Europe and the volume of plant material very competitively priced coming from Russia and the former Eastern Bloc countries.

The questions that have been asked by the management of the venture are:

- Should Scotland try to compete in the volume business for products from birch for instance, on the basis that the species grows like a weed in the Highlands?

- Should the venture go for unique products, which may provide a marketing advantage but have high entry costs?

- Should it try to develop niche markets for products from the more commonly traded species?

- Is there, particularly in the cosmetic sector, a premium for Scottish ingredients. Is there a marketing advantage to end users in this?

The solutions to these are being found but prospects are looking interesting.

6 CONCLUSIONS

So the final part of the paper completes the circle by identifying which of those plants on the original list are still used widely today and comparing the two lists. This shows that the early Celtic healers were not far out in their evaluation of the benefits of the plants that they and their successors were using from 1500 to 100 years ago. Some 15 0f the 20 are still used today and 7 are found on both lists. A number of examples are discussed below.

6.1 Bog Myrtle

This is a good example of an answer to the question, is there a future in the past, posed earlier. While not having a wide application by the Celts it was used as a food and beverage flavouring and for keeping summer insects at bay. Instances are also recorded of it being used for worming children.

Recent research initiated in Scotland has shown that a combination of some 10 compounds in the essential oil, of which terpenes are the key ones, are responsible for its insect repellent activity. Clinical and field trials have shown that it is as effective and in some cases better than diethyl toluamide against midge (*Culicoides* spp) and the mosquito (*Aedes aegypti*). Bog myrtle oil has the advantage of having a pleasant aroma and actually deters the insects from landing unlike DEET which only prevents the insects from biting. The oil has the potential to be used in creams including sun creams and does not need additives to mask its aroma.

It has also been found to contain antibacterial flavonoids which are responsible for the beer preserving activity. One of these, myrigalone B (MyB), is assigned to a rare class of flavonoids, C-methylated dihydrochalcones. As far as is known no further work has been done to identify the potential of these flavonoids and their antibacterial properties.

An extract of bog myrtle has been shown to have a suppressant effect on influenza A virus and a bacteriophage of Pseudomonas pyocyanea.

In a veterinary context the essential oil has been shown to be effective as a repellent against blowfly on sheep and mites on chickens

Research has also shown that it has anti fungal properties due to the presence of a flavonol glycoside. It was tested against the fungal pathogen *Trichopyton interdigitale* which causes infection of the feet and groin in humans and it compared well with nystatin, a compound already used against this pathogen.

6.2 Other Examples

There are also considered to be others and one or two are given below.

Another species is bogbean, which was used traditionally used as a poultice and more commonly against stomach pains. Its current use is against rheumatism and rheumatoid arthritis due partly to the presence of coumarins. It also is reported to have saliva inducing properties which provides potential for its use after operations.

Meadowsweet was used widely as a headache remedy. It is considered that just walking through it clears the head. Analysis of the compounds in the plant shows that it contains flavonoids, salicylates, tannins, volatile oils and other compounds such as coumarin, mucilage and ascorbic acid. It is currently used as a food flavouring, in beverages. It is also reported to be used for dyspepsia, muscular pains and peptic ulcers.

Thus, this project in the Highlands is still its early days, but with a combination of research into the past and the use of modern analytical techniques in is hoped that progress will be made in providing interesting and unique ingredients for industry from the Highlands of Scotland. In turn, it is hoped that new uses will be found for land that currently has little value and yet will preserve its landscape value.

5 Biosynthesis

IN VIVO AND *IN VITRO* BIOSYNTHETIC STUDIES : UNDERSTANDING AND EXPLOITING NATURAL PATHWAYS

Thomas J Simpson

School of Chemistry, University of Bristol, Cantock's Close, Bristol, BS8 1TS, UK

1 INTRODUCTION

Polyketides are naturally occurring compounds, most often produced by microorganisms such as fungi and the filamentous bacteria (the actinomycetes). The polyketide pathway is responsible for the formation of a huge diversity of natural products with structures varying from simple aromatic compounds such as 6-methylsalicylic acid to the gigantic polyether maitotoxin, whose molecular weight of 3422Da makes it the largest known secondary metabolite.[1] In addition to the inherent structural diversity shown by polyketide metabolites, many of them display potent biological activity, and it can be said that of all the pathways of secondary metabolism, it is the polyketide pathway that gives the greatest structural diversity and produces the widest range of biological activities. Furthermore, it is becoming clear that it is also the pathway that is most accessible to manipulation to increase the diversity of structures, either variations of existing skeletal types or totally novel structural classes.[2] In this chapter, *in vivo* and *in vitro* studies aimed at understanding the basic pathways for the assembly of polyketide metabolites, with a view to producing novel compounds, are described.

2 PRECURSOR-DIRECTED BIOSYNTHESIS OF FUNGAL METABOLITES

The manipulation of fermentations to facilitate the production of new compounds by incorporation of alternate precursor molecules has been well reviewed.[3] In the case of polyketide metabolites, a potentially useful approach is to look at the uptake of alternate starter acids which prime the polyketide chain assembly process. Norsolorinic acid (NSA, **1**) is the first isolable intermediate on the biosynthetic pathway leading to the potent hepatocarcinogenic mycotoxin, aflatoxin B_1 (**2**).[4] NSA is the first enzyme-free intermediate produced by the corresponding polyketide synthase (PKS), norsolorinic acid synthase (NSAS). The NSAS gene has been isolated from both *Aspergillus parasiticus*[5] and *Aspergillus nidulans*.[6] It has been shown to contain ketosynthase (KS), acyl carrier protein (ACP), acyl transferase (AT) and thioesterase (TE) domains only and, as expected, has none of the domains associated with reductive modifications of polyketides. It is associated with a dedicated fatty acid synthase (FAS) which appears to be responsible for the production of the hexanoate starter that primes formation of the octaketide chain that, after cyclisation and aromatisation, gives rise to NSA.[7] In the course of *in vivo* feeding studies to establish the role of hexanoate as a separately formed starter unit, we found that feeding [2-^{2}H$_2$]hexanoate in the form of its *N*-

acetylcysteamine (NAC) thioester resulted in a high level (*ca.* 40%) of intact incorporation.[8] This contrasted with feeding [2-^{2}H$_2$]hexanoate as the free acid which gave a low level of incorporation of label, ^{2}H NMR analysis of the isolated NSAS showed that most of the incorporation had occurred *via* prior degradation of the hexanoate to produce labelled acetyl CoA. Encouraged by this observation, we examined the incorporation of a number of alternate starter units to explore the specificity of NSAS for its chain-priming unit.

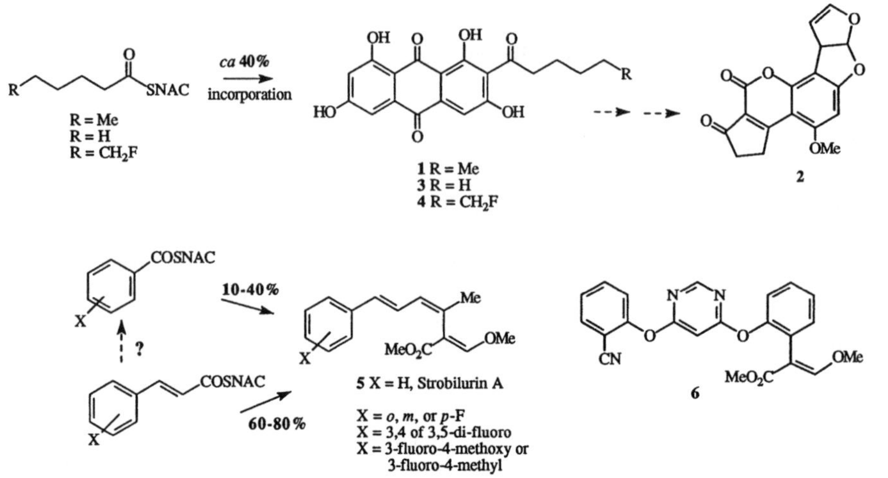

Scheme 1 *Examples of precursor-directed biosynthesis (PDB) of fungal metabolite analogues*

The studies, summarised in Scheme 1, showed that the NAC thioesters of pentanoic and fluorohexanoic acids are as efficient as the natural hexanoate starter and result in a high level of production of the nor-analogue (**3**) and the fluoro-analogue (**4**). While this ability to accept other starter units needs to be fully explored, it is noteworthy that shorter chains, e.g. butyrate, or longer chains, e.g. hexanoate, failed to show any incorporation.

The strobilurins, e.g. (**5**), are powerful natural anti-fungal agents produced mainly by a wide range of basidiomycetes.[9] While their inherent photolability renders them unsuitable as agrochemical reagents in their own right, their isolation has led to the development of the potent agrochemical agent azoxystrobin (**6**).[10] Their biosynthesis is relatively poorly understood but it has been shown that they are derived from a polyketide intermediate in which benzoate acts as the chain-initiator.[11] Encouraged by the successful incorporation of alternate starter units by feeding them as their NAC thioesters, we have studied the production of novel analogues of strobilurin A by incorporation of a range of fluorinated benzoates and cinnamates. In addition to the inherent interest in the fluoro-analogues, this has the advantage that the incorporation can be monitored directly by ^{19}F NMR. Some of the results are summarised in Scheme 1 which shows that a wide range of fluorobenzoates are incorporated with efficiencies between 10 and 40%. Interestingly, when the corresponding cinnamates were fed, the incorporation rates were substantially higher (60-80%) with the majority of the isolated metabolite being derived from the exogenous precursor.[12] Whether this reflects the fact that cinnamate is a later intermediate on the pathway, or that it is efficiently degraded to provide some of the fluorinated benzoyl CoA, remains to be determined. What it does show is a very loose substrate specificity of the strobilurin PKS.

3   BIOSYNTHESIS OF PYRROLE-CONTAINING POLYKETIDE METABOLITES

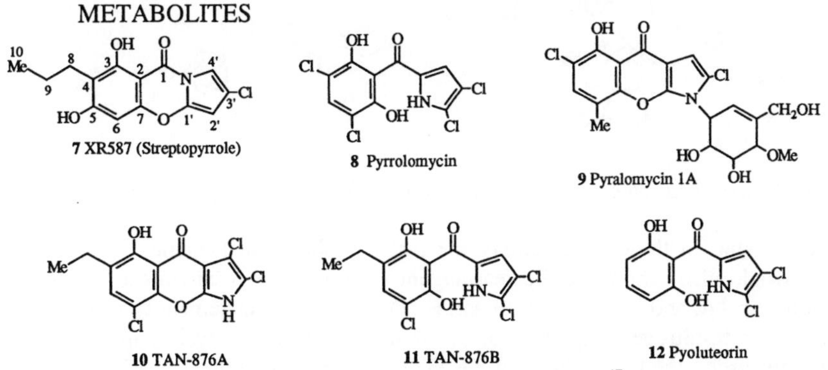

Figure 1 *Pyrrole-containing polyketide* Streptomyces *and other bacterial metabolites*

XR587 (**7**) is a pyrrole-containing antibiotic that was isolated from *Streptomyces rimosus* during a screening programme for antibacterial agents.[13] Its isolation from *Streptomyces armeniacus*, as streptopyrrole, was also recently reported.[14] It has an unusual amide-linkage to a phloroglucinol ring to which the pyrrole is also linked *via* an ether. A number of similar pyrrole-containing phenolic metabolites have been reported (Figure 1). Despite their superficial similarity, it is evident that they have significantly different biosyntheses. Pyrrolomycin (**8**)[15] is formed by straightforward cyclisation and aromatisation of a tetraketide intermediate in which a presumably proline-derived starter unit is elongated by three successive condensations with malonyl CoA. Recently reported labelling studies are consistent with a similar derivatisation of the pyralomycins, e.g. (**9**), *via* a tetraketide with the only difference being the introduction of the aromatic methyl group *via* methylmalonate and a rearrangement of the carbonyl group from C-2 to C-3 of the pyrrole.[16] Comparison of the structure of XR587 with the ochratoxin mycotoxins (Figure 2)[17] suggested that the amide linkage could be formed by condensation of proline with an aromatic carboxylic acid moiety formed by oxidation of the corresponding methyl group which could be introduced as in pyralomycin 1A (**9**) from propionate *via* methylmalonate or from *S*-adenosylmethionine.

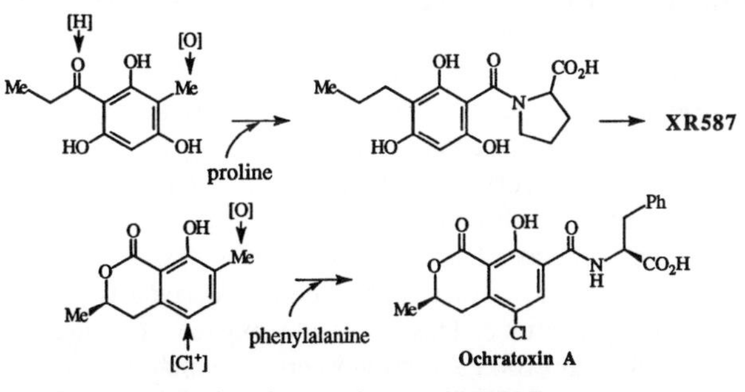

Figure 2 *Proposed ochratoxin-like biosynthesis of XR587*

However, this was soon ruled out by incorporation studies with ^{13}C-labelled acetate, propionate and proline, which gave the labelling pattern summarised in Figure 3, which showed that all four carbons of the pyrrole ring and the carboxyl group of the amide were derived by incorporation of an intact proline molecule.

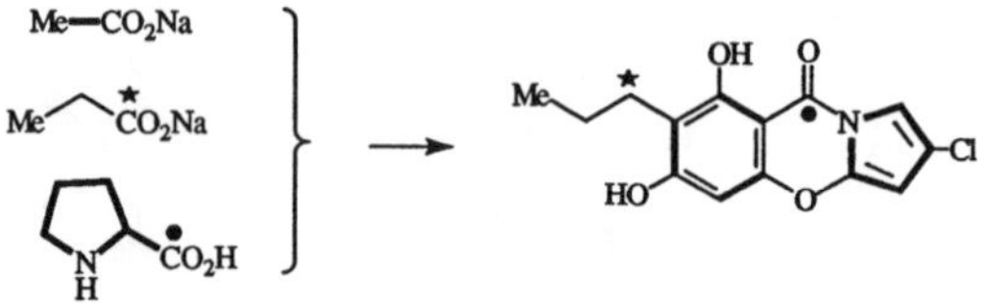

Figure 3 *Incorporation of ^{13}C-labelled acetate, priopionate and proline into XR587*

To account for these observations, the pathway shown in Scheme 2 is proposed,[18] which would also rationalise the rearrangements that must occur in the formation of pyralomycin 1A, and Tan-876A (**10**) and B (**11**).[19] The rearrangement to form an amide by a carbon-to-nitrogen migration of the proline-derived moiety is unprecedented. A further feature of the biosynthesis of XR587 (and presumably the Tan-876 metabolites) is that the single acetate incorporation pattern in the phenolic ring makes any pathway proceeding *via* a symmetrically substituted phloroglucinol ring unlikely[20] and so rules out a derivative *via* acylation of a simple polyketide-derived intermediate. It thus appears that the phenolic ring in these compounds must be derived, unusually, *via* condensation of two separately formed polyketide chains, as indicated in Scheme 2. Interestingly, very little is known about the timing and the mechanism of conversion of proline into pyrrole in these and a large number of pyrrole-containing metabolites. It is noteworthy that in these metabolites the pyrrole is usually variably halogenated to provide another source of structural diversity.

Pyoluteorin (**12**) is a further related pyrrole–phenol metabolite of *Pseudomonas fluorescens*. The biosynthetic gene cluster has been isolated and sequenced[21] and interestingly is shown to contain a Type I modular PKS. This is the first aromatic bacterial metabolite that has been demonstrated to be formed *via* a Type I rather than a Type II PKS (*vida infra*).

4　*IN VITRO* STUDIES OF TYPE II POLYKETIDE SYNTHASES

It has been shown that a large number of polycyclic aromatic metabolites are biosynthesised *via* Type II PKSs.[22] These metabolites include the important antibiotic oxytetracycline, and the antitumour agents tetracenomycin and daunomycin. Indeed, it was the isolation and sequencing of part of the gene cluster for actinorhodin (**19**) biosynthesis in *Streptomyces coelicolor*, and the demonstration that this was composed of a number of discrete open reading frames encoding essentially monofunctional proteins, by Hopwood and co-workers in 1984 that led to the recent burgeoning interest in polyketide biosynthesis.[23] Subsequent extensive *in vivo* and *in vitro* studies have helped to delineate the pathway of actinorhodin biosynthesis. The overall pathway is shown in Scheme 3, which shows that actinorhodin is derived *via* an octaketide chain which undergoes two cyclisations, loss of ketide oxygen and further modifications. Seminal work by Khosla, Hopwood and co-workers, in which they removed the biosynthetic gene cluster from *S. coelicolor* and derived an expression system that enabled the

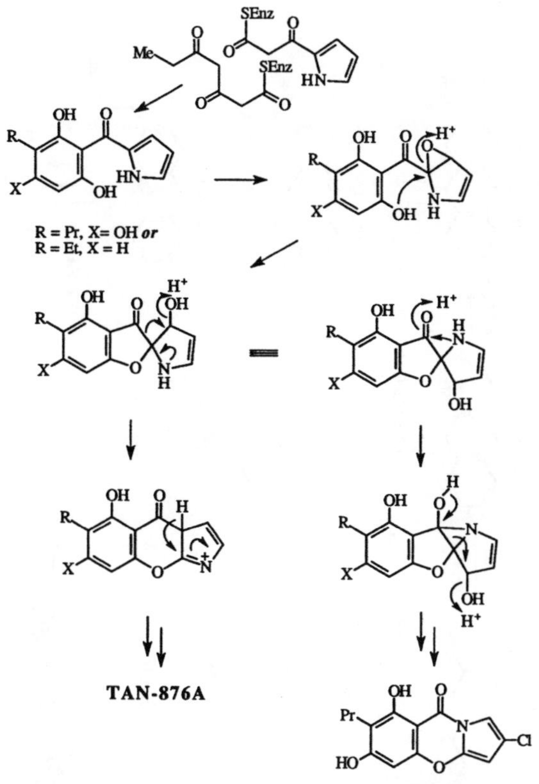

Scheme 2 *Proposed mechanism for rearrangement of the proline-derived pyrrole moiety in XR587 and TAN-876A*

re-insertion of one or more genes *in vivo*, led to the concept of the "minimal PKS", the minimum set of biosynthetic genes required for the production of an isolable polyketide metabolite.[24] The minimal PKS was shown, *in vivo*, to consist of the ACP, KS and so-called chain length factor (CLF), and this minimal PKS allowed the production of SEK-4 (**15**) which has the correct chain length and correct initial cyclisation for actinorhodin biosynthesis. Subsequent addition of other genes, notably the KR and aromatase, led to the production of further compounds such as mutactin (**16**), SEK-34 (**17**), aloesaponarin (**18**) and subsequently the full restoration of actino-rhodin production.[24]

A major aim of our work has been to overexpress and isolate, in pure form, the individual components of the minimal PKS and to see if these could be used to reconstitute, *in vitro*, the minimal PKS activity. *In vitro* studies would then allow us to dissect and study in detail the mechanisms involved in polyketide chain assembly, cyclisation and aromatisation.

Our initial studies focussed on the ACP component, as we believed that it could have a central role in stabilisation and cyclisation of the inherently unstable polyketide intermediates necessarily involved in chain assembly. The proposed pathway shown in Scheme 3 differs from the generally accepted pathway in a number of important aspects. First, we believe that the assembly pathway is a processive pathway, as in Type I polyketide biosynthesis,[25] and that the necessary modifications occur during the

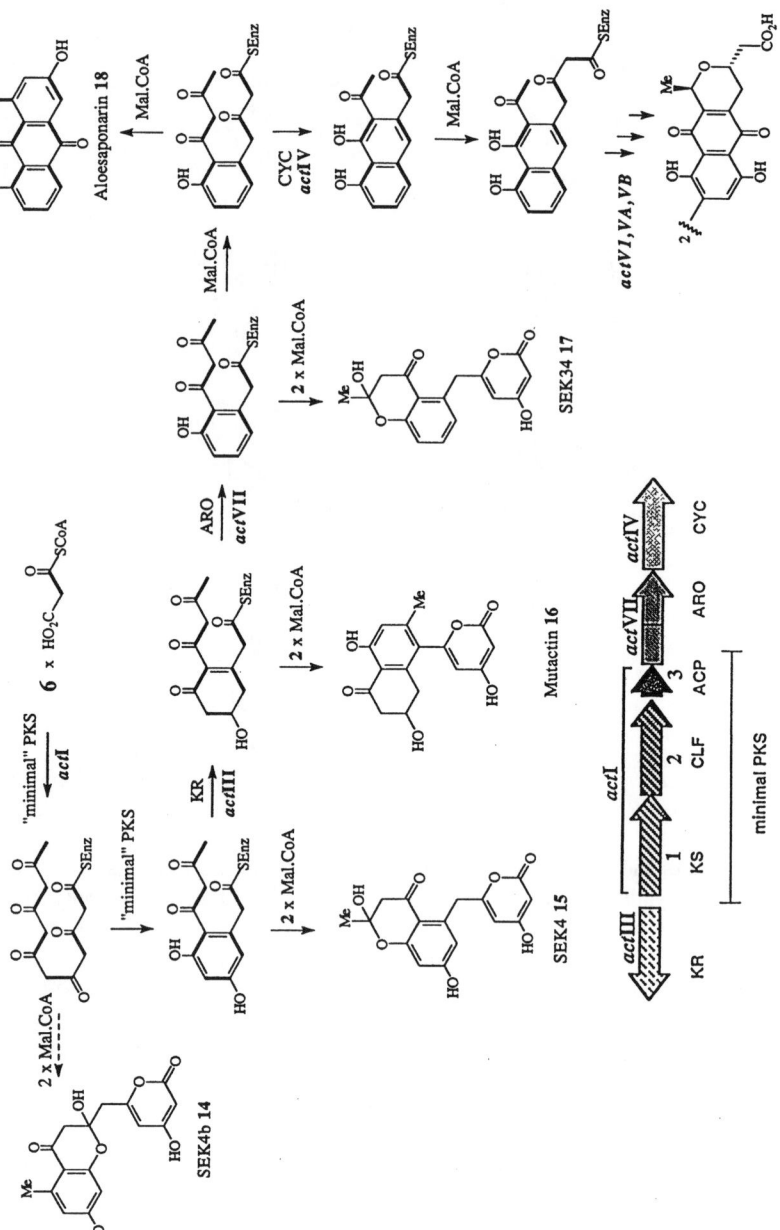

Scheme 3 *The actinorhodin PKS gene cluster and proposed "processive" assembly and cyclisation pathway catalysed by "minimal" PKS and further components*

assembly process. This cyclisation will occur as soon as the chain has reached the minimum length required for the key initial cyclisation. This would be at the hexaketide stage. This would then aromatise and the reduction and elimination processes would take place by reduction and dehydration processes analogous to those involved in e.g. the fungal melanin pathway in which 1,3,6,8-tetrahydroxynaphthalane is converted *via* the 1,3,8-trihydroxynaphthalene to 1,8-dihydroxynaphthalene.[26] Consistent with this proposal are sequence comparisons which indicate that the actIII, the actinorhodin "KR" gene, groups more closely with known fungal phenol reductase sequences than with typical FAS bacterial Type I ketoreductases.[27] SEK-4, mutactin and SEK-34 would then form by further chain elongation and cyclisations of the ACP-bound hexaketide intermediates as indicated. The formation of SEK-4 and SEK-34 involved hemiacetal formation between the remaining phenol hydroxyl and the first ketonic carbonyl in the side chain. In the non-aromatic intermediate formed in the reduction step, this hemiacetal formation is not feasible and so an alternate cyclisation pathway, leading to mutactin, occurs on further chain elongation.

Our initial efforts were directed at purification of a number of Type II PKS ACPs including those from the actinorhodin and oxytetracycline pathways.[28] The ACP genes encode the *apo*-form of the protein and this is converted to the active *holo*-form by transfer of the phosphopantetheine side chain from coenzyme A to the conserved serine-42 residue by the endogenous *E. coli holo*-ACP synthase. To ensure maximum expression of the *holo*-form of the ACP, the ACP gene is co-expressed with a copy of the ACPS gene as shown in Figure 4.[29] Electrospray mass spectrometry (ESMS) confirmed that the purified protein did, in fact, have the correct molecular weight, in contrast to the protein produced in the absence of the extra ACPS gene product.

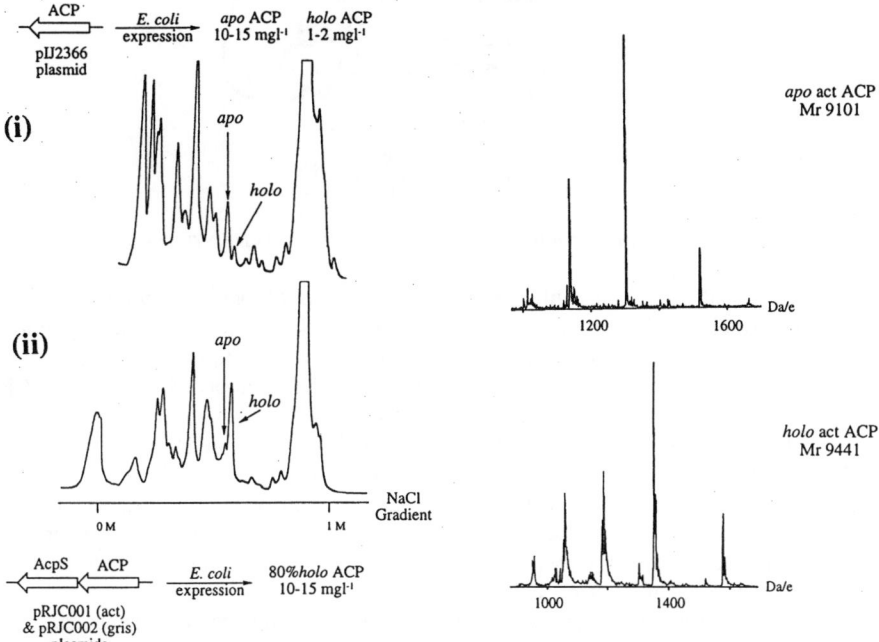

Figure 4 *FPLC purification and ESMS analysis of actinorhodin ACP expressed in i) the absence, and ii) the presence of ACP* holo-*synthase*

The three-dimensional structure of the ACP has been determined[30,31] by 2D NMR methods and this indicated a structure (Figure 5) analogous to that previously shown for the *E. coli* FAS ACP. The structure consists of four α-helical regions that define a potential binding cleft. A similar cleft is found in the *E. coli* ACP.[32] In the latter case, the cleft contains mainly hydrophobic residues, which is consistent with the binding of the growing saturated fatty acid side chain. However, the act ACP has a number of polar residues, e.g. arginine 72 and arginine 11, within the cleft, and these are exactly the type of residues that one might expect to participate in binding and stabilisation of the highly polar polyketide backbone of the actinorhodin precursor. Significantly, a number of these residues are highly conserved in Type II ACPs but are not present in FAS ACPs. To date, this is the only polyketide synthase component for which any 3D structure has been obtained.

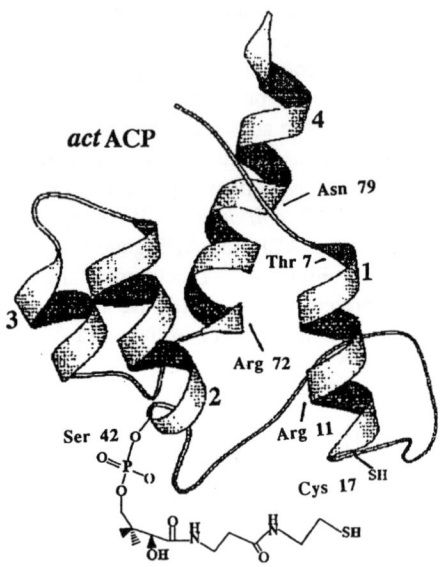

Figure 5 *Three-dimensional structure of the actinorhodin* (act) *PKS acyl carrier protein*

Our next goal was to establish conditions that would permit acylation of the phosphopantetheinyl thiol with acetate, malonate and further more advanced polyketide intermediates. The reasons for this are twofold. First, we wanted to look for evidence of specific binding of these acyl substituents consistent with the proposed role of the ACP in binding and stabilisation. Secondly, we wished to use these acyl ACPs in *in vitro* mechanistic studies of the assembly process. As is evident from the structure of the act ACP, there is an exposed cysteine residue at position 17, and preliminary studies showed, as expected, that selective chemical derivatisation of the phosphopantetheine thiol in the presence of cysteine 17 was not possible. This was overcome by mutagenesis to replace cysteine 17 with a serine, and the C17S mutant has been used in many of our subsequent *in vitro* studies.

The C17S mutant reacts cleanly with a number of thiol-specific agents, e.g. p-nitrophenyldisulphide, and a number of saturated and functionalised acyl residues can be cleanly added to the protein *via* the corresponding acyl imidazolides. In all cases the extent of fidelity of acylation can be monitored by ESMS.[29]

We then turned our attention to enzymatic methods for the preparation of acyl-ACPs. An important feature of most of the known Type II PKS gene clusters is the total absence of any gene corresponding to a malonyl transferase (MT or MCAT). This is surprising, as an essential feature of fatty acid and polyketide assembly is the transfer of malonate from malonyl CoA to the ACP where it is used for the essential chain-elongating condensation reaction. To account for this anomaly, it has been proposed that Type II PKSs use the endogenous MT associated with the essential FAS present in these bacteria, and the MT has been isolated and characterised from the FAS of both *S. coelicolor*[33] and tetracenomycin-producing *Streptomyces glaucescens*.[34] It has been proposed that the dual role of the MT in both FA and PK biosynthesis provides an unusual example of "cross-talk" between corresponding 1° and 2° metabolic pathways.[35,36]

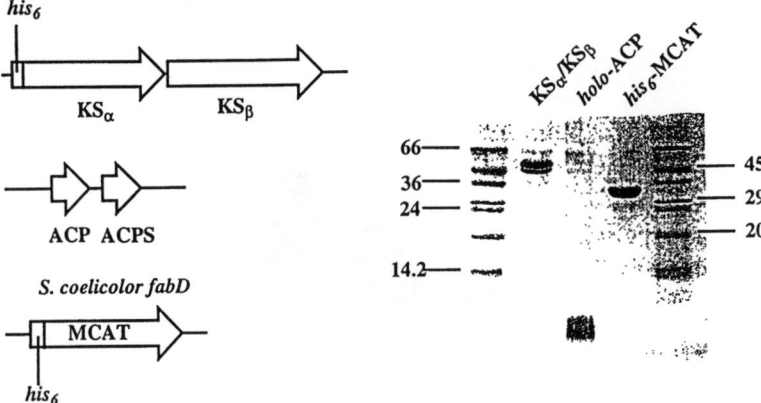

Figure 6 *Expression and purification of the components of the* act *"minimal" PKS: holo-ACP, KS and CLF (KSα and KSβ) and the* S. coelicolor *malonyl transferase*

With the need to prepare malonyl-ACP as an essential substrate for *in vivo* studies of actinorhodin biosynthesis, we overexpressed the *S. coelicolor* MT (MCAT) in his-tagged form to facilitate rapid purification. The purification is shown in Figure 6, and again ESMS confirmed the molecular weight of the protein. The activity of the protein was confirmed by its ability to rapidly malonylate the *S. coelicolor* FAS ACP. However, when we carried out corresponding experiments with the PKS ACP, we made the very surprising observation that the ACP reacted with malonyl CoA on its own and that addition of MT had no significant effect on the rate of this self-catalysed malonylation (Figure 7). Further studies showed that this was a general property of PKS but *not* FAS ACPs. Furthermore, no corresponding acylation was observed with acetyl CoA.[37] On the basis of this completely unexpected but intriguing observation, we proposed that the reason for the apparent lack of a dedicated MT activity in Type II PKS clusters could be a unique capacity of ACPs to catalyse their own malonylation and not use the FAS-associated MT. While the relevance of these *in vitro* observations to the true physiological roles of the proteins must be treated with caution, we were pleased with this early vindication of our belief in the value of *in vitro* studies of highly purified proteins in studying the details of polyketide assembly. Although the validity of this result was questioned by some others in the field,[38,39] it is now generally accepted to be correct.

The mechanism of ACP self-malonylation remains to be firmly established but it can be proposed that an initial binding of the malonate carboxylate to arginine 11 may have an important role. Support for this comes from the analogous self-acylation reaction also occurring with methylmalonyl CoA and the slow reaction with succinyl CoA. More surprisingly, both acetoacetyl CoA and the corresponding *N*-acetylcysteamine thioester are also efficient substrates for the reactions, resulting in the rapid and complete formation of the acetoacetyl derivative of the act-ACP.[37]

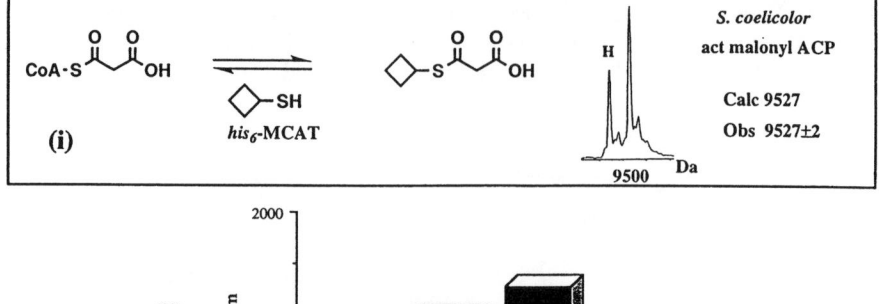

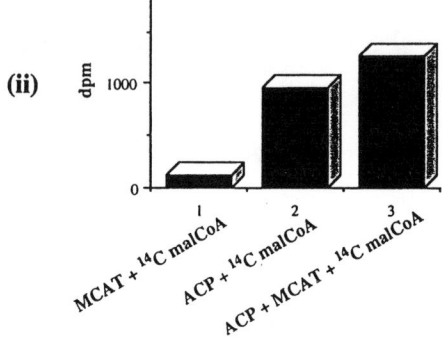

Figure 7 *Assay of the activity of his6-MCAT and the self malonylation of* act *holo-ACP, i) by ESMS analysis, and ii) by precipitation and counting of labelled protein*

The self-malonylation of the ACP suggested that the minimal PKS did indeed require only the KS, CLF and ACP. However, two *in vitro* experiments were reported that cast doubt on this. Khosla and co-workers reported that the act-KS and CLF, when incubated with malonyl CoA, produced no polyketide metabolite (SEK-4 or SEK-4b) unless the incubation was supplemented with malonyl transferase purified from *S. coelicolor*.[38] Hutchinson and co-workers, working with an analogous tetracenomycin minimal system, also found that production of tetracenomycin F_2 required the addition of the *S. glaucescens* MT. While they did observe self-malonylation of the *tcm* ACP, they suggested that the rate was not sufficient to account for more than 10% of the total polyketides produced. Interestingly, however, they did clearly show that malonyl CoA was the sole precursor and that acetyl CoA was not used as the starter unit.[39] The results of these experiments are summarised in Figure 8.

The results of our own experiments[40] are summarised in Figure 9. As the KS and CLF genes are translationally coupled, the proteins are expressed and purified as a heterodimer (see Figure 6). Two parallel sets of experiments were carried out. First, KS and CLF (each 1.5µM) were incubated with malonyl CoA and increasing concentrations of act *holo*-ACP. The incubations were analysed by HPLC which provides a very sensitive assay for production of polyketide metabolites. The lowest concentration of ACP (as in the previous experiments) was 2.5µM and detectable amounts of both SEK-4 and SEK-4b were observed. As the concentration of ACP was increased, the amount of total polyketide product increased. In the parallel experiments, MT was added. At the lowest concentrations of ACP, the presence of MT had a significant effect on the amount of polyketide but this difference rapidly decreased and at 20µM ACP there was no difference. Thus it is clear that in these *in vitro* experiments there was no absolute requirement for a MT and that the rate of polyketide production depended linearly on the concentration of ACP up to the highest concentrations (250µM) that we added.

i. *S.coelicolor* CH999/pSEK24

- cell-free extracts + malonyl CoA

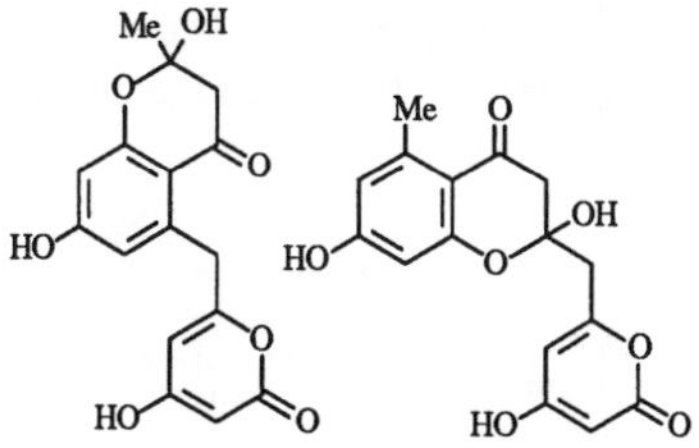

SEK4   **SEK4b**

- purified act KS/CLF + "fren *holo*-ACP" + malonyl CoA   - no polyketide synthesis

- ditto + MCAT - SEK4+ SEK4b

ii. *S.glaucescens*

- cell-free extracts + malonyl CoA

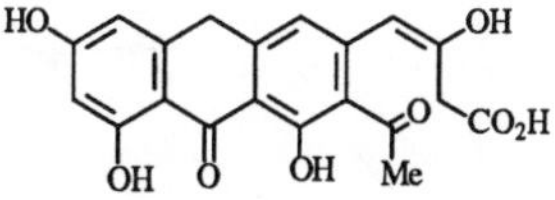

Tcm F2

- purified tcm KL + tcmM + tcmN - + malonyl CoA   no polyketide synthesis

- ditto + MCAT - Tcm F2

<u>Conclusions</u> - MCAT "essential" for polyketide synthesis

- little or no *holo*-ACP "autocatalytic" malonylation

Figure 8   *Summary of* in vitro *studies with cell-free extracts and purified "minimal" PKS systems from i)* S. coelicolor, *and ii)* S. glaucescens[38,39]

Thus we have achieved our initial aim which was to obtain a functional *in vitro* minimal PKS system that we can use to probe the fine details of the assembly process. This can be illustrated by recent experiments carried out in collaboration with the Cambridge group which have shown a possible role for the CLF component of the minimal PKS.[41] As stated above, acetyl CoA is not the starter unit for most Type II PKSs where the acetate starter unit is derived from malonyl CoA. The CLF and KS components have very similar sequences and differ mainly in the replacement of the essential active site cysteine in the KS with a glutamine in the CLF. This parallels an observation made for a number of Type I PKSs which include a KS analogue, KSQ, in their loading domains. Again, these have a cys→glu replacement. This is significant because it is well known that treatment of the essential active-site cysteine of the yeast FAS with iodoacetamide, converts it into a carboxamidomethylcysteine that is very similar in size, shape and polarity to glutamine.[42] Moreover, the treated enzyme is converted from a ketosynthase into a malonyl CoA decarboxylase. These observations strongly suggest that the CLF may function as a decarboxylase which converts malonyl ACP to acetyl ACP to provide the essential primer for polyketide assembly. This is summarised in Scheme 4.

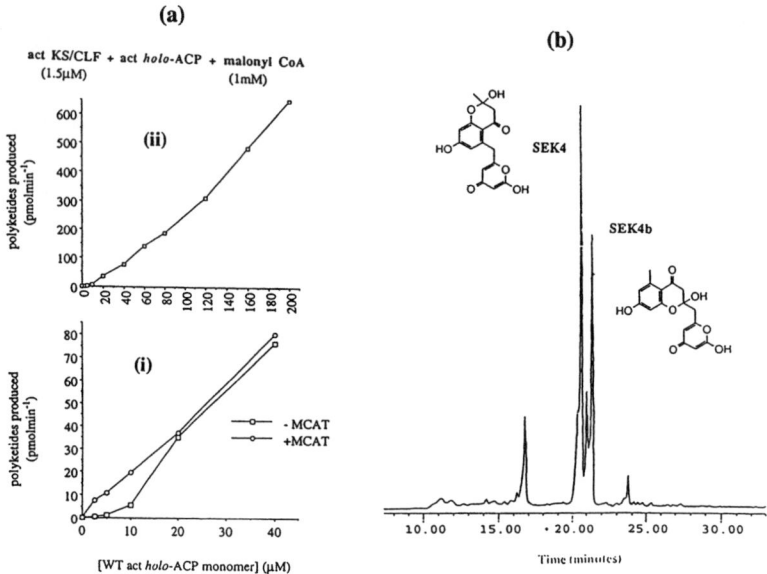

Figure 9 *(a) Variation in the rate of polyketide (SEK-4 and SEK-4a) production with ACP concentration: i) in the presence or absence of MT, and ii) increasing ACP concentration in the absence of MT. (b) HPLC trace of SEK-4 and SEK-4b production.*

To test this hypothesis, a number of KS-CLF mutants have been prepared and their ability to decarboxylate malonyl ACP determined. The results are summarised in Figure 10. In Figure 10a, we have the ESMS transform of the initial malonyl ACP. When this is incubated with KS(ala)/CLF(glu) in which the KS cysteine has been mutated so that no polyketide production is possible, we see slow conversion of malonyl ACP to acetyl ACP as predicted. The time course for the decarboxylation is shown below and this indicates *ca.* 40% decarboxylation after 1 hour. Intriguingly, when the malonyl ACP is incubated with a KS(glu)/CLF(ala) mutant, we see very rapid and complete decarboxylation. A similar result has been reported by Smith and co-workers with the mammalian FAS KS.[43]

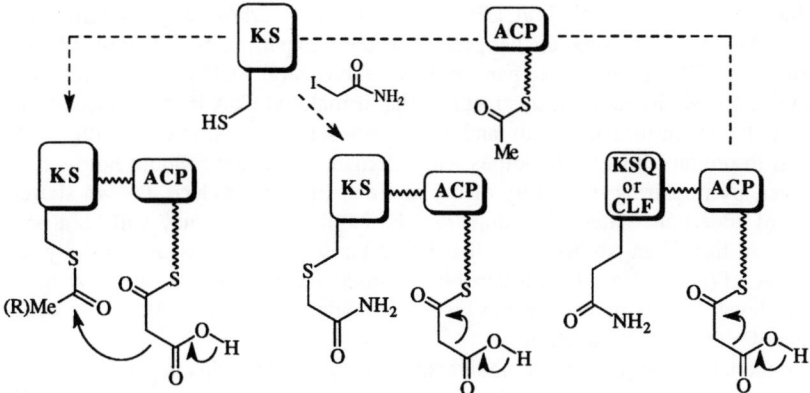

Scheme 4 *Proposed mechanism for decarboxylation of malonyl-ACP by iodoacetamide-modified KS and "CLF" domains. Acetyl-ACP from "CLF"-mediated decarboxylation of malonyl-ACP is transferred to KS for subsequent chain elongation by decarboxylative condensation. Reaction of KS with iodoacetamide blocks the condensation reaction and enhances the inherent decarboxylase activity of KS.*

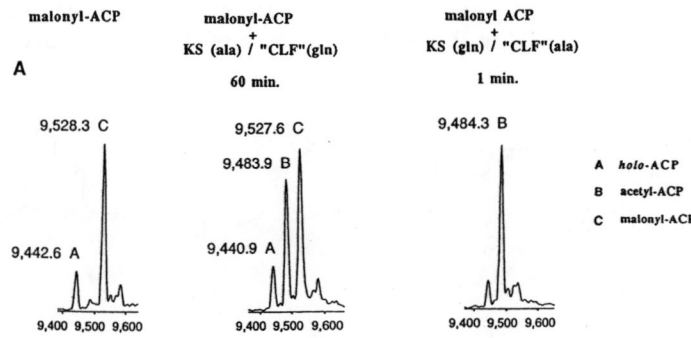

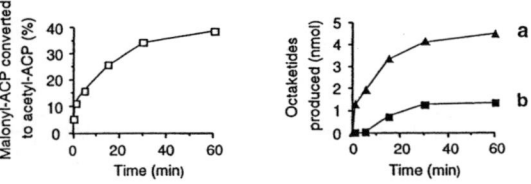

Figure 10 *Conversion of malonyl-ACP to acetyl-ACP and polyketide production by KS/CLF mutants. (a) ESMS analysis of initial sample of malonyl-ACP before decarboxylation and after incubation with either (middle) KS(ala)/CLF(gln) (60 min), or (right) KS(gln)/CLF(ala) (1 min). (b) Time course of decarboxylation of malonyl CoA by KS(ala)/CLF(gln), and production of octaketides (SEK-4 + SEK-4b) from malonyl ACP by KS(cys)/CLF(ala), both in the presence (triangles) and absence (squares) of acetyl-ACP.*

When treated with a KS(ala)/CLF(ala) mutant, no decarboxylation occurred and the malonyl ACP was recovered unchanged. Thus these experiments appear to indicate that the role of the CLF is, at least in part, to provide the acetyl ACP needed to initiate chain assembly. The significance of this result is that if malonyl CoA is the immediate source of both the chain-initiating unit and the chain starter, experiments with wild-type proteins to produce novel metabolites with alternate starter units would not be feasible. However, by impairing the ability of the minimal PKS to produce its own starter, the uptake of added alternate starter units will be facilitated. This may still be a problem because it has been shown that the KS retains a decarboxylase activity and a KS(cys)/CLF(ala) mutant is still capable of producing polyketides, albeit at a greatly reduced level. However, as shown in Figure 10b, addition of acetyl ACP to this restores the rate of polyketide production.

The theme of this meeting is biodiversity and to conclude this paper I would like to present a further intriguing example of nature's diversity. As mentioned above, 1,3,6,8-tetrahydroxyaphthalene (THN) is the essential polyketide precursor to melanin, which is involved in the pathogenic mechanisms of a number of fungal plant pathogens. In these organisms, THN has been shown to be produced by a Type I PKS containing a KS and two ACP domains within a single polypeptide of molecular weight 150K and encoded by a single gene (Figure 11).[44] It has been shown recently that in *Streptomyces griseus* and other bacteria, the same compound is produced by a protein of 42kDa which has no ACP and accepts malonyl CoA as its substrate.[45] The gene has high sequence similarity to the chalcone synthase gene found in higher plants. As more organisms are investigated, I am confident that further examples of this type of diversity will be discovered.

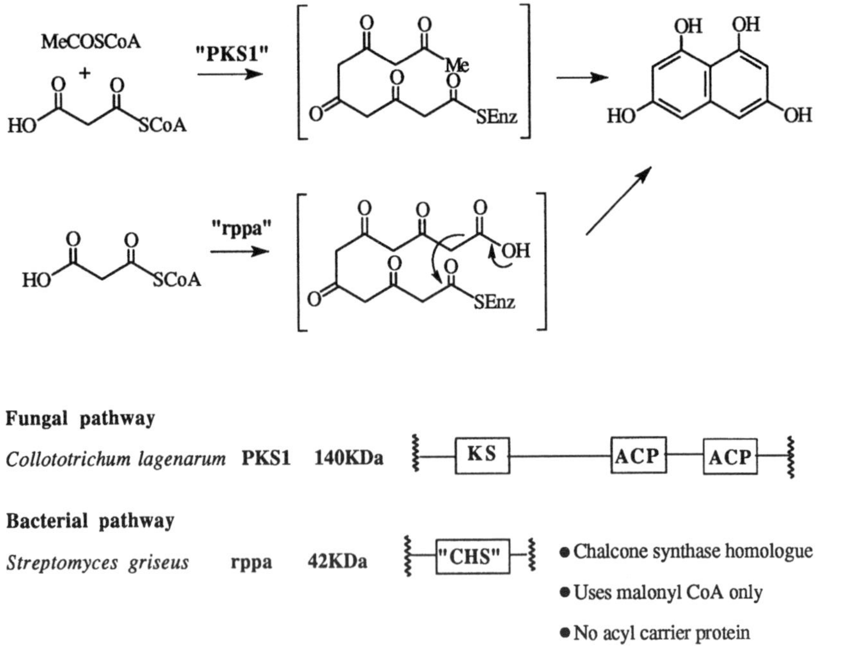

Figure 11 *Diverse pathways for tetrahydroxynaphthalene biosynthesis in fungi and bacteria*

Acknowledgements

The contribution of a number of senior collaborators, Professor P F Leadlay, Drs R J Cox and J Crosby, are acknowledged, in addition to the past and present postdoctoral and postgraduate researchers who carried out most of the work described above. These are Drs K J Byrom, M P Crump, T S Hitchman, M E Raggatt, A-L Matharu, N J Willett and A M Soares-Sello and Mr J Westcott. Financial support from EPSRC, BBSRC, the Commonwealth Scholarships Commission, University of Bristol and Xenova (now TerraGen) Discovery Ltd are gratefully acknowledged.

References

1. M. Murata, H. Naoki, T. Iwashita, S. Matsunaga, M. Sasaki, A. Yokoyama and T. Yasumoto, *J. Am. Chem. Soc.*, 1993, **115**, 2060.
2. P.F. Leadlay, *Curr. Opin. Chem. Biol.*, 1997, **1**, 162; J. Staunton, *Curr. Opin. Biol. Sci.*, 1998, **2**, 339.
3. R. Thiericke and J. Rohr, *Nat. Prod. Rep.*, 1993, **10**, 265.
4. R.E. Minto and C.A. Townsend, *Chem. Rev.*, 1997, **97**, 2537.
5. G.H. Feng and T.J. Leonard, *J. Bacteriol.*, 1995, **177**, 6246.
6. P-K. Chang, J.W. Cary, J. Yu, D. Bhatangar and T.E. Cleveland, *Mol. Gen. Genet.*, 1995, **248**, 270.
7. C.M.H. Watanabe, D. Wilson, J.E. Linz and C.A. Townsend, *Chem. Biol.*, 1996, **3**, 463.
8. D.S.J. McKeown, C. McNicholas, T.J. Simpson and N.J. Willett, *J. Chem. Soc., Chem. Commun.*, 1996, 301.
9. J.M. Clough and C.R.A. Godfrey, "The Strobilurin Fungicides", in *Fungal Activity*, Eds. E.H. Hutson and J. Miyamoto, John Wiley & Sons Ltd, 1998, pp109-148.
10. K. Beautemont, J.M. Clough, P.J. de Fraine and C.R.A. Godfrey, *Pestic. Sci.*, 1991, **31**, 499.
11. M. Vondarácek, J. Vondrácková, P. Sedmora and V. Musílek, *Collect. Czech. Chem. Commun.*, 1983, **48**, 1508.
12. A. Soares-Sello, PhD Thesis, University of Bristol, 1998.
13. E. Olsen, S.J. Trew, S.K. Wrigley, L. Pairet, M.A. Hayes, S. Martin and D.A. Kau, *Int. Pat.*, WO 98/25931.
14. J. Breinholt, H. Gürtler, A. Kjaer, S.E. Nielsen and C.E. Olsen, *Acta. Chem. Scand.*, 1998, **52**, 1040.
15. G.T. Carter, J.A. Nietsche, J.J. Goodman, M.J. Torrey, T.S. Dunne, M.M. Siegel and D.B. Borders, *J. Chem. Soc., Chem. Commun.*, 1989, 1271.
16. N. Kawamura, R. Sawa, Y. Takahashi, T. Sawa, H. Naganawa and T. Takeuchi, *J. Antibiot.*, 1996, **49**, 657.
17. A.E. de Jesus, P.S. Steyn, R. Vleggaer and P.L. Wessels, *J. Chem. Soc., Perkin Trans. 1*, 1980, 52.
18. M.E. Raggatt, T.J. Simpson and S.K. Wrigley, *J. Chem. Soc., Chem. Commun.*, 1999, 1039.
19. Y. Funabashi, M. Takizawa, S. Tsubotani, S. Tanida and S. Harada, *Takeda Kenkyushosho*, 1992, **51**, 73; *Chem. Abstr.*, 1993, **118**, 55722x.
20. T.J. Simpson, *Top. Curr. Chem.*, 1997, **195**, 1.

21. B. Nowark-Thompson, N. Chaney, J.S. Wing, S.J. Gould and J.E. Loper, *J. Bacteriol.*, 1999, **181**, 2166.
22. For a comprehensive review, see: B.J. Rawlings, *Nat. Prod. Rep.*, 1999, **16**, 425.
23. F. Malpartida and D.A. Hopwood, *Nature*, 1984, **309**, 462.
24. R. McDaniel, S. Ebert-Khosla, H. Fu, D.A. Hopwood and C. Khosla, *Proc. Nat. Acad. Sci. USA*, 1994, **91**, 11542.
25. T.J. Simpson, *Chem. Ind.*, 1995, 407.
26. N.S. Perpetua, Y. Kubo, N. Yusada, Y. Takeno and I. Furusawa, *Mol. Plant. Microb. Inter.*, 1996, **9**, 323.
27. T.P. Nicholson, PhD Thesis, University of Bristol, 2000.
28. J. Crosby, D.H. Sherman, M. Bibb, T.J. Simpson and D.A. Hopwood, *Biochim. Biophys. Acta,* 1995, **1251**, 32.
29. R.J. Cox, T.S. Hitchman, K.J. Byrom, S.C. Findlow, J.A. Tanner, J. Crosby and T.J. Simpson, *FEBS Letters*, 1997, **405**, 267.
30. M.P. Crump, J. Crosby, C.E. Dempsey, M. Murray, D.A. Hopwood and T.J. Simpson, *FEBS Letters*, 1996, **391**, 302.
31. M.P. Crump, J. Crosby, C.E. Dempsey, J.A. Parkinson, M. Murray, D.A. Hopwood and T.J. Simpson, *Biochemistry,* 1997, **36**, 6000.
32. T.A. Holak, J.H. Prestegard, K.J. Kearsley and Y. Kim, *Biochemistry*, 1988, **27**, 6135.
33. W.P. Revill, M.J. Bibb and D.A. Hopwood, *J. Bacteriol.*, 1995, **177**, 3946.
34. R.G. Summers, A. Ali, B. Shen, W.A. Wessal and C.R. Hutchinson, *Biochemistry*, 1995, **34**, 9389.
35. W.P. Revill, M.J. Bibb and D.A. Hopwood, *J. Bacteriol.*, 1996, **178**, 5000.
36. P. Zhou, G. Florova and K.A. Reynolds, *Chem. Biol.*, 1999, **6**, 577.
37. T.S. Hitchman, J. Crosby, K.J. Byrom, R.J. Cox and T.J. Simpson, *Chem. Biol.*, 1998, **5**, 35.
38. C.W. Carreras and C. Khosla, *Biochemistry*, 1998, **37**, 2084.
39. W. Bao, E. Wendt-Perkowski and C.R. Hutchinson, *Biochemistry*, 1998, **37**, 8132.
40. A.-L. Matharu, R.J. Cox, J. Crosby, K.J. Byrom and T.J. Simpson, *Chem. Biol.*, 1998, **5**, 699.
41. C. Bisang, P.F. Long, J. Cortés, J. Westcott, J. Crosby, A.-L. Matharu, R.J. Cox, T.J. Simpson, J. Staunton and P.F. Leadlay, *Nature*, 1999, **401**, 502.
42. G.-B. Kresze, L. Steber, D. Oesterhelt and F. Lynen, *Eur. J. Biochem.*, 1977, **79**, 191.
43. A. Witkowski, A.K. Joshi, Y. Lindquist and S. Smith, *Biochemistry*, 1999, **38**, 11643.
44. Y. Takano, Y. Kubo, K. Shimizu, K. Mise, T. Okuno and I. Furasawa, *Mol. Gen. Genet.*, 1995, **249**, 162.
45. N. Funa, Y. Ohnishi, I. Fujii, M. Shibuya, Y. Ebizuka and S. Horinouchi, *Nature*, 1999, **400**, 897.

FUSED RING AROMATIC POLYKETIDES ARE FORMED BY DIFFERENT CYCLISATION PATHWAYS IN FUNGI AND STREPTOMYCETES

Robert Thomas

Biotics Limited, School of Chemistry, Physics and Environmental Science, University of Sussex, Falmer, Brighton BN1 9QJ, United Kingdom

Dedicated to the late Professor Harold Raistrick F.R.S. in recognition of his seminal contributions to the chemistry of fungi

1 INTRODUCTION AND HISTORICAL PERSPECTIVE

At the dawn of the twentieth century, only a few simple fermentation products such as oxalic acid and citric acid had been isolated from the lower fungi.[1] Soon after World War I, Harold Raistrick initiated the first systematic studies of the chemistry of mould metabolites and in the course of the following four decades made a seminal contribution to the recognition of fungi as a major source of natural products.[2]

The mid-century clinical and industrial development of penicillin triggered a search for new antibiotics, an unanticipated bonus of which was the discovery that streptomycetes are an equally fertile source of novel structures. Perhaps not surprisingly, the taxonomic gulf between the streptomycetes, which are prokaryotic filamentous bacteria and the eukaryotic fungi is reflected in the structural and biosynthetic individuality of their respective secondary metabolites. While both form a wide variety of structures, pre-eminent among which are the polyketides, they have rarely been shown to yield identical products.

The chemical diversity of microbial constituents and continuing demand for improved antibiotics served to stimulate progressive improvements in the methodology of structure determination and total synthesis,[3] while the advent of isotopically labelled compounds provided a new dynamic for the investigation of biosynthetic pathways.[4] Landmark developments in this area include the classic speculative and experimental contributions of Arthur Birch dating from the 1950s, which have proved pivotal to our understanding of polyketide biosynthesis.[5] A further landmark in the closing two decades of the century has been the pioneering development by David Hopwood and his collaborators of a facile procedure for the controlled genetic manipulation of polyketide pathways in streptomycetes.[6]

Today, despite recent developments in combinatorial and other chemosynthetic strategies, fungi and streptomycetes remain the most prolific sources of new candidate drugs and agrochemicals. Both phyla elaborate bicyclic, tricyclic and tetracyclic fused ring polyketides, however, a preliminary survey of isotopically labelled precursor incorporation studies has revealed a consistent difference in the modes of cyclisation by which their characteristic polybenzenoid metabolites are formed.[7] These and subsequent observations (*vide infra*) provide the basis for a novel biosynthetic classification of microbial fused ring polyketides.

2 INCORPORATION OF INTACT C$_2$-UNITS IN FUNGAL AND STREPTOMYCETE FUSED RING POLYKETIDES

The distribution of intact C$_2$-units in cyclic polyketides and hence the geometry of folding of linear intermediates, can be determined through precursor feeding studies using [^{13}C$_2$]acetate.[8,9] This requires the unambiguous assignment of paired ^{13}C-^{13}C couplings in the ^{13}C-NMR spectra of biosynthetically labelled metabolites.

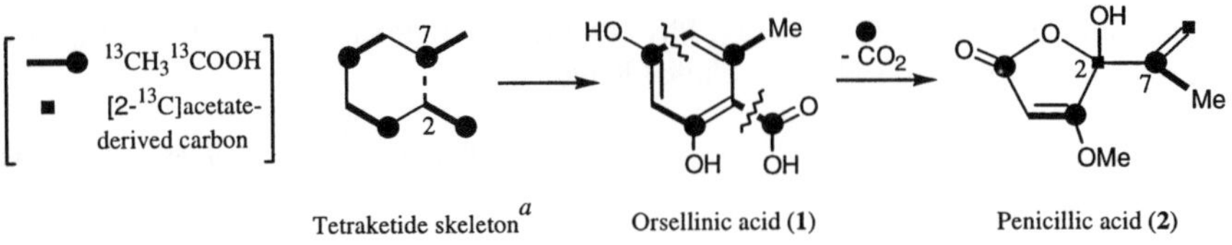

Tetraketide skeleton^a Orsellinic acid (1)   Penicillic acid (2)

Scheme 1 *Incorporation of [^{13}C$_2$]-labelled acetate into penicillic acid*
          ^{a}*Unless otherwise indicated, the numbering of the carbon atoms of all structures corresponds to that of the parent polyketide*

In one of the earliest applications of this methodology, Seto, Carey and Tanabe[10] determined the mode of cleavage of the fungal tetraketide orsellinic acid (1) which gives rise to penicillic acid (PA, 2). The observed couplings require a regiospecific ring cleavage in agreement with earlier studies based on the systematic chemical degradation of [^{14}C]acetate-labelled PA[11,12] and on the ^{3}H-NMR assignments of [^{3}H]acetate-derived PA[13] (Scheme 1).

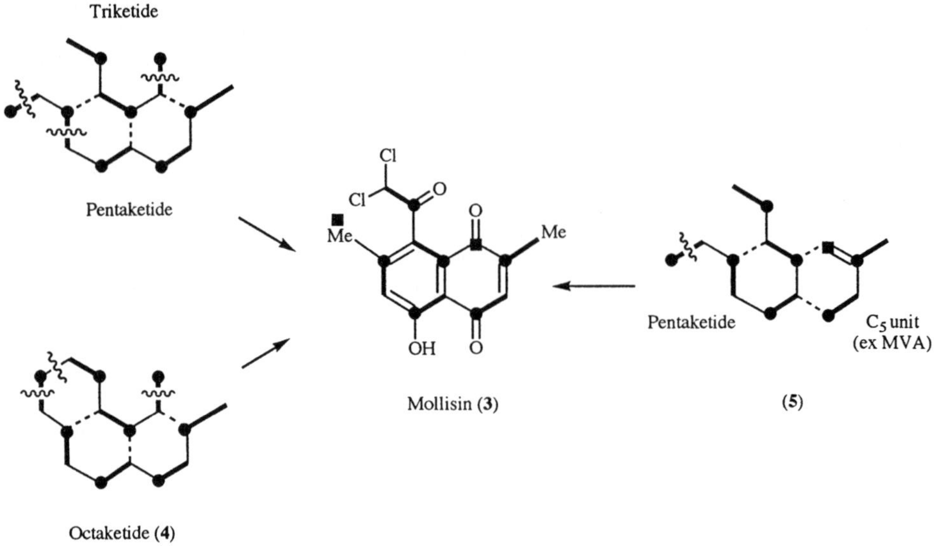

Mollisin (3)   (5)   Octaketide (4)

Scheme 2 *Incorporation of [^{13}C$_2$]-labelled acetate into mollisin*

These authors also described the first application of this methodology to the investigation of a fused ring fungal metabolite, namely the structurally unusual naphthoquinone mollisin (3).[8] [13]C-^{13}C couplings were interpreted as indicating a possible branched chain origin involving one of three options requiring two separate oligoketides (Scheme 2). This however does not exclude alternative pathways, either through ring cleavage of a linear octaketide (4),[9,14] or *via* an isopentenylpentaketide (cf. 5) derived from a linear pentaketide and mevalonate (MVA).

The rare tricyclic phenalenone fused ring system of the fungal heptaketide deoxyherqueinone (6)[15] could in theory form by three different modes of folding of a linear intermediate, but only one of these is consistent with the couplings observed in [[13]C₂]acetate-derived 6[16] (Scheme 3).

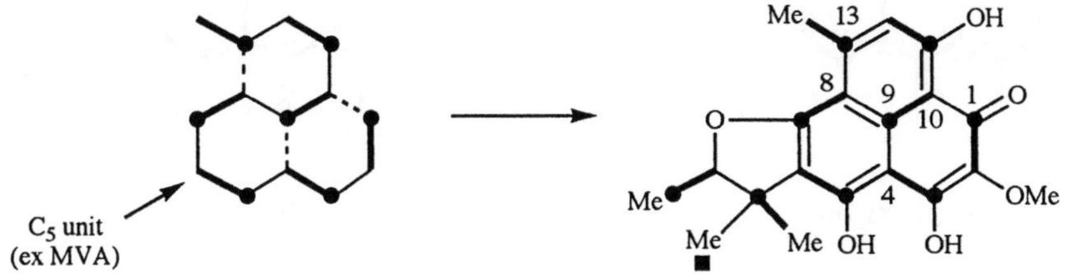

Deoxyherqueinone (6)

Scheme 3 *The heptaketide folding pattern of fungal phenalenones*

Structurally complex fused ring fungal polyketides studied in this manner include the potent carcinogen aflatoxin B₁ (7)[17] and its precursors averufin (8)[18] and sterigmatocystin (9)[17] (Scheme 4). The resulting [13]C-^{13}C couplings were consistent with the proposed formation of the heterocyclic skeleton of 7 via oxidative ring cleavages of an anthraquinone intermediate[19] and derivation of the C₄-furanoid moiety through oxidative rearrangement of the linear C₆-substituent of averufin.[20] Subsequent extensive investigations have clarified many details of the overall pathway, including the key finding that this C₆-substituent is derived from a *n*-hexanoyl starter unit.[21]

3 FUNGI AND STREPTOMYCETES UTILISE TWO DIFFERENT POLYKETIDE FOLDING STRATEGIES

Streptomycetes also produce a variety of fused ring aromatic polyketides, the carbon skeletons of some of which closely parallel those of their fungal counterparts. However, with very few exceptions, such as the melanin spore pigment precursor 1,3,6,8-tetrahydroxynaphthalene (10, Scheme 5) of *Alternaria alternata*[22] and *Streptomyces griseus*,[23] fungi and streptomycetes have not been found to produce identical fused ring structures.

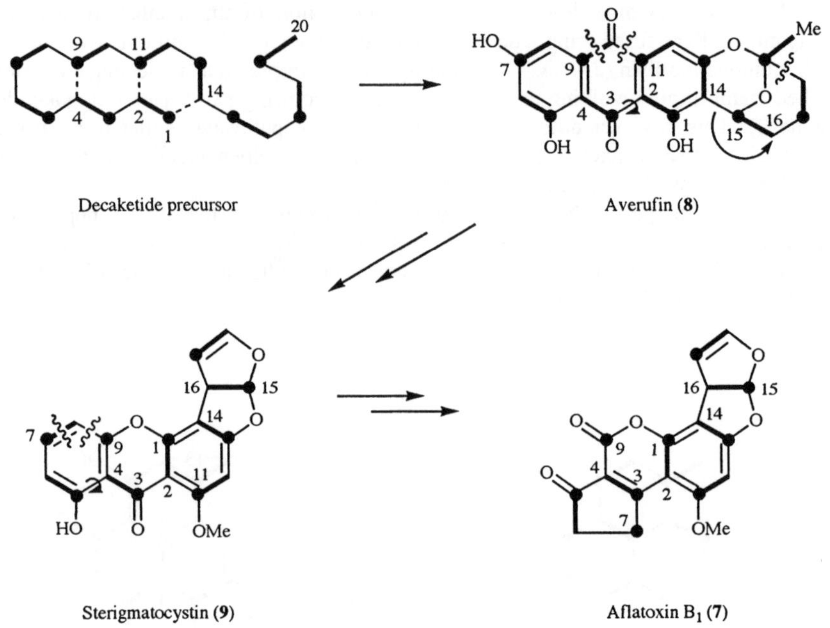

Decaketide precursor

Averufin (**8**)

Sterigmatocystin (**9**)

Aflatoxin B$_1$ (**7**)

Scheme 4 *Incorporation of [$^{13}C_2$]-labelled acetate into aflatoxin-related metabolites*

[$^{13}C_2$]acetate feeding studies have established the modes of cyclisation of the majority of known fused ring systems, which involve head to tail linkage of C$_2$-units with regiospecific folding and intramolecular aldol and occasional Claisen condensations of linear intermediates. Parallel investigations of the incorporation of labelled acetate into the *Streptomyces rimosus* antibiotic oxytetracycline (**11**)[24] and the structurally related *Penicillium viridicatum* tetracyclic amide viridicatumtoxin (**12**),[25] unexpectedly revealed two different folding strategies (Figure 1). Similarly contrasting folding patterns are evident in the corresponding acetyldecarboxamidotetracyclines cetocycline (**13**),[26] a product of *Nocardia sulphurea* and hypomycetin (**14**),[27] a metabolite of the fungus *Hypomyces aurentius*. Particularly noteworthy is the derivation of the initial cyclohexane ring of both of these bacterial polyketides through aldol crosslinking involving three intact C$_2$-units (hereafter designated polyketide folding mode S), whereas only two intact C$_2$-units are incorporated in the corresponding rings of the fungal tetracyclines (folding mode F).

A preliminary survey of [$^{13}C_2$]acetate incorporation studies indicated the consistent utilisation of mode F and S cyclisation pathways in the formation of the corresponding fused ring polyketides of fungi and streptomycetes.[7]

Other fungal metabolites shown in Figure 1 include the naphthalene derivatives dihydrofusarubin (**15**),[28] rubrofusarin (**16**)[29] and the anthraquinone islandicin (**17**),[30] all of which exhibit mode F folding. However, the reported labelling assignments of the fungal naphthopyrone fonsecin (**18**)[31] necessitate mode S folding, which consequently differs from that of its anhydro derivative rubrofusarin (**16**).

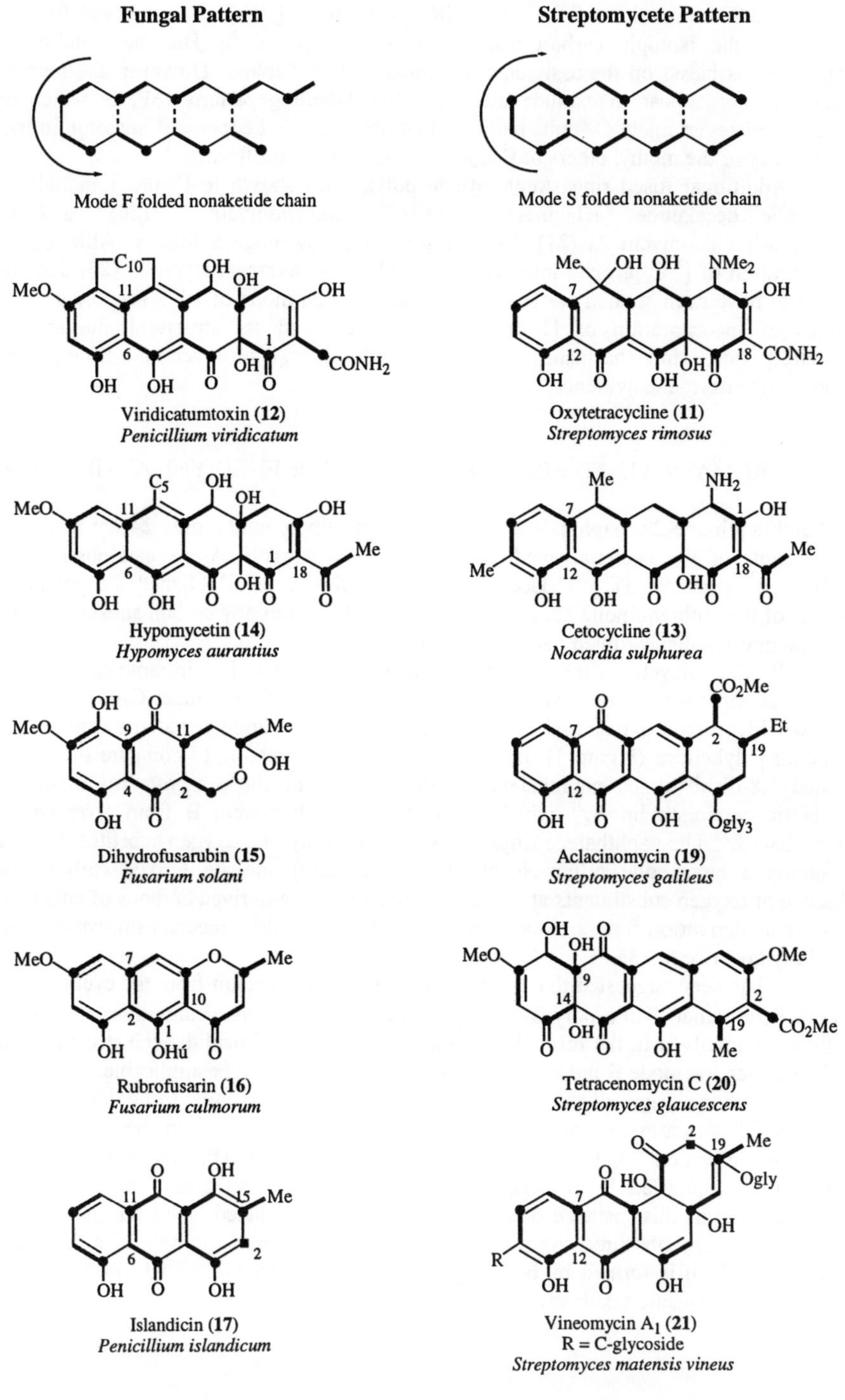

Fungal Pattern

Streptomycete Pattern

Mode F folded nonaketide chain

Mode S folded nonaketide chain

Viridicatumtoxin (**12**)
Penicillium viridicatum

Oxytetracycline (**11**)
Streptomyces rimosus

Hypomycetin (**14**)
Hypomyces aurantius

Cetocycline (**13**)
Nocardia sulphurea

Dihydrofusarubin (**15**)
Fusarium solani

Aclacinomycin (**19**)
Streptomyces galileus

Rubrofusarin (**16**)
Fusarium culmorum

Tetracenomycin C (**20**)
Streptomyces glaucescens

Islandicin (**17**)
Penicillium islandicum

Vineomycin A$_1$ (**21**)
R = C-glycoside
Streptomyces matensis vineus

Figure 1 *Distribution of C$_2$-units in microbial fused ring aromatic polyketides*

The interpretation of the ^{13}C-NMR spectrum of [^{13}C$_2$]acetate-derived fonsecin leading to the isotopic carbon distribution **18S** proposed by Bloomer, Smith and Caggiano[31] is based on the assignments summarised in Table 1. However, the reported values do not appear to exclude the alternative labelling pattern **18F**, in which the sequence of seven intact C$_2$-units parallels that observed by Leeper and Staunton in their NMR study of the methyl ethers of [^{13}C$_2$]acetate-derived rubrofusarin.[29]

Additional fused ring streptomycete polyketides shown in Figure 1 include the tetracyclic decaketides aclacinomycin (**19**),[32] tetracenomycin C (**20**)[33] and the angucycline vineomycin A$_1$ (**21**),[34] which are formed by mode S folding. Although the incorporation of [^{13}C$_2$]acetate into cetocycline (**13**) and tetracenomycin C (**20**) does not appear to have been studied *per se*, the proposed folding mode of **13** is supported by the modes of incorporation of [1 and 2-^{13}C]acetate[26] and its structural similarity to oxytetracycline, while the incorporation of C$_2$-units into **20** is in accord with extensive genetic and enzymatic evidence.[33]

4   AROMATIC FUSED RING CONSTITUENTS OF ENEDIYNE ANTIBIOTICS

The anthraquinone **22**, naphthalene **23a** and the naphthoic acid amide **24** are respective constituents of the bacterial enediyne antibiotics dynemycin A, neocarzinostatin and kedarcidin[35] (Scheme 5). [^{13}C$_2$]acetate feeding studies have established the polyketide origins of the anthraquinone nucleus **22**[36] which is devoid of any carbon substituents and also the rare 1,8-disubstituted naphthalene **23a**.[37]

If the bicyclic ring system of **23a** is formed by intramolecular aldol condensations, then the observed mode of incorporation of two intact C$_2$-units in each ring would provide a potential exception to the general mode S pattern exhibited by bacterial polyketides (Figure 1). In this context, it is interesting to compare the closely related 1,8-disubstituted naphthalene constituent **23b** of the *Streptomyces sahachiroi* antibiotic carzinophylin A,[38] which is identical to azinomycin B from *Streptomyces griseofuscus*.[39] The naphthalene rings of **23b** contain only one oxygen substituent which, assuming a hexaketide origin, is attached to a [2-C]malonate-derived carbon. The absence of oxygen substituents at any of the [1-C]malonate-derived carbons of either ring favours its derivation from a reduced polyketide by a non-aldol mechanism, which could similarly give rise to **23a**.

It has been suggested that the anthraquinone **22** may result from the cyclisation of an enediyne rather than a poly-β-ketoacyl heptaketide.[36] If this or any other a non-aldol pathway is involved in the formation of the 1,8-disubstituted naphthalene structures **23a** and **23b**, then the mode F and S classification criteria would not be applicable.

While the origin of the 2-naphthoic acid moiety **24** remains to be determined, it has an amide structure reminiscent of oxytetracycline, which has been shown to utilise a malonate starter unit[40] and to undergo typical S mode folding. The *Streptomyces griseus* pentaketide 1,3,6,8-tetrahydroxynaphthalene (**10**) synthase also uses a malonate starter unit, although in this instance the carboxyl group is eliminated[23] and the geometry of folding is thereby obscured. As previously noted, **10** is a rare example of a fused ring metabolite which is formed by both fungi and streptomycetes, although not necessarily *via* the same cyclisation pathway.

Table 1. ^{13}C NMR assignments of $[^{13}C_2]$acetate-derived fonsecin consistent with folding modes **18S** and **18F** and comparison with $[^{13}C_2]$acetate-derived rubrofusarin dimethyl ether ^{a,b}

Fonsecin **18S**
(mode S folding)

Fonsecin **18 F**
(mode F folding)

Rubrofusarin dimethyl ether
(mode F folding)

Fonsecin folding mode **18S** assignments[31]

Coupled carbons	1	2	3	4	5	6	7	8	9	10	11	12	13	14
δ (ppm)	25.5	98.1	45.7	195.6	104.1	151.8	100.9	160.0	94.6	159.1	100.0	141.7	99.6	162.9
J (Hz)	-	46	40		59		70		72		56		66	

Fonsecin folding mode **18F** assignments

Coupled carbons	1	2	3	4	5	6	7	8	9	8	13	12	13	14
δ (ppm)	25.5	98.1	45.7	195.6	104.1	151.8	100.9	160.0	94.6	159.1	100.0	141.7	99.6	162.9
J (Hz)	-	46	40		59		70		72		56		66	

Rubrofusarin dimethyl ether[28]:

Intact C$_2$-units	1	2	3	4	5	6	7	6	9	8	13	12	11	10
δ (ppm)	19.7	163.7	110.5	177.9	114.0	154.8	114.4	159.6	98.9	159.6	108.3	139.8	97.5	160.7
J (Hz)	52	52	57	57	61		70		75		59		59	61

^{a} Carbon atoms in **18S**, **18F** and rubrofusarin are numbered as in reference 31. ^{b} Broken bonds show second ring cyclisation sites.

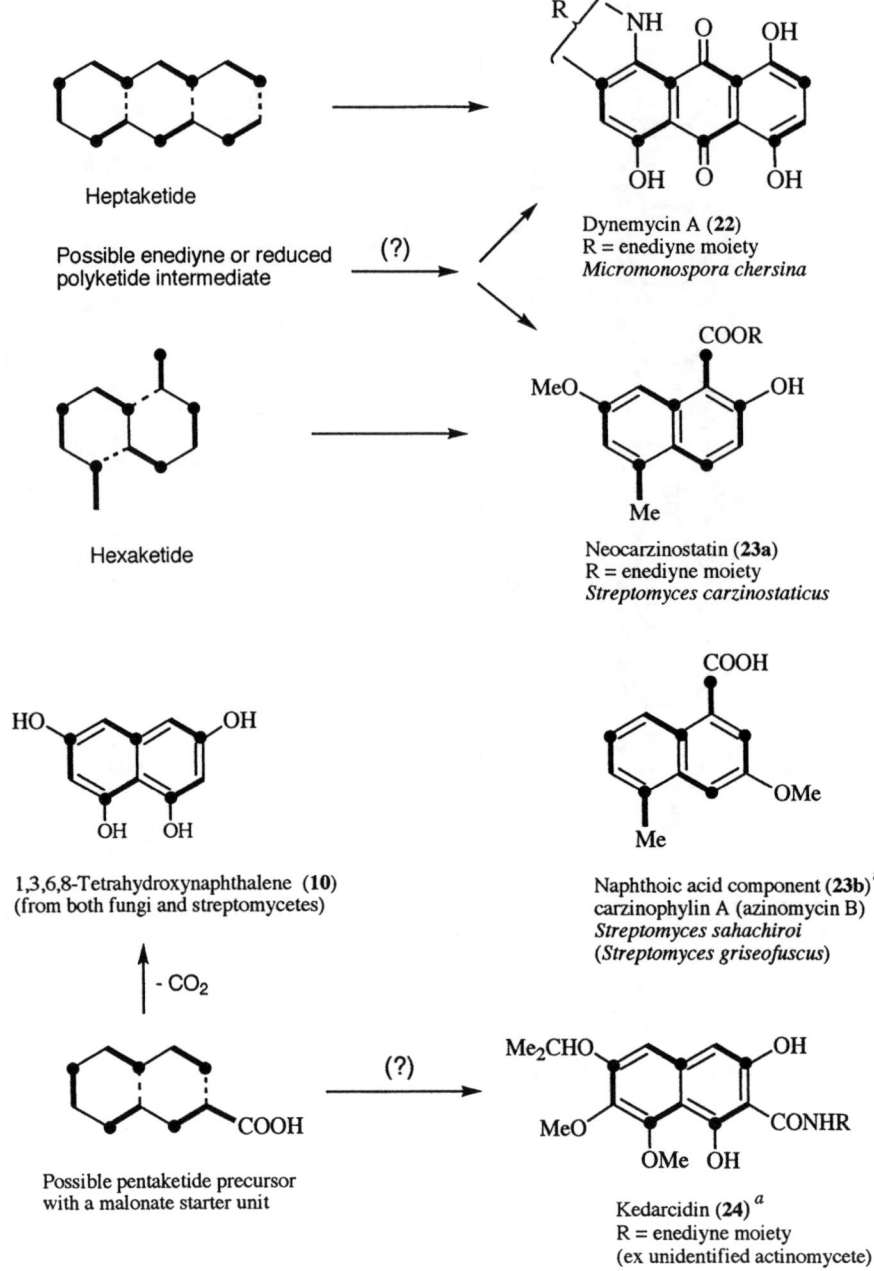

Heptaketide

Possible enediyne or reduced polyketide intermediate   (?)

Dynemycin A (**22**)
R = enediyne moiety
Micromonospora chersina

Hexaketide

Neocarzinostatin (**23a**)
R = enediyne moiety
Streptomyces carzinostaticus

1,3,6,8-Tetrahydroxynaphthalene (**10**)
(from both fungi and streptomycetes)

Naphthoic acid component (**23b**)[a]
carzinophylin A (azinomycin B)
Streptomyces sahachiroi
(*Streptomyces griseofuscus*)

- CO₂

Possible pentaketide precursor
with a malonate starter unit   (?)

Kedarcidin (**24**)[a]
R = enediyne moiety
(ex unidentified actinomycete)

Scheme 5 *Fused ring constituents of enediyne antibiotics*
      [a]*The incorporation of [¹³C₂]acetate into **23b** and **24** has not been investigated*

5   A GENERAL RULE FOR THE BIOSYNTHETIC CLASSIFICATION OF FUNGAL AND STREPTOMYCETE FUSED RING AROMATIC POLYKETIDES

Figure 2 illustrates the established distribution of intact C_2-units in the $[^{13}C_2]$acetate-derived carbon skeletons of the fungal polyketides dihydrofusarubin (**25**), rubrofusarin (**26**), averufin (**28**), islandicin (**29**) and hypomycetin (**30**); also the probable distribution of C_2-units in the streptomycete products tetracenomycin C (**35**) and cetocycline (**36**). The full structures of these compounds are shown in Figure 1 and Scheme 4. The carbon skeletons of other products for which $[^{13}C_2]$acetate incorporation patterns have been determined include the fungal metabolite bikaverin (**27**)[41] and the streptomycete products actinorhodin (**31**),[42] spectomycin A (**33**),[43] RM18 (**32**)[44] and aloesaponarin II (**34**)[45] (RM18 and aloesaponarin II are formed by streptomycete recombinant strains).

The above $[^{13}C_2]$acetate incorporation results are collectively representative of a wide range of carbocyclic fused ring products of microbial origin. It is evident from the observed labelling patterns that all of the fungal metabolites exhibit mode F folding, with the possible exception of the naphthopyrone fonsecin (**18**).[31] On the other hand, all of the streptomycete metabolites with one or two carbon side chains are mode S folded, other than the unusual 1,8-disubstituted naphthalene component of the enediyne neocarzinostatin (**23a**). These observations are consistent with a general rule for the biosynthetic classification of fused ring aromatic polyketides of fungi and streptomycetes, based on the regiospecificity of aldol crosslinking leading to the initial carbocyclic ring.[7] The archetypal mode F and mode S-derived carbon skeletons shown in Figure 3, illustrate the incorporation patterns of acetate-derived intact C_2-units in the corresponding initial rings and the two *ortho*-oriented carbon side chain substituents.

Thus, in fungal metabolites, mode F folding of a linear polyketide chain yields an initial cyclohexane ring containing *two* intact C_2-units with two adjacent side chains, each containing an *odd* numbers of carbons. The corresponding streptomycete polyketides exhibit mode S folding, leading to an initial ring derived from *three* intact C_2-units and also having two adjacent side chains, but with *even* numbers of carbon atoms in each. Streptomycete polyketides frequently utilise propionate starter units, which consequently give rise to a primer side chain with an odd number of carbon atoms, as illustrated in the example of aclacinomycin (Figure 3).

In the collective structures shown above, the first ring of each of the fungal fused ring products involves cross-linking of folded polyketide chains between carbon atoms 2 and 7, 4 and 9 or 6 and 11, whereas in the streptomycete products this ring is formed by cross-linking of carbons 7 and 12 or 9 and 14. In these instances, the U-turn fold leading to the first ring of the fungal products is not initiated beyond carbon-6 of the nascent polyketide chain, whereas in the streptomycete products it is only initiated at C-7 or C-9. This disparity in the minimum chain length required to initiate cylisation, may be a further distinguishing characteristic of the respective cyclase enzymes involved in the formation of fungal and streptomycete fused ring metabolites.

Fungal polyketides
(mode F folding)

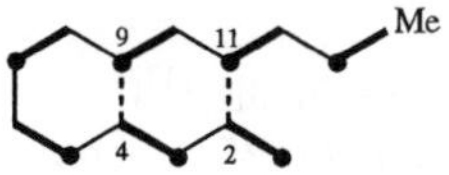

Streptomycete polyketides
(mode S folding)

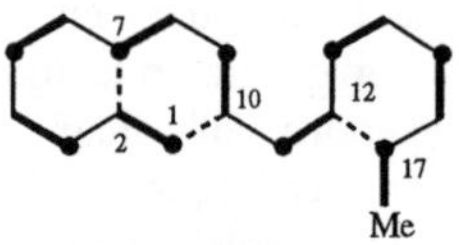

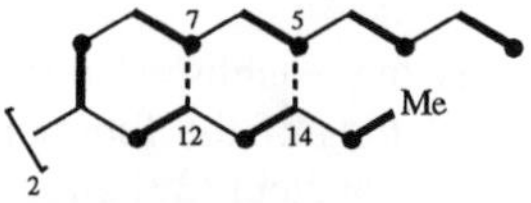

Dihydrofusarubin (**25**)

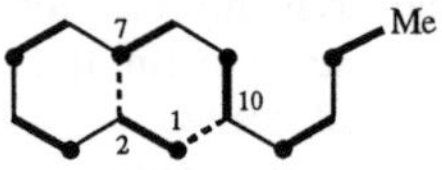

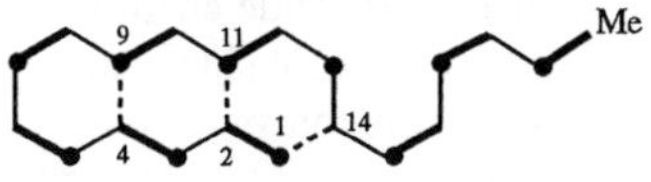

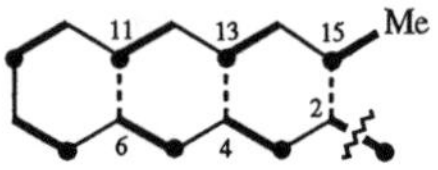

Actinorhodin (**31**)

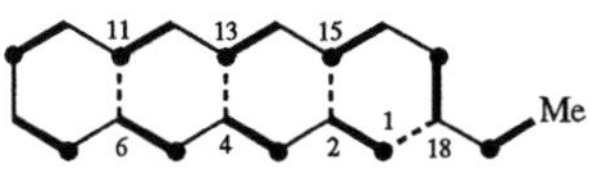

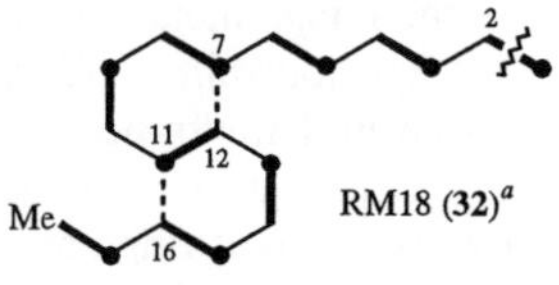

Rubrofusarin (**26**)

RM18 (**32**)[a]

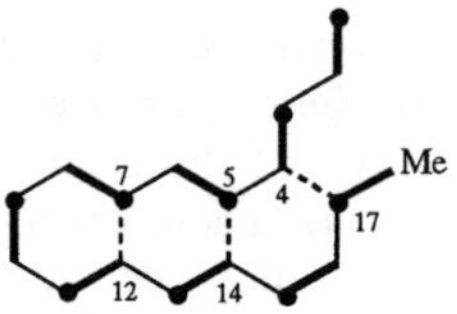

Bikaverin (**27**)

Spectomycin A (**33**)

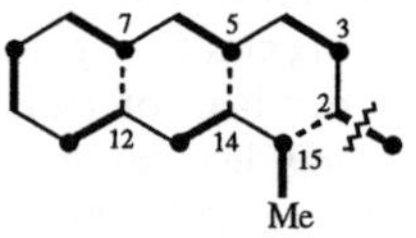

Averufin (**28**)

Aloesaponarin II (**34**)[a]

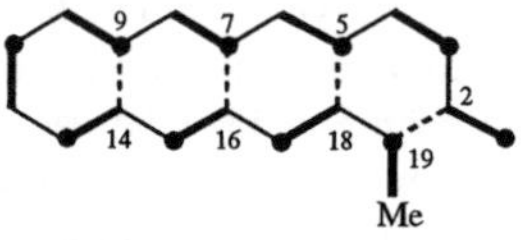

Islandicin (**29**)[a]

Tetracenomycin C (**35**)

Hypomycetin (**30**)

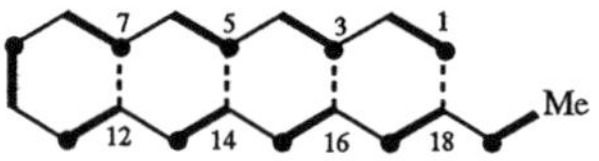

Cetocycline (**36**)

Figure 2. *Distribution of C₂-units in the carbon skeletons of microbial fused ring*
polyketides
[a]*The named product is the decarboxylated polyketide*

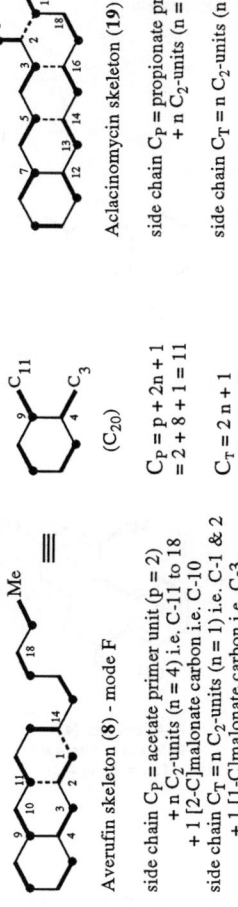

FUNGAL POLYKETIDES
(Mode F folding)

C_P (p + 2n + 1)

C_T (2n + 1)

STREPTOMYCETE POLYKETIDES
(Mode S folding)

C_T (2n)

C_P (p + 2n)

C_P = primer unit side chain, C_T = terminal unit side chain, p = primer unit carbons, n = intact C_2-units (i.e. 2n carbons)
The initial ring has two *ortho* substituents comprising a primer unit (e.g. acetate or propionate) and malonate-derived extender units

FUNGAL POLYKETIDES: C_2-UNIT DISTRIBUTION (MODE F)
(i)   The initial benzene ring contains 2 malonate-derived intact C_2-units
(ii) Primer side chain = p (primer unit) + n intact-C_2 units + 1 [2-C]malonate-derived benzylic carbon = p + 2n + 1 carbons
(iii) Terminal side chain = n intact C_2-units + 1 [1-C]malonate-derived benzylic carbon = 2n + 1 carbons
Where the primer unit is acetate (p = 2) both substituents have an *odd* number of carbons

STREPTOMYCETE POLYKETIDES: C_2-UNIT DISTRIBUTION (MODE S)
(i)   The initial benzene ring contains 3 malonate-derived intact C_2-units
(ii) Primer substituent side chain = p (primer unit) + n intact C_2-units = p + 2n carbons
(iii) Terminal substituent side chain = n intact C_2-units = 2n carbons
Where the primer unit is acetate (p = 2) both substituents have an *even* number of carbons

Representative fungal (mode F) and streptomycete (mode S) fused ring polyketides:

Me $\equiv$ C_{11}, C_3 (C_{20})

Averufin skeleton (**8**) - mode F

side chain C_P = acetate primer unit (p = 2)
+ n C_2-units (n = 4) i.e. C-11 to 18
+ 1 [2-C]malonate carbon i.e. C-10
C_P = p + 2n + 1 = 2 + 8 + 1 = 11
side chain C_T = n C_2-units (n = 1) i.e. C-1 & 2
+ 1 [1-C]malonate carbon i.e. C-3
C_T = 2n + 1 = 2 + 1 = 3

Et $\equiv$ C_6, C_9 (C_{21})

Aclacinomycin skeleton (**19**) - mode S

side chain C_P = propionate primer unit (p = 3)
+ n C_2-units (n = 3) i.e. C-13 to 18
C_P = p + 2n = 3 + 6 = 9
side chain C_T = n C_2-units (n = 3) i.e. C-1 to 6
C_T = 2n = 6

Figure 3. *Classification of fused ring polyketides of streptomycetes and fungi based on the mode of cyclisation of the initial carbocyclic ring*

Bacterial metabolites

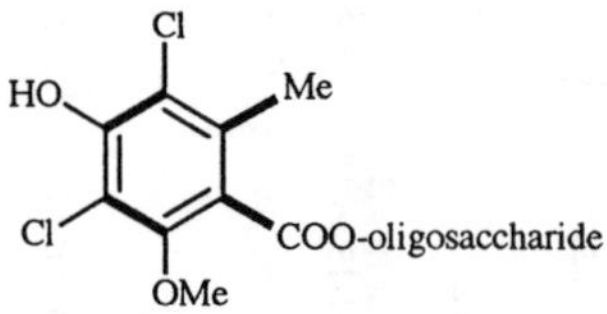

Everninomicin B (**39**)
Micromonospora carbonaceae
(B = 2)

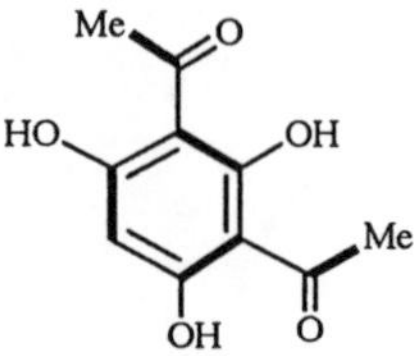

Diacetylphloroglucinol (**40**)
Pseudomonas fluorescens
(B = 3)

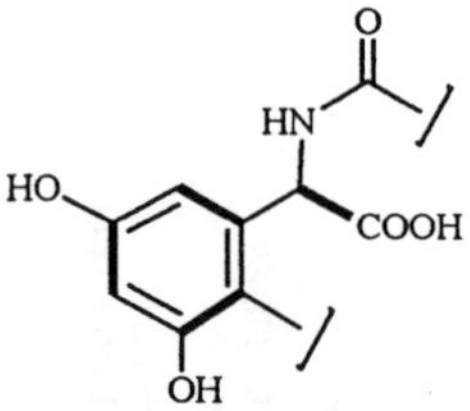

Vancomycin tetraketide unit (**41**)
Streptomyces orientalis
(B = 3)

Fungal metabolites

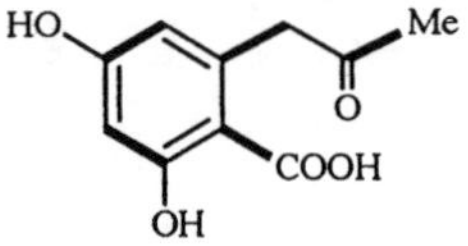

α-Acetylorsellinic acid (**37**)
Penicillium brevi-compactum
(B = 2)

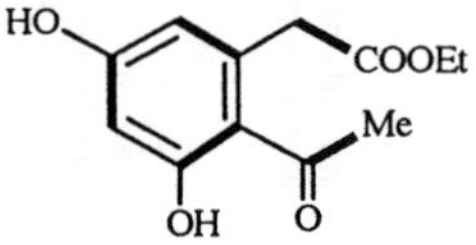

Curvulin (**38**)
Curvularia siddiqui
(B = 3)

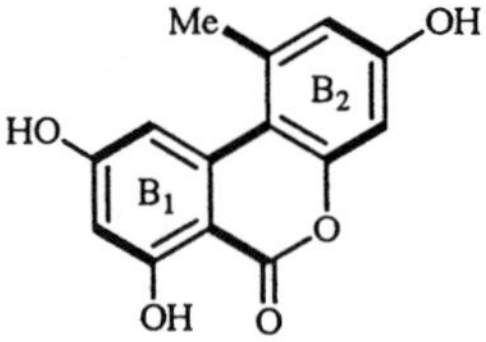

Alternariol (**42**)
Alternaria alternata
(B₁ = 2, B₂ = 2)

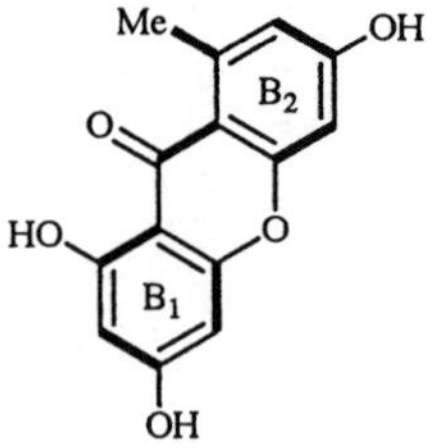

Norlichexanthone (**43**)
Alternaria alternata
(B₁ = 3, B₂ = 2)

Figure 4 *Fungal and bacterial non-fused ring carbocyclic polyketides (B = number of intact C₂-units in each benzene ring)*

The proposed classification only applies to fused ring polyketides, since both fungi and bacteria are known to produce monocyclic products with benzene rings derived from either two or three intact C₂-units. Both cyclisation modes are apparent in the fungal pentaketides α-acetylorsellinic acid (**37**)[46] and curvulin (**38**),[47] which are respectively metabolites of *Penicillium brevi-compactum* and *Curvularia siddiqui* (Figure 4). Analogous examples among streptomycete and other bacterial polyketides are the dichloroorsellinic acid ester constituent (39) of the *Micromonospora carbonaceae* everninomicin oligosaccharide antibiotics,[48] diacetylphloroglucinol (40) which is formed by various *Pseudomonas* species[49] and the 3,5-dihydroxyphenylacetate-derived tetraketide moiety of the antibiotic vancomycin (41) from *Streptomyces orientalis.*[50]

Fungi are prolific sources of monocyclic and bicyclic non-fused ring aromatic metabolites. Interesting examples of the latter are the dibenzo-α-pyrone alternariol (42), one of the earliest biosynthetically investigated polyketides[51] and norlichexanthone (43), which are heptaketide congeners of *Alternaria alternata*. While both of the carbocyclic rings of 42 and one ring of 43 incorporate two intact C_2-units, the second benzene ring of the 43 is derived from three C_2-units. The suggested precursor-product relationship of 42 and 43[52] was excluded by the results of a specifically targeted ^{13}C-labelling study.[53]

As previously emphasised, this general rule for the biosynthesis of fused ring polyketides is only applicable to compounds formed through intramolecular aldol condensations and does not apply to the products of olefinic or other non-aldol cyclisation pathways involving mechanisms such as the Diels-Alder cycloaddition or epoxide and peroxide-induced crosslinking. For example, the initial ring of the bicyclic fungal metabolite hamigerone (44) has been shown to incorporate three intact acetate-derived C_2-units[54] and *a priori* might appear to undergo an anomalous mode S cyclisation. However, this ring could alternatively could be formed in concert with the second ring by the Diels-Alder cyclisation of a partially reduced polyketide, as may also be involved in the biosynthesis of mevinolin (45) (Scheme 6). [55]

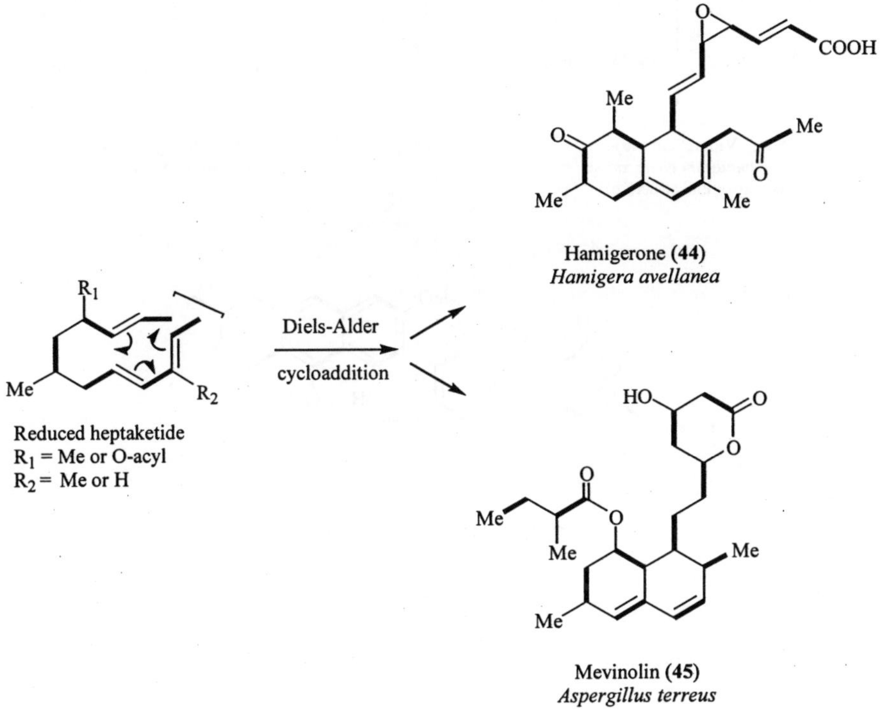

Hamigerone (44)
Hamigera avellanea

Reduced heptaketide
R_1 = Me or O-acyl
R_2 = Me or H

Diels-Alder
cycloaddition

Mevinolin (45)
Aspergillus terreus

Scheme 6 *Putative Diels-Alder cyclised fused ring polyketides*

   This biosynthetic classification facilitates the structural elucidation of microbial fused ring polyketides by restricting the selection of candidate structures to those which satisfy the predicted F or S modes of cyclisation. It similarly allows the identification of structures that do not conform to the expected cyclisation patterns and which consequently may warrant further examination. For example, the fungal polyketide viomellein was originally assigned the dimeric bisnaphthalene structure 46,[56] in which one of the monomeric components was a previously unprecedented naphthoquinone requiring the atypical mode S folding pattern, whereas the second naphthalene unit arises by the normal mode F folding. This structure was subsequently revised to 47 following X-ray crystallographic[57] and ^{13}C-NMR studies[58] of the closely related fungal metabolite xanthomegnin (48) in which both rings exhibit mode F folding (Figure 5).

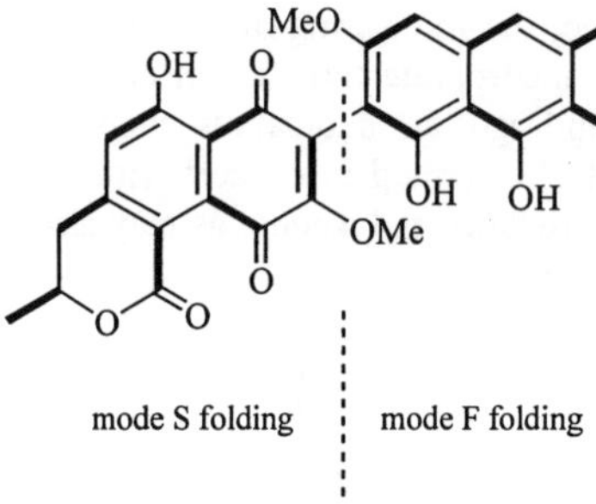

mode S folding ┊ mode F folding

Viomellein (46)
Penicillium viridicatum.
(original structure)

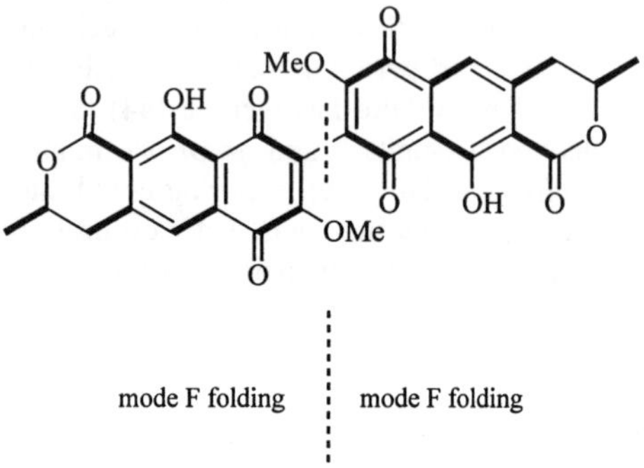

mode F folding ┊ mode F folding

Xanthomegnin (48)
Penicillium viridicatum

mode F folding ┊ mode F folding

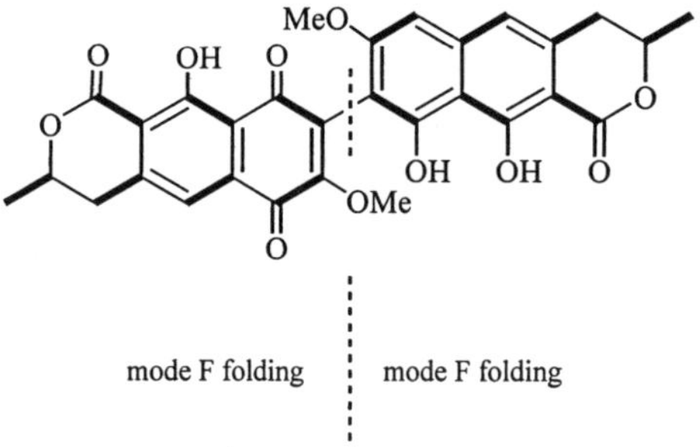

Viomellein (47)
(revised structure)

Figure 5   *Dimeric naphthalene polyketides from fungi*

6 REARRANGEMENT OF LINEAR PRECURSORS LEADING TO FUSED RING POLYKETIDES

While it is likely that some polyketides such as ochrephilone (**49**)[59] are derived from more than one chain, the majority of known fungal and streptomycete fused ring polyketides are formed through folding of single chain intermediates.

Ochrephilone (**49**)
Penicillium multicolor

This may be evident on inspection of the carbon skeletons of cyclic polyketides, although the linear origin of some metabolites can be obscured by intermediate rearrangements. For example, the anomalous labelling of the spirocyclopentane rings of the [$^{13}C_2$]acetate-derived streptomycete metabolite fredericamycin A (**50**) appears to favour its formation from two polyketide chains.[60] However, a linear polyketide pathway may be envisaged *via* oxidative rearrangement of an intermediate mode S folded pentacarbocyclic polyketide related to benastatin (**51**)[61] involving two sequential 1,2-shifts (Scheme 7)[7].

Other examples of the apparent incorporation of [$^{13}C_2$]acetate *via* oxidative rearrangements of fused ring intermediates were reported by Gould and co-workers, following extensive biosynthetic investigations of the streptomycete polyketides angucycline PD 116198 (**52**)[62] and the phenanthrene murayaquinone (**53**)[63] (Scheme 8). These authors proposed mechanistically feasible routes to **52** via the anthracycline (**54**) (path A1) and to **52** via the phenanthrene **55** (path B1). Alternative epoxide-driven rearrangements are possible via spiro intermediates analogous to the hypothetical fredericamycin A pathway (Scheme 7), involving the mode S folded angucycline (**56**) and anthraquinone (**57**) epoxides, as outlined in path A2 and path B2 respectively.

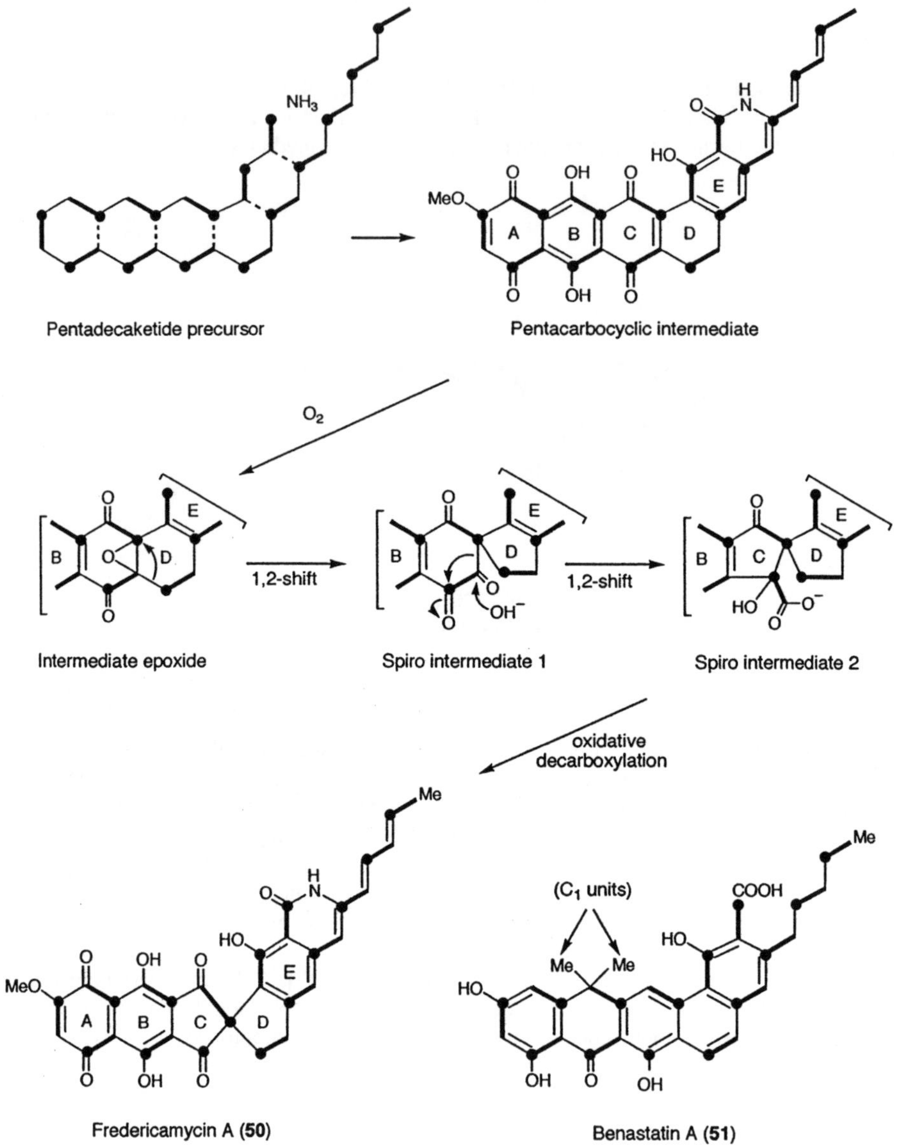

Scheme 7 *Possible formation of fredericamycin A via rearrangement of a mode S folded linear polyketide*

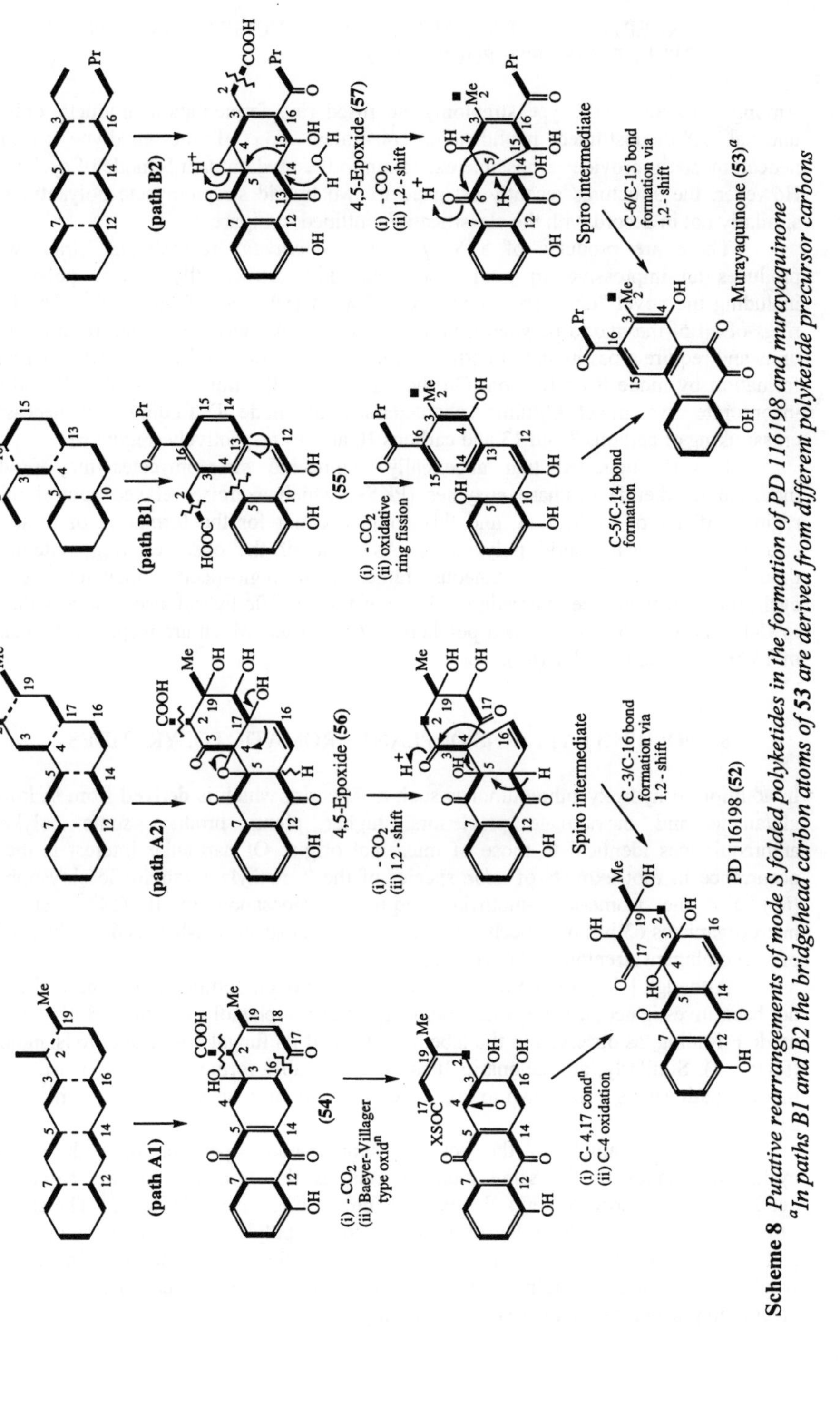

Scheme 8 *Putative rearrangements of mode S folded polyketides in the formation of PD 116198 and murayaquinone*
[a]*In paths B1 and B2 the bridgehead carbon atoms of 53 are derived from different polyketide precursor carbons*

7 EXCEPTIONS TO MODE S CYCLISATION IN HYBRID STREPTOMYCETE POLYKETIDES

Among the above wild type streptomycete fused ring fermentation products, only the unusual 1,8-disubstituted naphthalene constituent **23a** of the enediyne antibiotic neocarzinostatin provides a possible exception to the predominant S mode of cyclisation. However, the structures recently assigned to two hybrid streptomycete polyketides are similarly not in accord with the classification outlined in Figure 3.

These are products of a *Streptomyces coelicolor* recombinant strain which produces an impressive array of more than thirty structurally diverse polyketides, including the novel "unnatural compounds" TW93f (**59**) and TW93g (**60**).[64] The initial rings of all of the hybrid polyketides other than (**59**) and (**60**) are formed from three C_2-units and require crosslinking of carbon atoms 7 and 12, or 9 and 14, consistent with ring formation by mode S cyclisation. On the other hand, the initial rings of (**59**) and (**60**) incorporate two intact C_2-units, characteristic of mode F folding and necessitate crosslinking of carbons 8 and 13 and carbons 10 and 15 respectively (Figure 6).

It was suggested that genetically engineered streptomycetes may produce modified polyketide synthase enzymes (PKSs) which exercise relaxed control of the regiospecificity of cyclisation, and this could account for the formation of these two structurally atypical hybrid polyketides. As some of the observed ring systems are thought to arise from spontaneous rather than regiospecific enzyme catalysed cyclisations, it would be interesting to know if this prolific hybrid also produces the C-7 to C-12 and C-9 to C-14 bridged polyketides **61** and **62,** which are respectively isomers of TW93f (**59**) and TW93g (**60**).

8 FOLDING PATTERNS OF PLANT AROMATIC POLYKETIDES

In addition to hydroxyanthraquinones such as alizarin, which is derived from shikimate, glutamate and mevalonate precursors, higher plants produce some polyketide anthraquinones identical to those of microbial origin. Of particular interest is the co-occurrence in root extracts of *Aloe* species of the 2-methylanthraquinone chrysophanol (**63**) and the isomeric 1-methylanthraquinone aloesaponarin II (**64**),[65] since in microorganisms **63** has only been isolated from fungal species[2] whereas **64** is the product of a recombinant streptomycete[45] (Figure 7).

Although [$^{13}C_2$]acetate incorporation into the plant anthraquinones **63** and **64** has not been investigated, *a priori* one would expect chrysophanol to be formed in plants *via* mode F folding, as observed in the labelling study of its fungal co-metabolite islandicin[30] (Figure 1). Similarly, aloesaponarin II is most probably formed in plants *via* the same mode S pathway as was shown to take place in the hybrid streptomycete strain[45] (Figure 2).

Another example of the possible simultaneous operation of both F and S cyclisation modes in the same plant species, is the formation of the naphthalene octaketides eleutherin (**65**) and eleutherinol (**66**) in *Eleutherine bulbosa*.[66] The structure of eleutherin corresponds to that of a mode F folded polyketide, whereas the cyclisation pathway originally proposed by Birch and Donovan[67] for the biosynthesis of eleutherinol would require mode S folding, although its formation by rearrangement of an advanced eleutherin precursor is an alternative possibility.

Mode F Folding   Mode S Folding

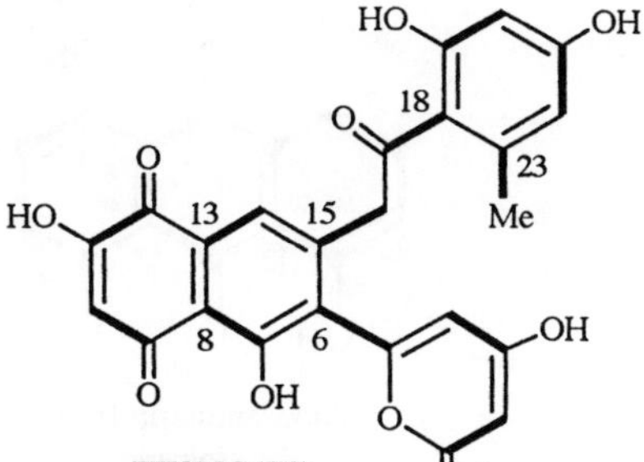

TW93f (**59**)   (**61**)
$C_{24}H_{16}O_{10}$ $C_{24}H_{16}O_{10}$

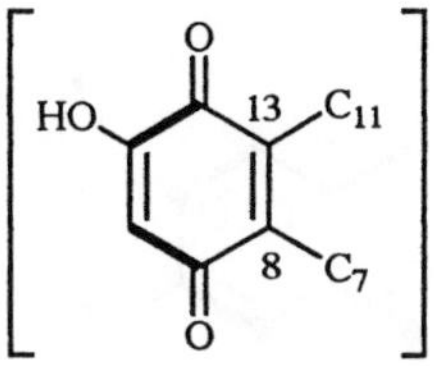

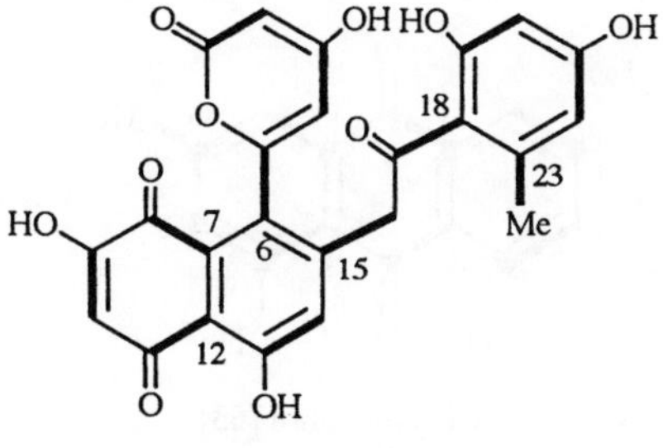

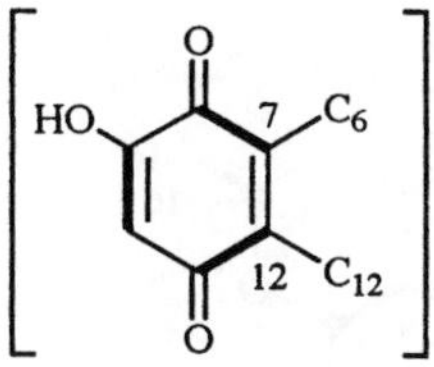

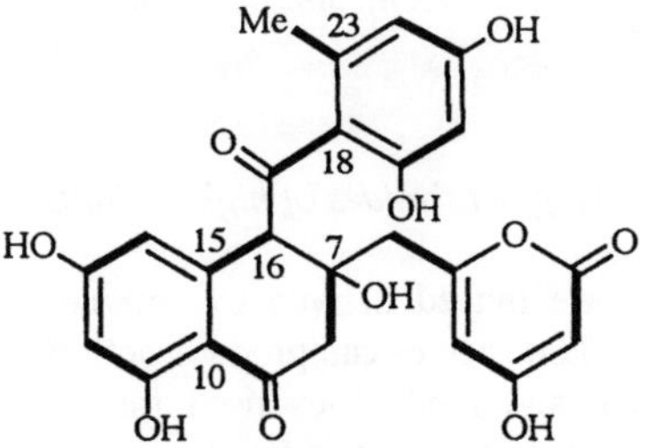

TW93g (**60**)   (**62**)
$C_{24}H_{20}O_{10}$ $C_{24}H_{20}O_{10}$

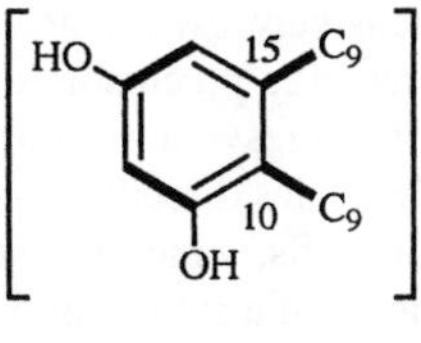

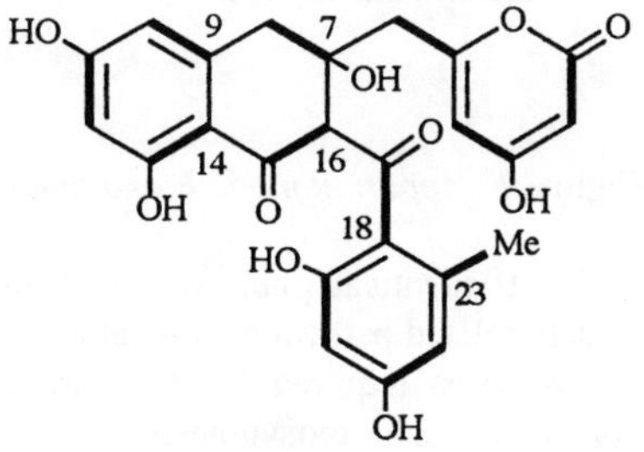

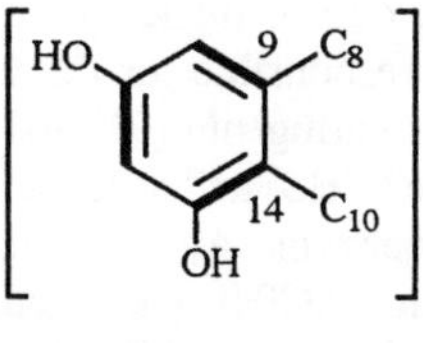

Figure 6 *Apparent exceptions to mode S folding in polyketides from hybrid streptomycetes*

Mode F Folding   Mode S Folding

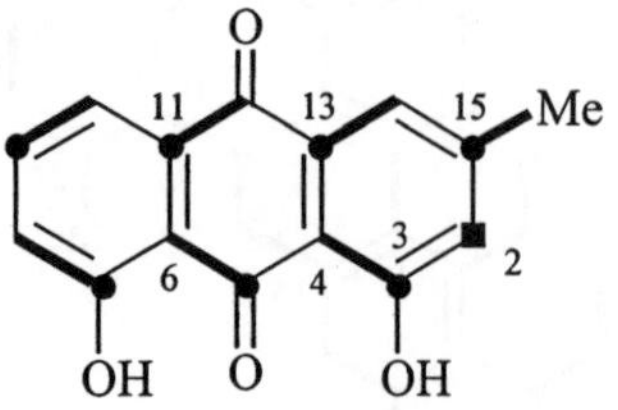

Chrysophanol [63]   Aloesaponarin II [64]
Aloe spp.   *Aloe spp.*

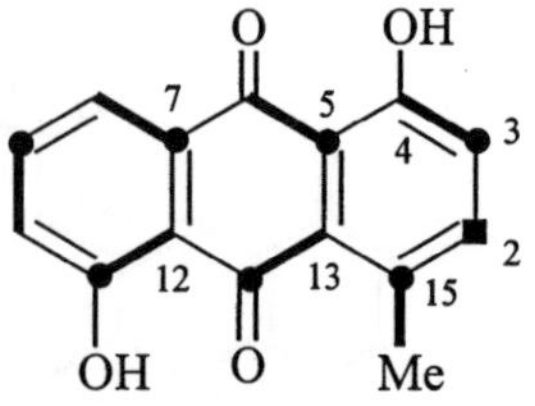

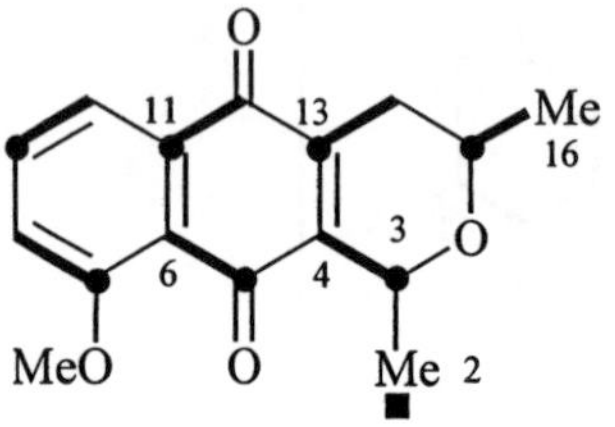

Eleutherin [65]   Eleutherinol [66]
Eleutherine bulbosa   *Eleutherine bulbosa*

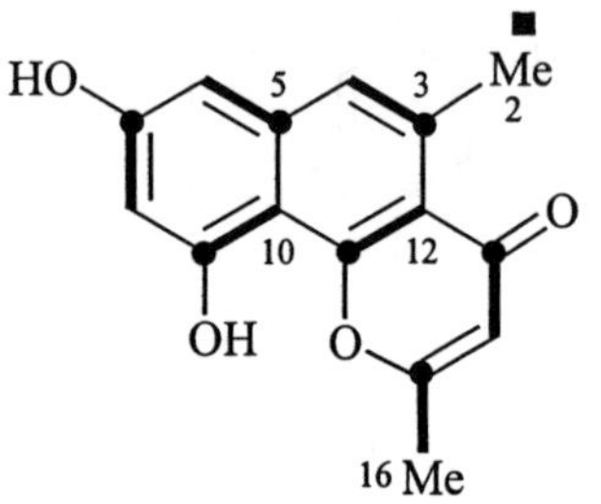

Figure 7 *Apparent mode F and mode S folded fused ring octaketides of higher plants*

If identical plant and microbial polyketides are indeed derived by common or closely related pathways, it would appear that higher plant species can produce both types of synthases required for the formation of mode F and mode S-cyclised fused ring polyketides. The biosynthesis of aloesaponarin II by steptomycetes is known to involve an iterative type II PKS and, by analogy with other fungal aromatic polyketide synthases, it is likely that a type I PKS is responsible for the formation of chrysophanol. However, at present little is known of the nature of PKSs responsible for the formation of fused ring polyketides in plants.

A major group of higher plant PKSs which have been extensively investigated, are the synthases responsible for the formation of the chalcone tetraketide precursors of flavonoids such as naringenin (**67**) and also the formation of the stilbene resveratrol (**68**). Chalcone synthase (CHS) and stilbene synthase (STS) assemble polyketides directly from acylthiosters of coenzyme A, in contrast to the fungal and bacterial PKSs, which utilise acyl carrier proteins (ACPs). These enzymes are thought to be members of a super family of polyketide synthases capable of producing diverse di, tri and tetraketides from a variety of starter units, as described in a recent review by Schröder.[68]

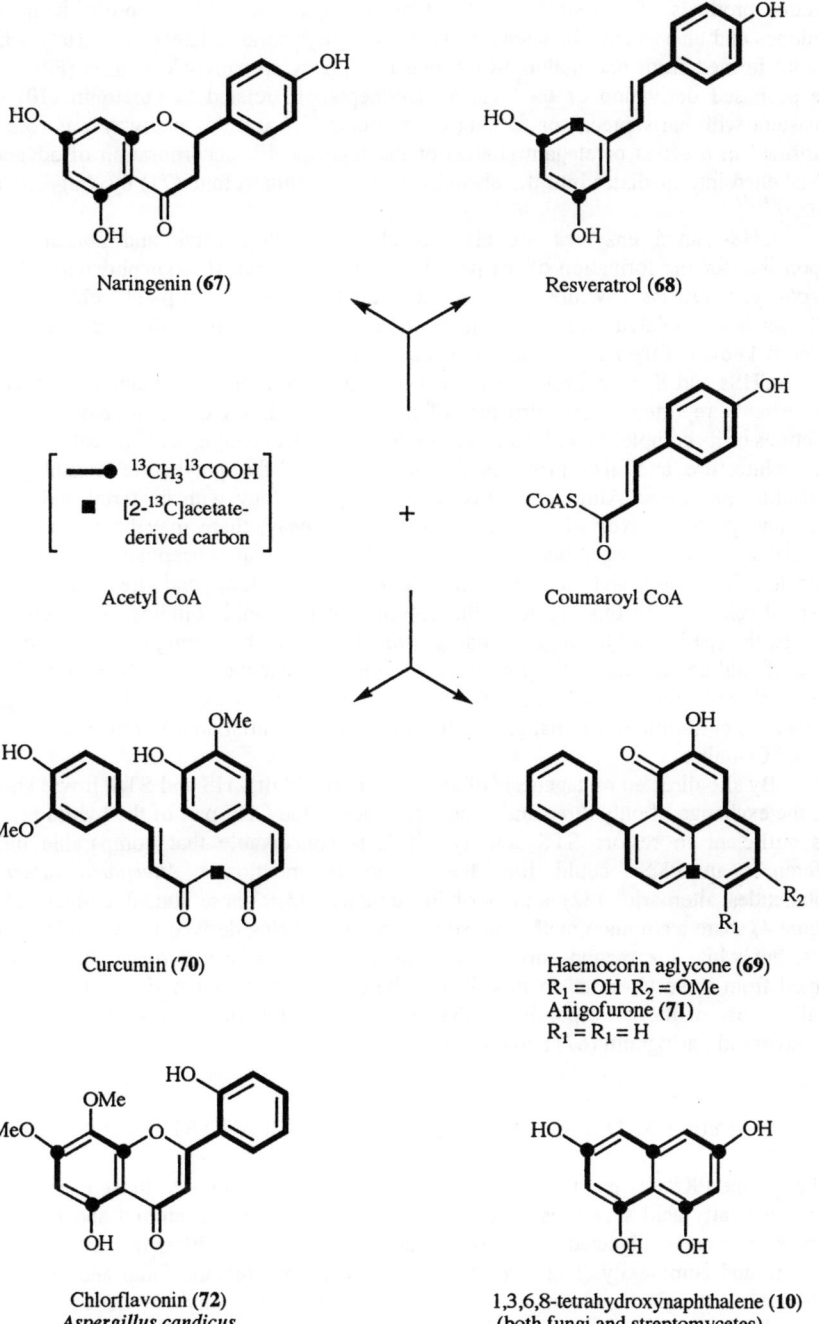

Figure 8 *Polyketides of the chalcone super family of chalcone and stilbene (The illustrated [$^{13}C_2$]acetate labeling is extrapolated from related incorporation studies)*

Diketide products of this super family of plant polyketide synthases probably include acridones and also phenylphenalenones such as the aglycone of haemocorin (**69**), which may be formed from one malonyl-CoA and two phenylpropanyl-CoA units (Figure 8). The proposed derivation of **69**[15] via a diarylheptanoid related to curcumin (**70**) was consistent with early precursor feeding experiments[69, 70] and this pathway was recently confirmed in a series of elegant studies of the regiospecific incorporation of advanced ^{13}C-labelled intermediates into the phenylphenalenone anigorufone (**71**) by *Anigozanthos preissii.*[71,72]

CHS-related enzymes are also found in microorganisms and appear to be responsible for the formation of the pentaketide 1,3,6,8-tetrahydroxynaphthalene (**10**) in *Streptomyces griseus.*[23] While flavonoids are widely distributed in plants, chlorflavonin (**72**) has been isolated from the fungus *Aspergillus candicus,*[73] although nothing is currently known of the nature of the associated synthase.

CHSs and STSs are relatively small homodimeric proteins (subunits 40-45 kDa). The recently reported crystal structure of recombinant alfalfa CHS2 together with the structures of its complexes with some substrate and product analogues,[74] reveals its active site architecture and also provides the first clues to the polyketide assembly and cyclisation processes. Although CHSs show little homology with bacterial and fungal PKSs and probably evolved *via* an independent pathway, there may be some general similarities in their assembly strategies which provide a conceptual basis for the characteristic mode F and mode S cyclisation pathways of fungi and streptomycetes. Of potential relevance in this regard is the presence of polyketide binding and cyclisation pockets, the spatial and hydrogen bonding geometry of which is thought to determine the mode of folding leading to the formation of either a chalcone or a stilbene tetraketide. The combined crystal structures indicate that the growing carbon skeleton undergoes progressive conformational changes in the course of sequential addition of three acetate-derived C_2-units.

By site-directed mutagenesis of an inactive hybrid of CHS and STS, it was shown that the exchange of only three amino acid residues in the CHS part of the hybrid protein was sufficient to restore STS activity.[68,75] It is conceivable that comparable minor differences in PKSs could link the previously mentioned *Alternaria alternata* heptaketides alternariol (**42**) and norlichexanthone (**43**). These fungal co-metabolites (Figure 4) share a common methyl substituted resorcinol ring derived from two intact C_2-units, but while the second carbocyclic ring of **42** is similarly derived, that of **43** is formed from three C_2-units. In this regard, the origins of the latter rings of **42** and **43** parallel those of the corresponding polyketide rings of the stilbene resveratrol (**68**) and the flavonoid naringenin (**67**) shown in Figure 8.

9   TYPE I AND II AROMATIC PKS'S OF FUNGI AND STREPTOMYCETES

Although the PKSs of plants have received little attention relative to those of microbial origin, the fatty acid synthases (FASs) of a wide range of plant, animal and microbial sources have been subjected to intensive study[76] and shown to differ in their molecular structure and complexity. Thus type I FASs produced by animals, fungi and yeasts are high molecular weight multienzyme complexes, in which the individual active sites may be covalently linked components of a single polypeptide chain. In the type II FASs of bacteria and plants, the constituent enzymes are freely dissociable proteins which may be organised *in vivo* as loose aggregates.

As previously mentioned, microbial PKSs also exist as type I and type II enzymes.[6,77] Both synthases produce aromatic polyketides through the iterative operation of single sets of active sites in successive rounds of chain assembly involving sequential addition of malonate units and intramolecular cyclisation. Although considerably less information is available regarding the properties of the aromatic PKSs of fungi relative to those of streptomycetes, examples of the former have been found to be type I whereas the latter are type II synthases. Few PKSs have been purified to homogeneity and little is known of their three dimensional structures, however, the solution structure of the actinorhodin ACP from the type II PKS of *Streptomyces coelicolor* has been determined by ^{1}H NMR and shown to contain pockets which may harbour growing polyketide chains.[78]

Conceivably, mode F and S folding patterns may reflect inherent characteristics of the respective aromatic type I and type II fused ring PKSs of fungi and bacteria and their parent gene clusters. An improved understanding of the enzymes responsible for F and S modes of cyclisation could facilitate the combinatorial biosynthesis of structurally novel fused ring polyketides, if viable hybrid clusters can be prepared by mixing and matching equivalent PKS genes of fungi and bacteria.

References

1. C. Wehmer, *Botan. Zeit.*, 1891, **49**, 233.
2. H. Raistrick, *Proc. Roy. Soc. London*, 1949, **A199**, 141;
   R. Bentley and R. Thomas, *The Biochemist*, 1990, **12**, 3.
3. K. C. Nicolau, D. Vourloumis, N. Winnsinger and P. S. Baran,
   Angew. Chem. Int. Ed., 2000, **39**, 45.
4. For recent reviews see: Comprehensive Natural Product Chemistry;
   D. H. R. Barton and K. Nakanishi, Eds; Elsevier: Oxford, 1999; vol. 1,
   (U. Sankawa, vol. ed.), Elsevier, Oxford, 1999.
5. A. J. Birch and F. W. Donovan, *Austr. J. Chem.*, 1953, **6**, 360;
   A. J. Birch, To See The Obvious; American Chemical Society,
   Washinton DC, 1995.
6. D. A. Hopwood, *Chem. Rev.*, 1997, **97**, 2465.
7. R. Thomas, *Folia Microbiol.*, 1995, **40**, 4.
8. H. Seto, L. W. Carey and M. Tanabe, *J. Chem. Soc., Chem. Commun.*,
   1973, 867.
9. A. G. McInnes and J. L. C. Wright, *Acc. Chem. Res.*, 1975, **8**, 313.
10. H. Seto, L. W. Carey and M. Tanabe, *J. Antibiotics*, 1974, **27**, 558.
11. K. Mosbach, *Acta Chem. Scand.*, 1960, **14**, 457.
12. R. Bentley and J. G. Kiel, *J. Biol. Chem.*, 1962, **237**, 867.
13. J. M. A. Al-Rawi, J. A. Elvidge, D. K. Jaiswal, J. R. Jones
    and R. Thomas, *J. Chem. Soc., Chem. Commun.*, 1974, 220.
14. M. L. Casey, R. C. Paulick and H. W. Whitlock,
    J. Amer. Chem. Soc., 1976, **98**, 2636.
15. R. Thomas, *Biochem. J.*, 1961, **78**, 807.
16. T. J. Simpson, *J. Chem. Soc., Chem. Commun.*, 1976, 258.
17. K. G. R. Pachler, P. S. Steyn, R. Vleggaar, P. L. Wessels and D. B.Scott,
    J. Chem. Soc., Perkin Trans.1, 1976, 1182.

18. C. P. Gorst-Allman, K. G. R. Pachler, P. S. Steyn, P. L. Wessels and
     D. B.Scott, *J. Chem. Soc., Perkin Trans.1*, 1977, 2181.
19. R. Thomas, in Biogenesis of Antibiotic Substances; eds. Z. Vanek and
     Z Hostalek; Czech. Acad. Sci., Prague , 1965, p. 155.
20. R. Thomas, personal communication to M. O. Moss, in Phytochemical
     Ecology; ed. J. B. Harborne; Academic Press, London, 1972, p. 140.
21. For recent reviews see R. E. Minto and C. A. Townsend, *Chem. Rev.*,
     1997, **97**, 2537; C. A. Townsend and R. E. Minto , in Comprehensive Natural
     Product Chemistry; D. H. R. Barton and K. Nakanishi, Eds; Elsevier: Oxford,
     1999; vol. 1, (vol. ed. U. Sankawa), p.443.
22. I. Fujii, in Comprehensive Natural Product Chemistry; D. H. R. Barton and
     K. Nakanishi, Eds; Elsevier: Oxford, 1999; vol. 1, (vol. ed. U. Sankawa), p.409.
23. N. Funa, Y. Ohnishi, I. Fujii, M. Shibuya, Y. Ebizoka and S. Horinouchi,
     Nature, London, 1999, **400,** 897.
24. R. Thomas and D. J. Williams, *J. Chem. Soc., Chem. Commun.*, 1983, 128.
25. A. E. de Jesus, W. E. Hull, P. J. Steyn, F. R. van Heyden and
     R. Vleggaar,*J. Chem. Soc., Chem. Commun.*, 1982, 902.
26. L. A. Mitscher, J. K. Swayze, T. Högberg, I. Khanna, G. S. R. Rao,
     R. J. Theriault, W. Kohl, C. Hanson and R. Egan, *J. Antibiotics*, 1983, **36**, 1405.
27. J. Breinholt, G. W. Jensen, A. Kjaer, C. E. Olsen and C. N. Rosendahl,
     Acta Chem. Scand., 1997, **51**, 855.
28. I. Kurobane, L. C. Vining, A. G. McInnes and J. A. Walter,
     *Can. J. Chem.,*1980, **58**, 1380.
29. F. J. Leeper and J. L. Staunton, *J. Chem. Soc., Perkin Trans. 1*, 1984, 2919.
30. M. L. Casey, R. C. Paulick and H. W. Whitlock, *J. Org. Chem.*,
     1978, **43**, 1627.
31. J. L. Bloomer, C. A. Smith and T. J. Caggiano, *J. Org. Chem.,*
     1984, **49**, 5027.
32. I. Kitamura, H. Tobe, A. Yoshimoto, T. Oki, H. Naganawa, T. Takeuchi
     and H. Umezawa, *J. Antibiotics*, 1981, **34**, 1498.
33. C. R. Hutchinson, *Chem. Rev.*, 1997, **97**, 2525.
34. N. Imamura, K. Kakinuma, N. Ikekawa, H. Tanaka and S. Omura,
     J. Antibiotics, 1982, **35**, 602.
35. S. Iwasaki, in Comprehensive Natural Product Chemistry; D. H. R. Barton
     and K. Nakanishi, Eds; Elsevier: Oxford, 1999; vol. 1,
     (vol. ed.U. Sankawa), p. 557.
36. Y. Tokiwa, M. Miyoshi-Saitoh, H. Kobayashi, R. Sunaga, M. Konishi,
     T. Oki and S. Iwasaki, *J. Amer. Chem. Soc.*, 1992, **114**, 4107.
37. O. D. Hensens, J-L. Giner, and I. H. Goldberg,
     J. Amer. Chem. Soc., 1989, **111**, 3295.
38. T. Hata, F. Koga, Y. Sano, K. Kanamori, A. Matsumae, R. Sugawara,
     T. Shima, S. Ito and S. Tomizawa, *J. Antibiotics, Ser. A*, 1954, **17**, 107.
39. K.Yohoi, K. Nagaoka and T. Nakashima, *Chem. Pharm.
     Bull.*, 1986, **34**, 4554.
40. R. Thomas and D. J. Williams, *J. Chem. Soc., Chem. Commun.*, 1983, 677.
41. A. G. McInnes, D. G. Smith, J. A. Walter, L. C. Vining and
     J. L. C. Wright, *J. Chem. Soc., Chem. Commun.*, 1975, 66.
42. C. P. Gorst-Allman, B. A. M. Rudd, C-j. Chang and H. G. Floss,
     J. Org. Chem., 1981, **46**, 455.

43. A. L. Staley and K. L. Rhinehart, *J. Antibiotics*, 1994, **47**, 1425.
44. R. McDaniel, S. Ebert-Khosla, D. A. Hopwood and C. Khosla, *J. Amer. Chem. Soc.*, 1993, **115**, 11671.
45. P. L. Bartel, C. B. Zhu, J. S. Lampel, D. C. Dosch, N. C. Connors, W. R. Strohl, J. M. Beal and H. G. Floss, *J. Bact.*, 1990, **172**, 4816.
46. A. E. Oxford and H. Raistrick, *Biochem. J.*, 1933, **27**, 634.
47. A. A. Qureshi, R. W. Rickards and A. Kamal, *Tetrahedron*, 1967, 23, 3801.
48 S. Gaisser, A. Trefzer, S. Stockert, A. Kirschning and A. Bechthold, *J. Bact.*, 1997, **179**, 6271.
49. M. G. Bangera and L. S. Thomashow, *J. Bact.*, 1999, **181**, 3155.
50. S. J. Hammond, M. P. Williamson, D. H. Williams, L. D. Boeck and G. G. Marconi, *J. Chem. Soc., Chem. Commun.*, 1982, 344.
51. R. Thomas, *Biochem. J.*, 1961, **78**, 748.
52. E. E. Stinson, W. B. Wise, R. A. Moreau, A. J. Jurewicz and P. E. Pfeffer, *Can. J. Chem.*, 1986, **64**, 1590.
53. J. Dasenbrock and T. J. Simpson, *J. Chem. Soc., Chem. Commun.*, 1987, 1235.
54. J. Breinholt, A. Kjaer, C. E. Olsen, B. R. Rassing and C. N. Rosendahl, *Acta Chem. Scand.*, 1997, **51**, 1241.
55. R. N. Moore, G. Bigam, J. K. Chan, A. M. Hogg, T. T. Nakashima and J. C. Vederas, *J. Amer. Chem. Soc.*, 1985, **107**, 3694.
56. T. J. Simpson, *J. Chem. Soc., Perkin Trans.1*, 1977, 592.
57. V. L. Himes, A. D. Mighell, S. W. Page and M. E. Stack, *Acta Cryst., Section B*, 1981, **37**, 1932.
58. G. Höfle and K. Röser, *J. Chem. Soc., Chem. Commun.*, 1978, 611.
59. H. Seto and M. Tanabe, *Tetrahedron Lett.*, 1974, 651.
60. K. M. Byrne, B. D. Hilton, R. J. White, R. Misra and R. C. Pandey, *Biochemistry*, 1985, **24**, 478.
61. T. Aoyama, H. Naganawa, Y. Muraoka, T. Aoyagi and T. Takeuchi, *J. Antibiotics*, 1992, **45**,1767.
62. S. J. Gould and X.-C. Cheng, *Tetrahedron*, 1993, **49**, 11135.
63. S. J. Gould, C. R. Melville, and J. Chen, *Tetrahedron*, 1997, **53**, 4561.
64. Y. M. Shen, P. Yoon, T. W. Yu, H. G. Floss, D. Hopwood and B. S. Moore, *Proc. Natl. Acad. Sci. USA*, 1999, **96**, 3622.
65. E. Dagne, A. Yenesew, S. Asmellash, S. Demissew and S. Mavi, *Phytochemistry*, 1994, **35**, 401.
66. H. Schmid and A. Ebnöther, *Helv. Chim. Acta*, 1951, **34**, 1041.
67. A. J. Birch and F. W. Donovan, *Austr. J. Chem.*, 1953, **6**, 373.
68. J. Schröder, in Comprehensive Natural Product Chemistry; D. H. R. Barton and K. Nakanishi, Eds; Elsevier: Oxford, 1999; vol. 1, (vol. ed. U. Sankawa), p. 409.
69. R. Thomas, *J. Chem. Soc., Chem. Commun.*, 1971, 739; R. Thomas, *Pure and Applied Chem.*, 1973, **34**, 515.
70. J. M Edwards, R. C. Schmitt and U. Weiss, *Phytochemistry*, 1972, **11**, 1717; R. G. Cooke and J. M. Edwards, *Fortsch. Chem. Org. Naturstoffe*, 1981, **40**, 153.
71. D. Holscher and B. Schneider, *Nat. Prod. Lett.*, 1995, **7**, 177.
72. B. Schmitt and B. Schneider, *Phytochemistry*, 1999, **52**, 45.
73. R. Marchelli and L. Vining, *J. Chem. Soc., Chem. Commun.*, 1973, 555.

74. J-L. Ferrer, J. M. Jez, M. E. Bowman, R. A. Dixon and J. P. Noel, *Nature Struct. Biol.*, 1999, **6**, 775.
75. S. Tropf, T. Lanz, S. A. Rensing, J. Schröder and G. Schröder, *J. Mol. Evol.*, 1994, **38**, 610.
76. Fatty Acid Metabolism and Its Regulation, Ed. S. Numa, 1984, Elsevier, Amsterdam. (New Comp. Biochem. Vol. 7 Gen. Eds. A. Neuberger & L. L. M. van Deenen).
77. P. M. Shooligan-Jordan and I. D. G. Campuzano, in Comprehensive Natural Product Chemistry; D. H. R. Barton and K. Nakanishi, Eds; Elsevier: Oxford, 1999; vol. 1, (vol. ed. U. Sankawa), p. 345.
78. M. P. Crump, J. Crosby, C. E. Dempsey, J. A. Parkinson, M. Murray, D. A. Hopwood and T. J. Simpson, *Biochemistry*, 1997, **36**, 6000.

6 Natural Products as Leads for Synthesis

THE STROBILURIN FUNGICIDES - FROM MUSHROOM TO MOLECULE TO MARKET

John M Clough

Zeneca Agrochemicals, Jealott's Hill Research Station, Bracknell, Berkshire, RG42 6ET

1 THE COMMERCIAL STROBILURIN FUNGICIDES

The strobilurin fungicides are a group of synthetic agrochemicals which were inspired by the naturally-occurring strobilurins, oudemansins and myxothiazols. There are two strobilurin fungicides on sale at present, azoxystrobin from Zeneca and kresoxim-methyl from BASF (Figure 1). Both were announced in 1992[1,2] and first sold in 1996. Azoxystrobin is now registered for use on 55 crops in 49 countries. It is sold, for example, under Zeneca's trademarks "Amistar" on cereals, "Quadris" on grape vines and "Heritage" on turf. Already it is one of the world's leading fungicides with sales in 1998 totalling about $290 million. BASF sells kresoxim-methyl mainly in mixture with other fungicides, in combination with fenpropimorph, for example, under the trademark "Brio", or with epoxiconazole under the trademark "Allegro", both for use on cereals.

In addition to these two products, other strobilurin fungicides are being developed by various agrochemical companies. Closest to the market are metominostrobin from Shionogi[3] and trifloxystrobin from Novartis[4] (Figure 1), both of which are expected to be on sale for the first time this year.

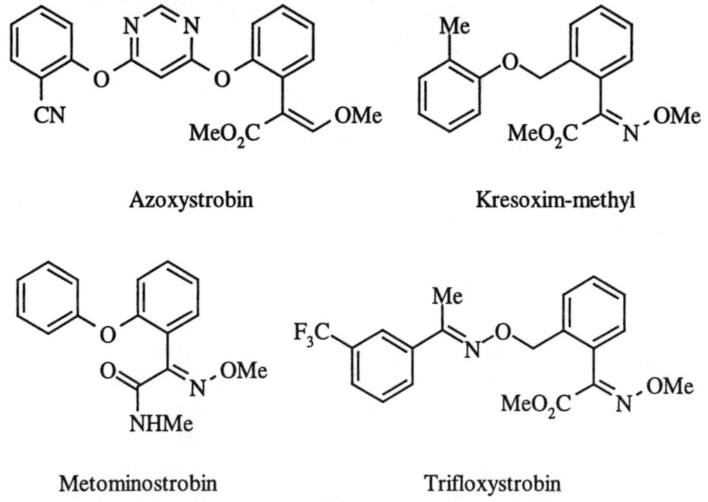

Figure 1. *Synthetic Strobilurin Fungicides*

The strobilurins continue to be an area of intense research within the agrochemical industry. Patent applications claiming strobilurins as fungicides have been filed by more than 20 companies, and a total of about 585 had been published by the end of July 1999.

2   THE NATURAL STROBILURIN FUNGICIDES AND THE DISCOVERY OF AZOXYSTROBIN

More than 30 strobilurins, oudemansins and myxothiazols are now known. The strobilurins and oudemansins have been isolated from several genera of small fungi which grow on decaying wood, such as *Oudemansiella mucida*. By contrast, the myxothiazols are produced by the gliding bacterium *Myxococcus fulvus*. The simplest of the strobilurins and oudemansins, together with the most abundant myxothiazol, are shown in Figure 2. We have already described how these natural products formed the starting point for a programme of synthesis which led ultimately to the discovery of azoxystrobin. The evolution of ideas is summarised in Figure 3 and the following paragraph, and full details can be found in reference 5 and other references cited therein.

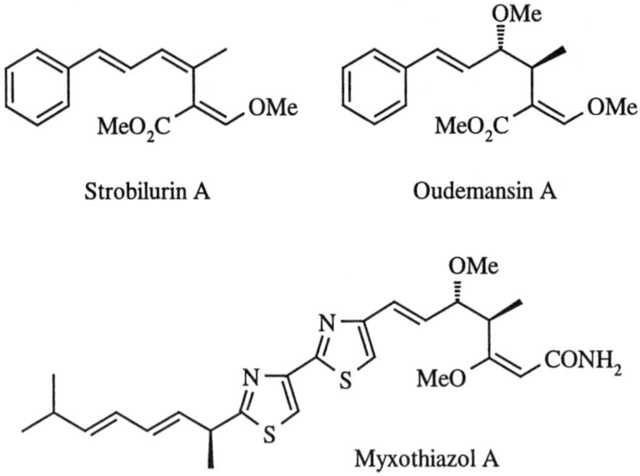

Strobilurin A   Oudemansin A

Myxothiazol A

Figure 2. *Natural Fungicidal Derivatives of β-Methoxyacrylic Acid*

Strobilurin A, though active against fungi growing on agar, does not control fungi growing on plants in the glasshouse because of its photochemical instability and volatility. The stilbene (1) is active in the glasshouse but still degrades too quickly in light to express high activity in the field. The diphenyl ether (2) is much more stable in light and is active in the field, albeit at rather high application rates. Furthermore, (2) is systemic in plants, an important property of many modern fungicides which improves field performance by redistribution of the compound within plant tissue after spraying. Modifications to (2) led to the analogues (3) and (4). The tricyclic derivative (3) has improved levels of fungicidal activity, but is too lipophilic to have systemic movement. The pyridine (4) retains the activity of its isostere (2) and, as expected, is highly mobile (probably too mobile) in plants. A combination of the ideas which led from (2) to (3) and (4), together with a huge amount of experimentation, led finally to the discovery of azoxystrobin, an outstanding systemic fungicide with excellent bioefficacy and safety in the environment.[5]

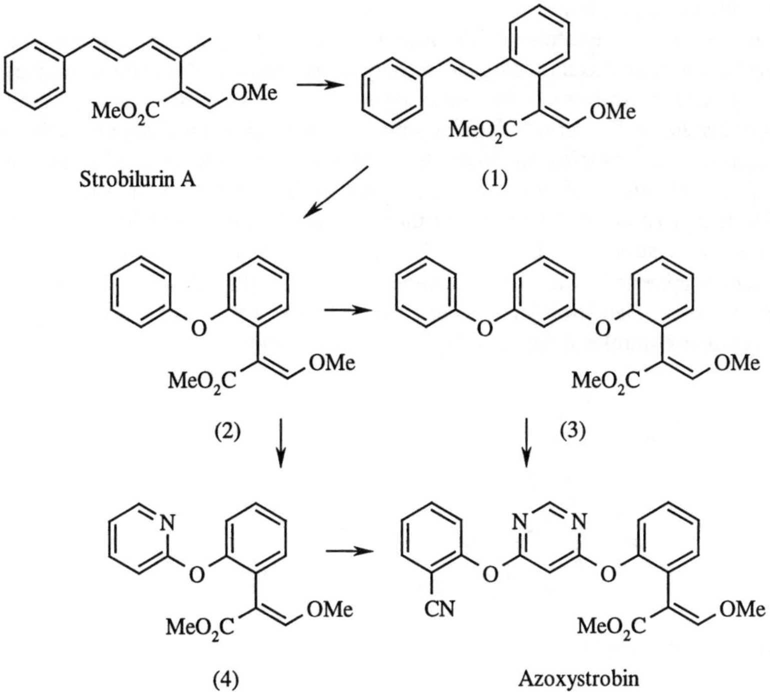

Figure 3. *Evolution of Ideas leading from the Natural Products (here represented by Strobilurin A) to Azoxystrobin*

3 LESSONS FROM THE RESEARCH PROGRAMME WHICH LED TO THE DISCOVERY OF AZOXYSTROBIN

The purpose of this final section is to draw some general lessons for current and future research programmes from the work leading to the discovery of azoxystrobin. These relate to (a) the use of natural products as starting points for the discovery of new agrochemicals; (b) targeting markets from which a healthy return on the research investment can be made; and (c) ways in which biological activity can most efficiently be optimised.

Nature is an excellent source of novel and diverse compounds with fungicidal activity. Such compounds can occasionally be isolated and used as agrochemicals in their own right. However, under the influence of evolutionary processes, nature has optimised the properties of its fungicides for its own particular purposes and conditions, and we should not expect these to be the same as those required by the farmer. Strobilurin A, for instance, is photochemically unstable and volatile. Photochemical instability is not a problem for the fungi which produce strobilurin A, since it is formed and does its job deep within a piece of rotting wood, and volatility is likely to help its distribution. However, important agricultural fungicides are applied in sunlight as foliar sprays. This means that good photostability is a key property, and volatility must not be too high either. So we can conclude that nature provides us with compounds with fungicidal activity but, in order for this activity to be expressed, we must expect to have to design new analogues with the physical chemical properties which we require.

Although there are exceptions, it is usual for many of the functional groups of a natural product to contribute to its biological activity. This means that a natural product with a relatively simple structure is the most promising starting point for a programme of analogue synthesis. The term "relatively simple" in this context is changing as the years go by, because more complex structures have become amenable as organic synthesis has developed, and this trend will continue. In the strobilurin project, it was fortunate that strobilurin A was isolated first, rather than one of the more complex members of the family such as strobilurin E (Figure 4). Strobilurin A has a simple structure and so immediately appealed as a lead for synthesis, with the β-methoxyacrylate group as the likely toxophore. Had strobilurin E been isolated first, we could have wrongly assumed that its complex spiroketal functionality was the toxophore.

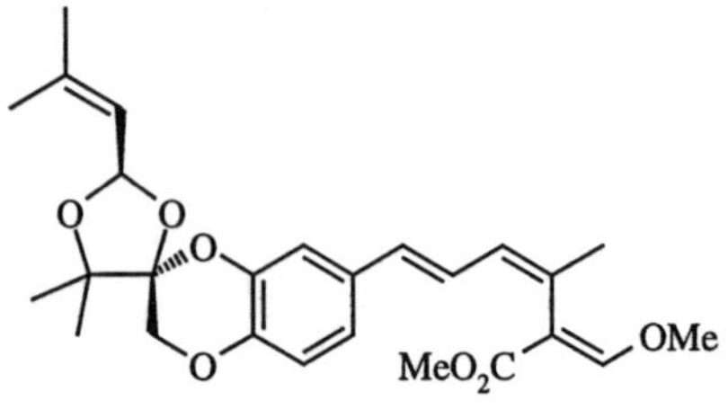

Figure 4. *Strobilurin E*

A family of fungicidal natural products is always a more attractive starting point than a single compound, because it indicates that there is at least some scope for structural modification without loss of activity. If the mode of action is known, that is also extremely useful.

If the scientific literature is to be used as the source of new fungicidal natural products, it is important to obtain a sample of the natural product itself for testing in-house with commercial fungicides as standards. This is because, in our experience, natural products claimed in the literature to be fungicidal very often have only low levels or a narrow spectrum of activity. To embark on a programme of synthesis before confirming the activity of the lead compound is therefore likely to result in disappointment.

It is also worth noting that, when using the scientific literature as the source of new fungicidal natural products, one needs to begin work as soon as the structure is published since the whole agrochemical community is reading the same information! In the strobilurin area, Zeneca and BASF, the latter working closely with Professors Timm Anke and Wolfgang Steglich, both independently initiated programmes of synthesis based on the natural products.[5,6] Scientists in the two companies worked along strikingly similar lines, discovering the limiting properties of the strobilurins and preparing several of the same synthetic analogues such as the stilbene (1). Only when Zeneca's first patent applications were published in April 1986 did BASF become aware of Zeneca's work in this area. And then, some seven months later when BASF's first patent applications were published, we in Zeneca also became aware that we had competitors. During the optimisation phases of the two programmes, different biological objectives carried Zeneca and BASF in different directions, resulting in the two products azoxystrobin and kresoxim-methyl with their distinctive properties.

Before embarking on a research programme, it is clearly important to understand the market for the product one is seeking. Wood Mackenzie provides the following figures.[7] In 1997, the fungicide market was valued at $5.5 billion at the end-user level, which is about 18% of the whole agrochemical market. The markets for insecticides and especially herbicides are substantially larger. Furthermore, the market for fungicides is fragmented, with the fruit and vegetables sector, itself embracing a wide variety of crops and diseases, constituting the largest part at 47%, and temperate cereals and rice forming other important sectors at 25% and 14%, respectively. Finally, systemic compounds constitute 69% of the fungicide market, a proportion which is rising year by year. One can conclude from these figures that a new fungicide needs to have a broad spectrum of activity as well as systemic movement in plants if it is to become a major product. Screens must then be devised accordingly because, of course, you only find what you screen for, or, conversely, you won't find a particular property like systemic activity if you don't screen for it.

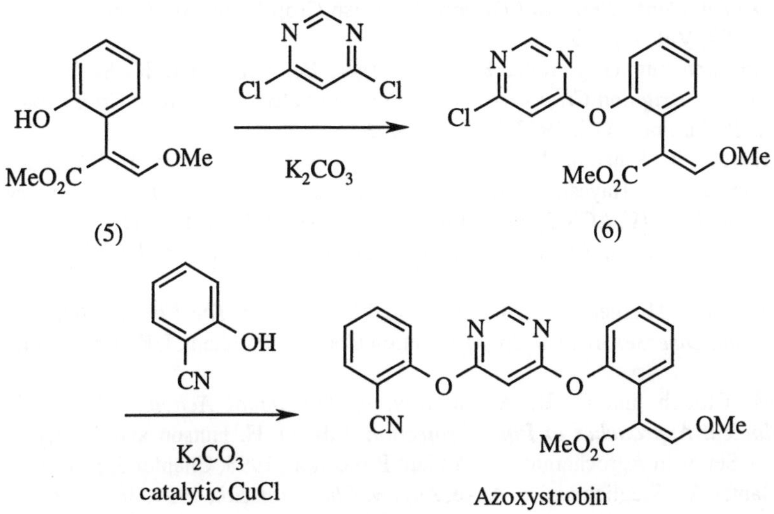

Figure 5. *Laboratory Synthesis of Azoxystrobin*

Finally, the optimisation of activity still requires the synthesis and testing of many series of compounds. A variety of properties, such as potent fungicidal activity, systemicity, and safety in the environment, must be combined in a commercial fungicide, and it is impossible to find these except by many cycles of synthesis and screening. Divergent intermediates, from which a variety of analogues can quickly be made, are very useful for streamlining synthesis. For example, the laboratory synthesis of azoxystrobin shown in Figure 5 follows a route with two branch points for the synthesis of related compounds. The phenol (5), for example, reacts with a variety of electrophilic heterocycles to give products containing the β-methoxyacrylate toxophore, some of which have useful fungicidal activity. Similarly, the chloropyrimidine (6) reacts with a variety of phenols to give products which are closely related to azoxystrobin.[8] Automated synthesis was not available to us in the 1980's when we conducted the research described here. These days we would be able to use robots to prepare libraries of related compounds.

4 CONCLUSIONS

The strobilurins are an important new class of agricultural fungicides. The best examples exhibit an extremely useful spectrum of activity against many plant diseases of commercial importance. The first two products of this class, azoxystrobin and kresoxim-methyl, first sold just three years ago in April 1996, have already made a major impact on the global fungicide market. The importance of the strobilurins will increase in the years ahead as the established products are registered for use on further crops and in further territories, and as new products, currently under development, are launched.

References

1. J. R. Godwin, V. M. Anthony, J. M. Clough and C. R. A. Godfrey, in *Brighton Crop Prot. Conf.: Pests and Diseases'*, British Crop Protection Council, Farnham, UK, 1992, Vol. 1, p. 435.
2. E. Ammermann, G. Lorenz, K. Schelberger, B. Wenderoth, H. Sauter and C. Rentzea, in *Brighton Crop Prot. Conf.: Pests and Diseases*, British Crop Protection Council, Farnham, UK, 1992, Vol. 1, p. 403.
3. Y. Hayase, T. Kataoka, M. Masuko, M. Niikawa, M. Ichinari, H. Takenaka, T. Takahashi, Y. Hayashi and R. Takeda, in *Synthesis and Chemistry of Agrochemicals IV*, ACS Symposium Series No. 584, Eds. D. R. Baker, J. G. Fenyes and G. S. Basarab, American Chemical Society, Washington, DC, USA, 1995, Chapter 30, p. 343.
4. P. Margot, F. Huggenberger, J. Amrein and B. Weiss, in *The 1998 Brighton Conf.: Pests and Diseases*, British Crop Protection Council, Farnham, UK, 1998, Vol. 2, p. 375.
5. J. M. Clough and C. R. A. Godfrey, in *Fungicidal Activity. Chemical and Biological Approaches to Plant Protection*, Eds. D. H. Hutson and J. Miyamoto, Wiley Series in Agrochemicals and Plant Protection, 1998, Chapter 5, p. 109.
6. H. Sauter, W. Steglich and T. Anke, *Angew. Chem. Int. Ed.*, 1999, **38**, 1328.
7. Wood Mackenzie Consultants Limited, Edinburgh and London, in *Agrochemical Service, Update of the Products Section*, May 1998, pp. 1-11 and 64-85.
8. J. M. Clough, D. A. Evans, P. J. de Fraine, T. E. M. Fraser, C. R. A. Godfrey and D. Youle, in *Natural and Engineered Pest Management Agents*, ACS Symposium Series No. 551, Eds. P. A. Hedin, J. J. Menn and R. M. Hollingworth, American Chemical Society, Washington, DC, USA, 1994, Chapter 4, p. 37.

A SYNTHESIS OF (+)-PRELOG-DJERASSI LACTONIC ACID

Steven D. Hiscock, Peter B. Hitchcock and Philip J. Parsons*

The Chemical Laboratories, Arundel Building, School of Chemistry, Physics & Environmental Science, University of Sussex, Falmer, Brighton, BN1 9QJ, UK.

A new approach to the synthesis of Prelog-Djerassi Lactonic acid (**1**) is reported. A key step in this synthesis involves an Ireland-Claisen rearrangement/silicon-mediated fragmentation sequence to provide the carbon framework in (**1**).

1 INTRODUCTION

The isolation of (+)-Prelog-Djerassi Lactonic acid (PDLA) (**1**) was reported in 1956 independently by Prelog and Djerassi.[1-3] Prelog *et al.* isolated the six-membered lactone (**1**) as a key oxidative degradation product of the macrolide antibiotics narbomycin and pikromycin.[1] Djerassi and Zderic initially reported the lactone as a degradation fragment of the macrolide antibiotic methymycin[2] and subsequently, as a degradation product of neomethymycin.[3] The fragment holds a prominent position in the field of macrolide antibiotic chemistry, not only having provided essential information for their structure elucidation,[4,5] but it has also served in their synthesis.[6] The structure of (+)-PDLA was determined in 1970 by Rickards and Smith[4] and shown to be (**1**).

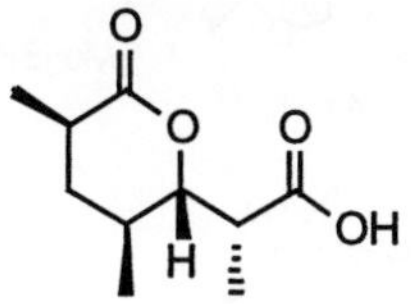

1

We now report our synthesis of (+)-PDLA based on the retrosynthetic analysis shown in Scheme 1.

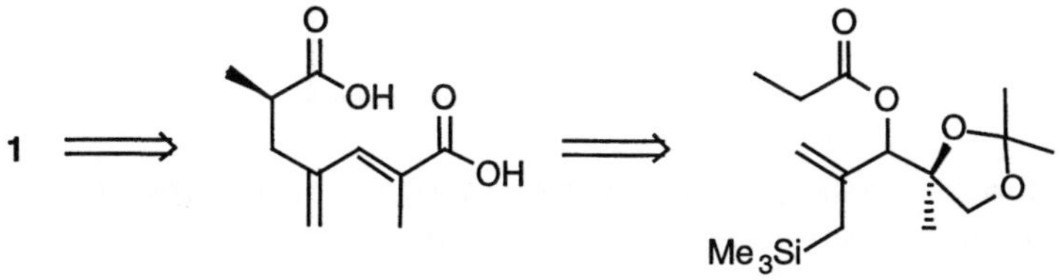

Scheme 1 *Retrosynthetic analysis of (+)-PDLA*

2 RESULTS & DISCUSSION

We have recently investigated the remote control of asymmetry by cascade rearrangement/silicon-mediated fragmentation of allylic epoxides,[7] and now report our work on the fragmentation of acetals. Our synthesis of (+)-PDLA relied on the construction of key acid (5), and our approach is shown in Scheme 2. Treatment of the homochiral aldehyde (2)[8] with 1-trimethylsilyl-2-propenyl-magnesium bromide[9] gave an easily separable mixture of allylic alcohols (3a) and (3b) in 80 % yield, with the stereochemical assignments determined *via* single crystal x-ray analysis of the corresponding *p*-nitrobenzoates. The *syn*- and *anti*-allylic alcohols were then each subjected to esterification in the presence of propionic anhydride, triethylamine and 4-(dimethylamino)pyridine,[10] to give the corresponding propionate esters (4a) and (4b) in excellent yield (98 %). Both the *syn*- and *anti*-propionate esters could be converted to a single diastereomeric acetonide-acid (5) in optically pure form by modifying the reaction conditions to obtain either the *E*- or the *Z*-silyl ketene acetal (Scheme 3) as intermediate for the key Ireland-Claisen rearrangement.[11]

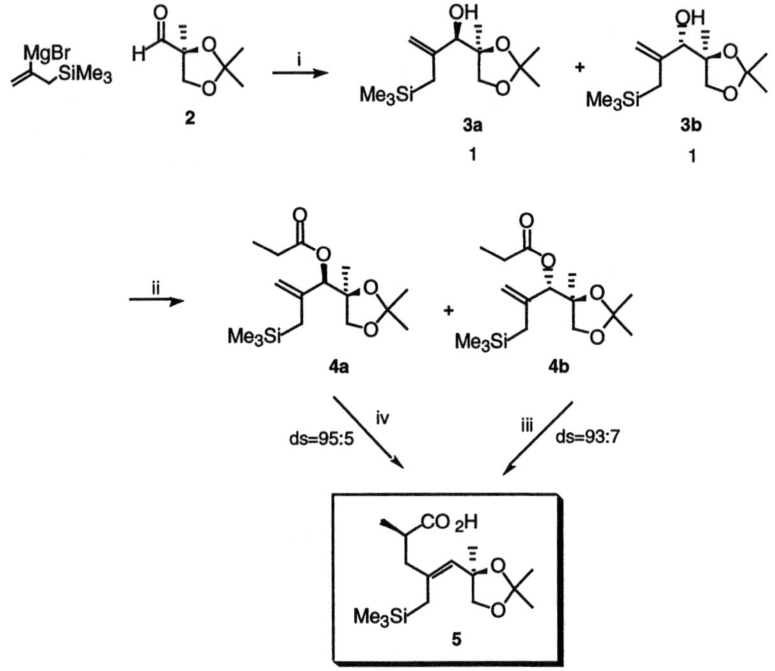

Reagents : (i) 0ºC/THF (82 %); (ii) (EtCO)$_2$O/Et$_3$N/DMAP/DCM (97 %); (iii) LDA/THF/TBDMSCl then DMPU (88 %); (iv) LDA/THF/DMPU then TBDMSCl (65 %)

Scheme 2 *Synthesis of acid intermediate (5) in synthesis of (+)-PDLA*

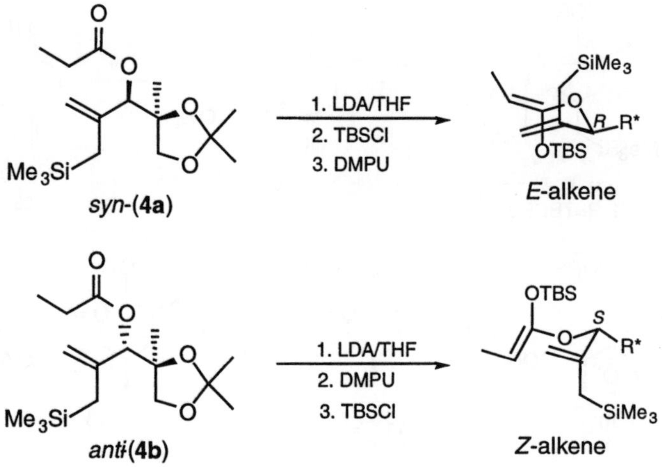

Scheme 3 *Synthesis of intermediate for the Ireland-Claisen rearrangement*

Scheme 4 depicts the remainder of our total synthesis. The acid (**5**) was subjected to RedAl® reduction[12] to give the hydroxymethyl intermediate (**6**) in 99 % yield, which was then protected as the *tert*-butyldiphenylsilyl ether (**7**) under standard conditions (96 %).[13] Silicon-mediated fragmentation of the acetonide (**7**) was carried out under Lewis acid-mediated conditions (BF$_3$•OEt$_2$) to furnish allylic alcohol (**8**) in 96 % yield. Sharpless epoxidation provided (**9**),[14] followed by a diastereoselective diimide reduction of the remaining olefin,[15] to give the epoxyalcohol (**10**) in satisfactory yield, and as exclusively one diastereoisomer.

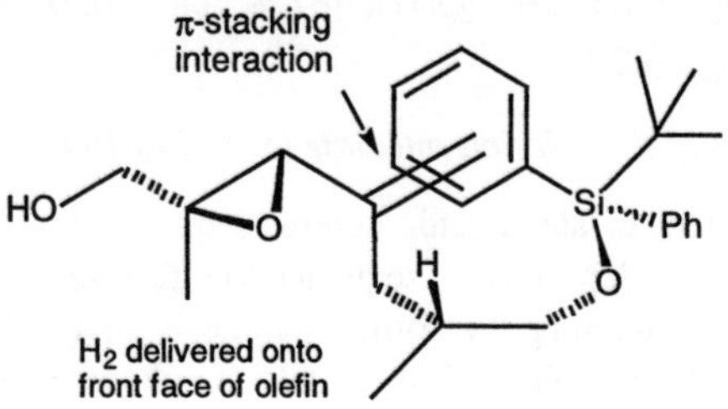

Figure 1 *Postulated conformation of the reactant during reduction by diimide*

We invoke π-stacking of the alkene with a phenyl moiety on the silicon protecting group (since this high degree of selectivity was only observed for TBDPS and not with TBS), in the precursor to explain this remarkable selectivity (Figure 1).[16] Lewis acid induced reduction of the epoxide with sodium cyanoborohydride led regioselectively to the 1,3-diol (**11**); the hydride attacks the more substituted position *via* an S$_N$2 mechanism.[17]

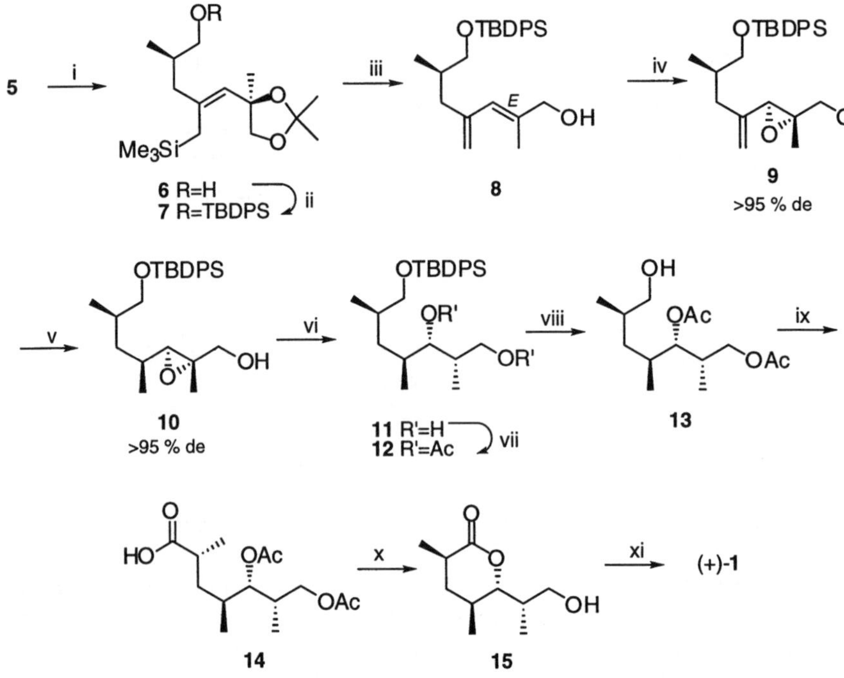

Reagents: (i) RedAl/THF/0°C (80 %); (ii) TBDPSCl/DMAP/DMF/imidazole (95 %); (iii) BF₃•OEt₂/THF/–78°C to rt/48 h (95 %); (iv) Ti(OⁱPr)₄/(+)-DET/–20°C (93 %); (v) (NH₂)₂/EtOH/CuSO₄/Δ (65 %); (vi) BF₃•OEt₂/NaCNBH₃/THF/Δ(90 %); (vii) (CH₃CO)₂O/Et₃N/DMAP (80 %); (viii) TBAF/THF (70 %); (ix) RuCl₃•3H₂O/ NaIO₄/CCl₄/MeCN/H₂O; (x) LiOH/H₂O/THF (67 % from **13**); (xi) RuCl₃•3H₂O/ NaIO₄/CCl₄/MeCN/H₂O (93 %)

Scheme 4 *Conversion of the acid intermediate (5) to (+)-PDLA*

With the full complement of stereogenic centres required, the 1,3-diol (**11**) was then taken through a protection/ deprotection sequence to afford the corresponding diacetoxy alcohol (**13**), thereby completing the formal synthesis of (+)-PDLA. The remaining steps in the total synthesis followed the route employed by Yamaguchi and co-workers.[18] Diacetoxy alcohol (**13**) was oxidised with RuCl₃/NaIO₄ to provide acid (**14**), which underwent concomitant lactonisation to (**15**) under saponification conditions. The primary alcohol (**15**) was then oxidised to afford (+)-Prelog-Djerassi lactonic acid in an overall 9 % yield, with all spectroscopic data in accord with the literature values.

Acknowledgments

We thank E.P.S.R.C. and Tocris-Cookson Ltd. for financial support. We also thank Dr Neil Edwards for helpful discussions and assistance in the preparation of the manuscript, and Dr C. S. Penkett for helpful discussions.

References

1. R. Anliker, D. Dvornik, K. Gubler, H. Heusser and V. Prelog, *Helv. Chim. Acta*, 1956, **39**, 1785.
2. C. Djerassi and J. A. Zderic, *J. Am. Chem. Soc.*, 1956, **78**, 6390; C. Djerassi, A. Bowers, R. Hodges and B. Rinker, *J. Am. Chem. Soc.*, 1956, **78**, 1733; C. Djerassi and J. A. Zderic, *J. Am. Chem. Soc.*, 1956, **78**, 2907.
3. C. Djerassi and O. J. Halpern, *J. Am. Chem. Soc.*, 1957, **79**, 2023.
4. R. W. Rickards and R. M. Smith, *Tetrahedron Lett.*, 1970, 1025; R. W. Rickards and R. M. Smith, *Tetrahedron Lett.*, 1970, 1029.
5. For a review concerning the synthesis of Prelog-Djerassi lactonic acid, see: S. F. Martin and D. E. Guinn, *Synthesis*, 1991, 245 and references cited therein.
6. I. Paterson and M. M. Mansuri, *Tetrahedron*, 1985, **41**, 3569.
7. J. J. Eshelby, P. J. Parsons, N. C. Sillars and P. J. Crowley, *J. Chem. Soc., Chem. Commun.*, 1995, 1497.
8. J. S. Dung, R. W. Armstrong, O. P. Anderson and R. M. Williams, *J. Org. Chem.*, 1983, **48**, 3592.
9. J. J. Eshelby, P. J. Parsons and P. J. Crowley, *J. Chem. Soc., Perkin Trans. 1*, 1996, 191.
10. W. Steglich and G. Hoefle, *Angew. Chem., Int. Ed. Engl.*, 1969, **8**, 981.
11. R. E. Ireland and R. H. Mueller, *J. Am. Chem. Soc.*, 1972, **94**, 5897; R. E. Ireland, R. H. Mueller and A. K. Willard, *J. Am. Chem. Soc.*, 1976, **98**, 2868.
12. M. Gugelchuk, *Encyclopedia of Reagents for Organic Synthesis*, Ed. L. A. Paquette, J. Wiley and Sons, Chichester, NY, 1995, **7**, 4518.
13. T. W. Greene and P. G. M. Wuts, *Protective Groups in Organic Synthesis*, 2nd Edn., Wiley-Interscience, New York, 1991, 39.
14. K. B. Sharpless and T. Katsuki, *J. Am. Chem. Soc.*, 1980, **102**, 5974; Y. Gao, R. M. Hanson, J. M. Klunder, S. Y. Ko and K. B. Sharpless, *J. Am. Chem. Soc.*, 1987, **109**, 5765.
15. D. J. Pastog, *Encyclopedia of Reagents for Organic Synthesis*, Ed. L. A. Paquette, J. Wiley and Sons, 1995, **3**, 1892.
16. J. d'Angelo and J. Maddaluno, *J. Am. Chem. Soc.*, 1986, **108**, 8112; A. B. Smith, N. J. Liverton, H. J. Hriab, H. Sivaramakrishnan and K. Winzenberg, *J. Am. Chem. Soc.*, 1986, **108**, 3040; J. Kallmerton and T. J. Gould, *J. Org. Chem.*, 1986, **51**, 1152.
17. D. F. Taber and J. B. Houze, *J. Org. Chem.*, 1981, **46**, 5214; H. Tone, T. Nishi, Y. Oikawa, M. Hikota and O. Yonemitsu, *Tetrahedron Lett.*, 1987, **28**, 4569.
18. M. Yamaguchi, T. Katsuki and M. Honda, *Tetrahedron Lett.*, 1984, **25**, 3857.
19. D. A. Evans and J. Bartroli, *Tetrahedron Lett.*, 1982, **23**, 807.
20. M. Miyashita, M. Hoshino, A. Yoshikoshi, K. Kawamine, K. Yoshihara and H. Irie, *Chem. Lett.*, 1992, 1101.
21. W. Oppolzer, E. Walther, C. P. Balado and J. De Brabander, *Tetrahedron Lett.* 1997, **38**, 809.

N-(α-AMINOACYL)-5'-O-SULFAMOYLADENOSINES: NATURAL PRODUCT BASED INHIBITORS OF AMINO ACYL tRNA SYNTHETASES

Kevin Beautement, Ewan J T Chrystal, Joy Howard and Stuart M Ridley

Zeneca Agrochemicals, Jealott's Hill Research Station, Bracknell, Berkshire, RG42 6ET.

1 INTRODUCTION

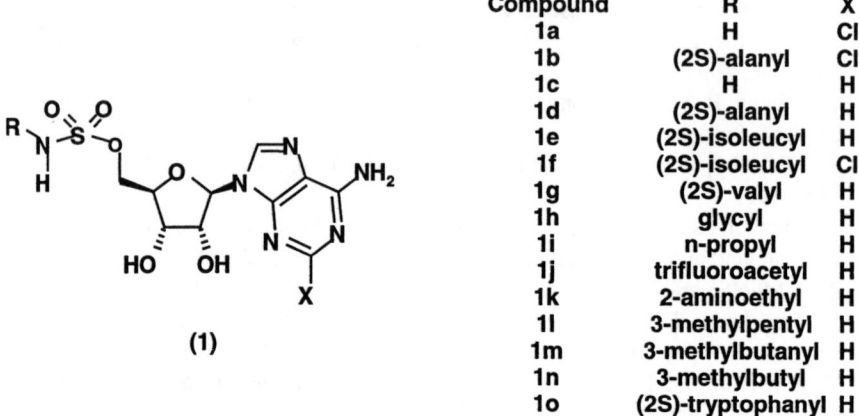

Compound	R	X
1a	H	Cl
1b	(2S)-alanyl	Cl
1c	H	H
1d	(2S)-alanyl	H
1e	(2S)-isoleucyl	H
1f	(2S)-isoleucyl	Cl
1g	(2S)-valyl	H
1h	glycyl	H
1i	n-propyl	H
1j	trifluoroacetyl	H
1k	2-aminoethyl	H
1l	3-methylpentyl	H
1m	3-methylbutanyl	H
1n	3-methylbutyl	H
1o	(2S)-tryptophanyl	H

Figure 1 *Sulfamoyladenosine and derivatives*

Natural products have a long history as a source of leads for both novel modes of action and novel chemistry in agrochemical research.[1] The *Streptomyces* natural products, dealanylascamycin, compound **1a**, and ascamycin, compound **1b** (Figure 1), have been reported to possess herbicidal activity.[2,3] Sulfamoylnucleosides (in particular 5'-O-sulfamoyl-adenosine, compound **1c**) are well documented as bioisosters of the corresponding nucleotides, their derivatives and analogues.[4] Such compounds are reported to possess a wide range of biological activities, e.g. as trypanocides,[5] antiviral agents,[6] and inhibitors of protein biosynthesis.[7, 8]

Suitable N-substituted sulfamoyladenosines (**1**) inhibit amino acyl tRNA synthetases (aaTRS's) by mimicking the enzyme bound reaction intermediate **2** (Scheme 1).[9] Thus the Structure Activity Relationship (SAR) obtained by varying the R group in structure **1** should probe the key interactions for amino acid selectivity in the aaTRS under investigation.

Prompted by the herbicidal activity of monate derivatives, which are inhibitors of prokaryotic isoleucyl-tRNA synthetase (ITRS),[10] several analogues of structure **1** were prepared and their biological activities characterised to establish if there was a common SAR between herbicidal activity and aaTRS inhibitor activity.

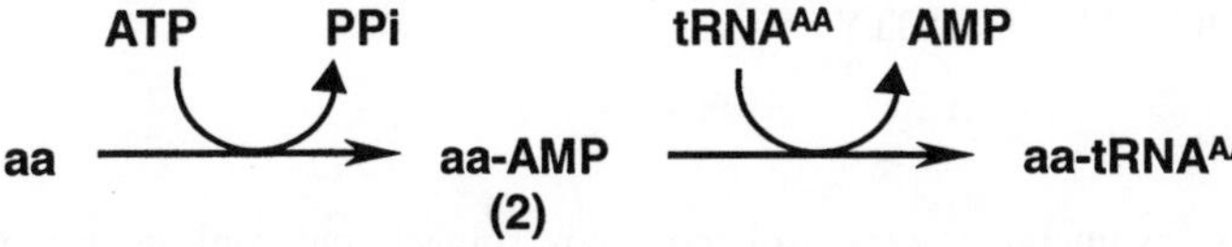

Scheme 1 *The reaction catalysed by aminoacyl-tRNA synthetases*

2 SYNTHESIS

N-Acyl derivatives (**1b**, **1d**, **1e**, **1f**, **1g**, **1h**, **1j**, **1m** and **1o**) were prepared from either the protected adenosine **3**, or its chloro analogue, by standard procedures (Scheme 2). N-alkyl derivatives (**1i**, **1k**, **1l** and **1n**) were prepared by the base mediated alkylation of compound **4** followed by acid catalysed deprotection, e.g. compound **1i** was prepared from compound **4** in a 22% overall yield. Dealanylascamycin **1a** was prepared by the hydrolysis of the chloro analogue of compound **4**.[11, 12]

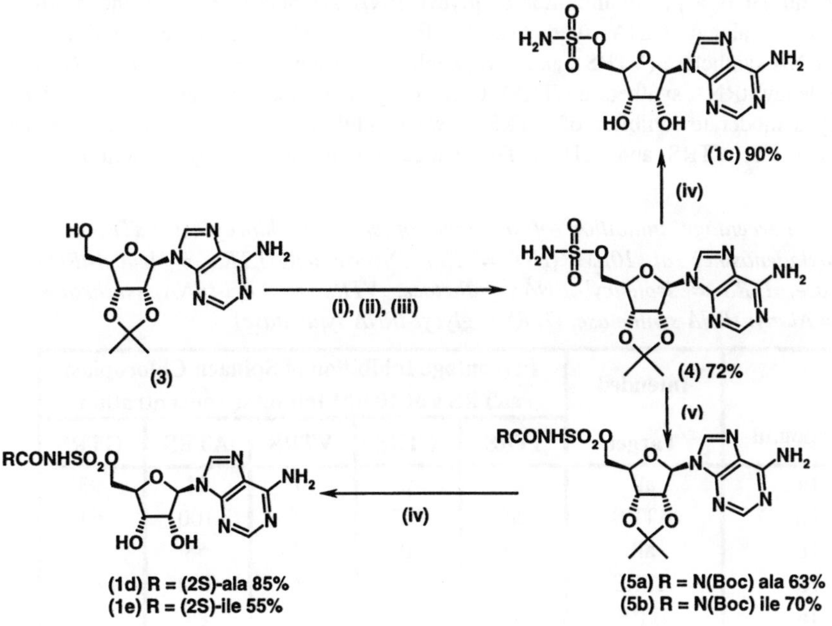

Scheme 2 *Synthetic route: (i) Toluene/Dean Stark/(Bu₃Sn)₂O/Δ/2h, (ii) Sulfamoylchloride/1,4-dioxane/0°C to 5°C, (iii) NH₄OH/MeOH, (iv) TFA:H₂O::5:2, (v) DBU/DMF/N-tertButoxycarbonyl-α-aminoalkanoic acid N-hydroxysuccinamide ester*

3 BIOLOGICAL ACTIVITY

3.1 Assays

Aminoacid-tRNA synthetases were isolated from spinach chloroplasts by methods developed at Jealott's Hill. Cytoplasmic aaTRS's from carrot were isolated by standard procedures. Compounds were assayed as inhibitors of the chloroplastic aaTRS's at a single rate, 10 μM. The reaction was followed by phosphate generation and the percentage inhibition determined using literature methods. The IC_{50} value for the spinach chloroplastic enzymes and inhibition data for carrot cytoplasmic aaTRS's were determined for compounds of interest. All compounds were tested for post-emergence herbicidal activity. at 0.125, or at 0.5, or at 2.0 kg/ha in standard screens.[11, 12]

3.2 Single Rate Chloroplast α-Aminoacyl-tRNA Synthetase Inhibition Data

Even with single concentration inhibition data, an SAR may be determined across a range of aaTRS's for features of amino acid selectivity (Table 1). For example, compound **1h** is a potent inhibitor of glycyl-tRNA synthetase (GTRS) and a strong inhibitor of alanayl-tRNA synthetase (ATRS), but only a moderate inhibitor of valyl-tRNA synthetase (VTRS), and a weak inhibitor of leucyl-tRNA synthetase (LTRS) and isoleucyl-tRNA synthetase (ITRS). Conversely the desoxo analogue, compound **1k**, is only a moderate inhibitor of GTRS, a weak inhibitor of LTRS and was inactive against ITRS, VTRS and ATRS. This suggests that the carbonyl function of the

Table 1 *Percentage inhibition of a range of spinach chloroplast aaTRS's by sulfamoyladenosines at 10μM (ND = Not Determined, LTRS = leucyl-tRNA synthetase, ITRS = isoleucyl-tRNA synthetase, VTRS = valyl-tRNA synthetase, ATRS = Alanyl-tRNA synthetase, GTRS = glycyl-tRNA synthetase)*

Compound	Intended Target	Percentage Inhibition of Spinach Chloroplast aaTRS's at 10 μM inhibitor concentration				
		ITRS	LTRS	VTRS	ATRS	GTRS
1a	all	61	0	0	11	93
1b	ATRS	35	ND	25	100	61
1c	all	11	0	11	58	86
1d	ATRS	58	61	96	100	46
1e	ITRS	100	88	88	89	10
1f	ITRS	100	98	84	71	71
1g	VTRS	99	100	95	100	82
1h	GTRS	25	19	40	95	100
1i	ATRS	15	ND	ND	0	ND
1j	ATRS	0	ND	0	58	89
1k	GTRS	0	17	0	0	48
1l	ITRS	0	ND	ND	ND	ND
1m	VTRS	25	13	0	0	ND
1n	VTRS	23	0	0	ND	ND

Table 2 *IC$_{50}$ (nM) values for selected α-aminoacylsulfamoyl-adenosines against plastidic and cytosolic aaTRS's (ND = Not Determined, LTRS = leucyl-tRNA synthetase, ITRS = isoleucyl-tRNA synthetase)*

Compound	Test aaTRS	IC$_{50}$ (nM)	
		Spinach Chloroplast	Carrot Cell Cytoplasm
1a	ITRS	4800	>10000
1e	ITRS	0.7	0.7
1f	ITRS	0.2	1.4
1g	ITRS	300	ND
1g	LTRS	1	8.1

α-aminoacyl substituent plays an important role in the binding of the N-(α-aminoacyl)-sulfamoyladenosines, either by direct interaction with the aaTRS, or by increasing the acidity of the amido group and making the sulfamoyl group a better mimic of the ionised phosphate in the intermediate **2**. However, sulfamoyladenosine **1c** is good inhibitor of GTRS, a moderate inhibitor of ATRS, a very weak inhibitor of ITRS and VTRS, and is inactive against LTRS. Thus the interactions between potential inhibitor and target aaTRS are more complex and may not be interpreted on the basis of class of aaTRS under study (ITRS, LTRS and VTRS are class I, and ATRS and GTRS are class II).

Compound **1o** was inactive against all aaTRS's tested. This unexpected result was attributed to its hydrolytic instability.

3.3 Comparison of Inhibition of Chloroplast and Cytoplasmic Enzymes.

IC$_{50}$ values were determined for selected compounds against the appropriate aaTRS from both spinach chloroplast and carrot cell cytoplasm (Table 2). No differentiation in potency was observed for inhibition of the chloroplastic, i.e. prokaryotic, and cytoplasmic, i.e. eukaryotic, enzymes. Enzymes from both tissues were equally susceptible to inhibition by the appropriate aminoacyl compound at nanomolar concentrations. Even closely related aminoacyl-sulfamoyladenosines were much weaker inhibitors, e.g. the leucyl derivative **1g** and dealanylascamycin, **1a**, were several orders of magnitude less potent than the appropriate aminoacyl-sulfamoyladenosines, compounds **1e** and **1f**, *versus* ITRS.

3.4 Herbicidal Activity

Compounds **1i**, **1l**, **1m**, and **1n** failed to show any herbicidal activity when tested in standard herbicide screens in the glass house. The remaining compounds showed similar activity as the natural products dealanylascamycin, **1a**, and ascamycin, **1b**. They exhibited only weak to moderate activity post-emergence and were usually more active against broad leaf plants than grasses (Table 3).

Table 3 *Herbicidal activity as mean percentage damage against broad leaf weeds and grasses in glass house testing*

Compound	Test Rate kg/ha	Herbicidal Activity as Mean % Damage	
		Grasses	Broad Leaf Weeds
1a	0.125	33	69
1b	0.125	35	68
1c	0.125	33	69
1d	0.125	14	48
1e	0.125	0	27
1f	2.0	90	100
1g	0.125	26	47
1h	0.125	2	14
1j	0.125	40	72
1k	0.125	9	25

4 CONCLUSIONS

The high potency of the specific ITRS inhibitor compound **1e** is not reflected in the observed herbicidal activity. Thus there are other factors that are severely limiting the herbicidal activity of these compounds. In contrast, monic acid derivatives are highly active as herbicides, e.g. methyl monate A, with an IC_{50} for ITRS from spinach chloroplasts of less than 1 nM, controls a variety of broad leaf weeds at a rate of 100 g/ha in the glasshouse.[10] Both ascamycin **1b** and dealanylascamycin **1a** are bactericidal. However, ascamycin is unable to enter the bacterial cells. For expression of its bactericidal activity, ascamycin is cleaved by a membrane peptidase to give dealanylascamycin, **1a**.[13, 14, 15] Delanylascamycin can then enter the bacteria on an AMP active transport system. The glasshouse test data is consistent with a similar limitation applying to *in planta* herbicidal activity. Thus compounds, which are not readily cleaved to either compound **1a**, or **1c** (e.g. straight N-alkyl derivatives), are not easily transported into cells and fail to express herbicidal activity. Further, any herbicidal activity associated with the active analogues probably results from non-specific interaction with AMP binding sites rather than specific inhibition of the target aaTRS. Thus for useful *in vivo* activity, any future sulfamoyl-adenosine derivatives should be capable of cellular uptake and be metabolically stable.

Monic acid derivatives possess high levels of intrinsic selectivity between prokaryotic and eukaryotic ITRS, which is highly desirable for a potentially benign mammalian toxicology. The sulfamoyladenosine derivatives, described above, lack a similar selectivity between the plastidic, i.e. prokaryotic, and cytoplasmic, i.e. eukaryotic forms of the same aaTRS. For the sulfamoyladenosine derivatives, there are tight SAR requirements for the aminoacyl functionality to achieve nanomolar IC_{50} values at the target aaTRS. Thus replacement of the aminoacyl residue of this class of inhibitors to achieve such intrinsic selectivity will be problematic and require very specific structural elements. Any attempts at enhancing the intrinsic selectivity for prokaryotic *versus* eukaryotic enzymes are more likely to come from either ribosyl sugar replacement, and or adenine base replacement. This latter approach has been successfully employed by Hill *et al.*, who achieved prokaryotic *versus* eukaryotic differentiation and *in vivo* bacterial protectant activity in mouse models.[16, 17]

Publication of the x-ray structures of the N-serinyl-sulfamoyladenosine and the N-glutaminyl-sulfamoyladenosine bound to serinyl-tRNA synthetase (STRS) and glutaminyl-tRNA synthetase (QTRS) respectively provide a clear insight into how this class of inhibitors bind.[18, 19] In particular the QTRS structure, shows the potential for hydrogen bonding between the α-amino group of the aminoacyl residue and the 3'-hydroxyl of a ribosyl ring of the terminal adenosine in the tRNA^{gln}. In this structure, the 2'-hydroxyl on the ribosyl ring is held at the appropriate position to bind the carbonyl functional group of the glutaminyl residue. These observations suggests that the binding of this class of inhibitor to their target aaTRS's may be more complex than just mimicking the intermediate mixed anhydride **2** and that the tightness of binding may depend on the presence or absence of tRNA.

These studies serve to illustrate how natural products can act as leads for both novel modes of action and for novel chemistry in agrochemical research. They also show that while such compounds are by definition "biologically compatible", this does not preclude their "useful biological activity" being limited by similar factors to those which limit the activity of synthetic chemicals, i.e. in this case uptake and metabolic stability. This is not surprising as the natural products will have evolved to meet a specific need or challenge faced by their producer organism in a specific environment. Commercial uses tend to be general and non-specific in needs and environment, e.g. controlling several pest/weed species in one or more crops robustly under a variety of climatic conditions. For both natural products and synthetic leads, an optimisation programme will probably be required to achieve a cost effective useful compound.

References

1.   L. G. Copping, in *Crop Protection Agents From Nature*, ed. L. G. Copping, Royal Society of Chemistry, Cambridge, 1996, Introduction, pp. xvi.

2.   K. Kristinsson, K. Nebel, A. O'Sullivan, J. P. Pachlatko and Y. Yasuchika, *ACS Symp. Ser.*, 1995, **584**(Synthesis and Chemistry of Agrochemicals IV), 206.

3.   A. Scacchi, R. Bortolo, G. Cassani and E. Nielsen, *Pesticid. Biochem. Physiol.*, 1994, **50**, 149.

4.   D. A. Shuman, M. J. Robins and R. K. Robins, *J. Amer. Chem. Soc.*, 1970, **92**, 3434.

5.   A. Alcina, M. Fresno and B. Alarcon, *Antimicrob. Agents Chemother.*, 1988, **32**, 1412.

6.   E. A. Meade, L. Wotring, J. C. Drach, L. B. Townsend, *J. Med. Chem.*, 1993, **36**, 3834.

7.   A. Bloch and C. Coutsogeorgopoulos, *Biochem.*, 1971, **10**, 4394.

8.   H. Osada and K. Isono, *Antimicrob. Agents Chemother.*, 1985, **27**, 230.

9.   H. Ueda; Y. Shoku, N. Hayashi, J. Mitsunaga, Y. In, M. Doi, M. Inoue, and T. Ishida, *Biochim. Biophys. Acta*, 1991, **1080**, 126.

10. P. Bellini, J. M. Clough and G Hatter, in *Book of Abstracts Volume 1 Topics 1-4, 9th International Congress Pesticide Chemistry, The Food-Environment Challenge*, Royal Society of Chemistry and The International Union of Pure and Applied Chemistry, 1998, Abstract 1A-032.
11. K. Beautement and E. J. T. Chrystal, GB 2284811, 1995.
12. K. Beautement and E. J. T. Chrystal, GB 2287464, 1995.
13. H. Osada and K. Isono, *J. Antibiot.*, 1986, **39**, 286.
14. H. Osada and K. Isono, *Biochem. J.*, 1986, **232**, 459.
15. M. Ubukata, H.Osada, J. Magae and K. Isono, *Agric. Biol. Chem.*, 1988, **52**, 1117.
16. P. Schimmel, T. Jiansshi and J.Hill, *FASEB J.*, 1998, **12**, 1599.
17. J. M. Hill, G. Yu, Y-K. Shue, T. M. Zydowsky and J. Rebek Jr., WO 9705132, 1997.
18. H. Belrhali, A. Yaremchuk, M. Tukalo, K. Larsen, C. Berthet-Colominas, R. Leberman, B. Beijer, B. Sproat, J. Als-Nielsen, G. Grubel, J-F. Legrand, M. Lehmann and S. Cusack, *Science*, 1994, **263**, 1432.
19. V. L. Rath, L. F. Silvian, B. Beijer, B. S. Sproat and T. A. Steitz, *Structure*, 1998, **6**, 439.

Subject Index